CAMBRIDGE STUDIES IN ADVANCED MATHEMATICS 214

*Editorial Board*
J. BERTOIN, B. BOLLOBÁS, W. FULTON, B. KRA, I. MOERDIJK,
C. PRAEGER, P. SARNAK, B. SIMON, B. TOTARO

# ALGEBRAIC VARIETIES: MINIMAL MODELS AND FINITE GENERATION

The finite generation theorem is a major achievement in modern algebraic geometry. Based on the minimal model theory, it states that the canonical ring of an algebraic variety defined over a field of characteristic 0 is a finitely generated graded ring.

This graduate-level text is the first to explain this proof. It covers the progress on the minimal model theory over the last 30 years, culminating in the landmark paper on finite generation by Birkar–Cascini–Hacon–M$^c$Kernan. Building up to this proof, the author presents important results and techniques that are now part of the standard toolbox of birational geometry, including Mori's bend-and-break method, vanishing theorems, positivity theorems, and Siu's analysis on multiplier ideal sheaves. Assuming only the basics in algebraic geometry, the text keeps prerequisites to a minimum with self-contained explanations of terminology and theorems.

**Yujiro Kawamata** is a University Professor at the University of Tokyo. He is the recipient of various prizes and awards, including the Mathematical Society of Japan Autumn award (1988), the Japan Academy of Sciences award (1990), ICM speaker (1990), and ISI Highly Cited Researcher (2001).

CAMBRIDGE STUDIES IN ADVANCED MATHEMATICS

*Editorial Board*
J. Bertoin, B. Bollobás, W. Fulton, B. Kra, I. Moerdijk, C. Praeger, P. Sarnak, B. Simon, B. Totaro

All the titles listed below can be obtained from good booksellers or from Cambridge University Press. For a complete series listing, visit www.cambridge.org/mathematics.

*Already Published*

174. P. Garrett *Modern Analysis of Automorphic Forms by Example, II*
175. G. Navarro *Character Theory and the McKay Conjecture*
176. P. Fleig, H. P. A. Gustafsson, A. Kleinschmidt & D. Persson *Eisenstein Series and Automorphic Representations*
177. E. Peterson *Formal Geometry and Bordism Operators*
178. A. Ogus *Lectures on Logarithmic Algebraic Geometry*
179. N. Nikolski *Hardy Spaces*
180. D.-C. Cisinski *Higher Categories and Homotopical Algebra*
181. A. Agrachev, D. Barilari & U. Boscain *A Comprehensive Introduction to Sub-Riemannian Geometry*
182. N. Nikolski *Toeplitz Matrices and Operators*
183. A. Yekutieli *Derived Categories*
184. C. Demeter *Fourier Restriction, Decoupling and Applications*
185. D. Barnes & C. Roitzheim *Foundations of Stable Homotopy Theory*
186. V. Vasyunin & A. Volberg *The Bellman Function Technique in Harmonic Analysis*
187. M. Geck & G. Malle *The Character Theory of Finite Groups of Lie Type*
188. B. Richter *Category Theory for Homotopy Theory*
189. R. Willett & G. Yu *Higher Index Theory*
190. A. Bobrowski *Generators of Markov Chains*
191. D. Cao, S. Peng & S. Yan *Singularly Perturbed Methods for Nonlinear Elliptic Problems*
192. E. Kowalski *An Introduction to Probabilistic Number Theory*
193. V. Gorin *Lectures on Random Lozenge Tilings*
194. E. Riehl & D. Verity *Elements of ∞-Category Theory*
195. H. Krause *Homological Theory of Representations*
196. F. Durand & D. Perrin *Dimension Groups and Dynamical Systems*
197. A. Sheffer *Polynomial Methods and Incidence Theory*
198. T. Dobson, A. Malnič & D. Marušič *Symmetry in Graphs*
199. K. S. Kedlaya *p-adic Differential Equations*
200. R. L. Frank, A. Laptev & T. Weidl *Schrödinger Operators: Eigenvalues and Lieb–Thirring Inequalities*
201. J. van Neerven *Functional Analysis*
202. A. Schmeding *An Introduction to Infinite-Dimensional Differential Geometry*
203. F. Cabello Sánchez & J. M. F. Castillo *Homological Methods in Banach Space Theory*
204. G. P. Paternain, M. Salo & G. Uhlmann *Geometric Inverse Problems*
205. V. Platonov, A. Rapinchuk & I. Rapinchuk *Algebraic Groups and Number Theory, I (2nd Edition)*
206. D. Huybrechts *The Geometry of Cubic Hypersurfaces*
207. F. Maggi *Optimal Mass Transport on Euclidean Spaces*
208. R. P. Stanley *Enumerative Combinatorics, II (2nd Edition)*
209. M. Kawakita *Complex Algebraic Threefolds*
210. D. Anderson & W. Fulton *Equivariant Cohomology in Algebraic Geometry*
211. G. Pineda Villavicencio *Polytopes and Graphs*
212. R. Pemantle, M. C. Wilson & S. Melczer *Analytic Combinatorics in Several Variables (2nd Edition)*
213. A. Yadin *Harmonic Functions and Random Walks on Groups*

# Algebraic Varieties: Minimal Models and Finite Generation

YUJIRO KAWAMATA
*University of Tokyo*

TRANSLATED BY CHEN JIANG
*Fudan University*

Shaftesbury Road, Cambridge CB2 8EA, United Kingdom

One Liberty Plaza, 20th Floor, New York, NY 10006, USA

477 Williamstown Road, Port Melbourne, VIC 3207, Australia

314–321, 3rd Floor, Plot 3, Splendor Forum, Jasola District Centre, New Delhi - 110025, India

103 Penang Road, #05–06/07, Visioncrest Commercial, Singapore 238467

Cambridge University Press is part of Cambridge University Press & Assessment,
a department of the University of Cambridge.

We share the University's mission to contribute to society through the pursuit of
education, learning and research at the highest international levels of excellence.

www.cambridge.org
Information on this title: www.cambridge.org/9781009344678

DOI: 10.1017/9781009344647

KOJIGENDAISUTAYOTAIRON (Higher Dimensional Algebraic Varieties)
by Yujiro Kawamata
© 2014 by Yujiro Kawamata
Originally published in 2014 by Iwanami Shoten, Publishers, Tokyo.
This English edition published 2024
by Cambridge University Press, Cambridge
by arrangement with Iwanami Shoten, Publishers, Tokyo

English translation © Cambridge University Press 2024

This publication is in copyright. Subject to statutory exception and to the provisions
of relevant collective licensing agreements, no reproduction of any part may take place
without the written permission of Cambridge University Press & Assessment.

When citing this work, please include a reference to the DOI 10.1017/9781009344647

First published 2024

*A catalogue record for this publication is available from the British Library.*

*Library of Congress Cataloging-in-Publication Data*
Names: Kawamata, Yūjirō, 1952– author.
Title: Algebraic varieties : minimal models and finite generation / Yujiro Kawamata ;
translated by Chen Jiang.
Other titles: Kōjigen daisū tayōtairon. English
Description: New York : Cambridge University Press, 2024. | Series: Cambridge studies in advanced
mathematics | Translation of: Kōjigen daisū tayōtairon by Iwanami Shoten in 2014. | Includes
bibliographical references and index.
Identifiers: LCCN 2023052216 (print) | LCCN 2023052217 (ebook) | ISBN 9781009344678
(hardback) | ISBN 9781009344647 (epub)
Subjects: LCSH: Algebraic varieties.
Classification: LCC QA564 .K3913 2024 (print) | LCC QA564 (ebook) | DDC 512–dc23/eng/20240326
LC record available at https://lccn.loc.gov/2023052216
LC ebook record available at https://lccn.loc.gov/2023052217

ISBN 978-1-009-34467-8 Hardback

Cambridge University Press & Assessment has no responsibility for the persistence
or accuracy of URLs for external or third-party internet websites referred to in this
publication and does not guarantee that any content on such websites is, or will
remain, accurate or appropriate.

# Contents

| | | |
|---|---|---:|
| | *Preface* | *page* ix |
| | **Introduction** | 1 |
| **1** | **Algebraic Varieties with Boundaries** | 6 |
| | 1.1  **Q**-divisors and **R**-divisors | 6 |
| | 1.2  Rational Maps and Birational Maps | 11 |
| | 1.3  Canonical Divisors | 17 |
| | 1.4  Intersection Numbers and Numerical Geometry | 20 |
| | 1.5  Cones of Curves and Cones of Divisors | 24 |
| | 1.5.1  Pseudo-Effective Cones and Nef Cones | 25 |
| | 1.5.2  Kleiman's Criterion and Kodaira's Lemma | 28 |
| | 1.6  The Hironaka Desingularization Theorem | 33 |
| | 1.7  The Kodaira Vanishing Theorem | 36 |
| | 1.8  The Covering Trick | 38 |
| | 1.9  Generalizations of the Kodaira Vanishing Theorem | 41 |
| | 1.10  KLT Singularities for Pairs | 46 |
| | 1.11  LC, DLT, and PLT Singularities for Pairs | 50 |
| | 1.11.1  Various Singularities | 50 |
| | 1.11.2  The Subadjunction Formula | 55 |
| | 1.11.3  Terminal and Canonical Singularities | 59 |
| | 1.12  Minimality and Log Minimality | 60 |
| | 1.13  The 1-Dimensional and 2-Dimensional Cases | 64 |
| | 1.13.1  The 1-Dimensional Case | 64 |
| | 1.13.2  Minimal Models in Dimension 2 | 65 |
| | 1.13.3  The Classification of Algebraic Surfaces | 68 |
| | 1.13.4  Rational Singularities | 69 |
| | 1.13.5  The Classification of DLT Surface Singularities I | 73 |

v

| | | | |
|---|---|---|---|
| | 1.13.6 | The Classification of DLT Surface Singularities II | 76 |
| | 1.13.7 | The Zariski Decomposition | 80 |
| 1.14 | The 3-Dimensional Case | | 82 |

**2 The Minimal Model Program** — 87

| | | | |
|---|---|---|---|
| 2.1 | The Basepoint-Free Theorem | | 87 |
| | 2.1.1 | Proof of the Basepoint-Free Theorem | 88 |
| | 2.1.2 | Paraphrasings and Generalizations | 93 |
| 2.2 | An Effective Version of the Basepoint-Free Theorem | | 96 |
| 2.3 | The Rationality Theorem | | 99 |
| 2.4 | The Cone Theorem | | 104 |
| | 2.4.1 | The Contraction Theorem | 104 |
| | 2.4.2 | The Cone Theorem | 106 |
| | 2.4.3 | Contraction Morphisms in Dimensions 2 and 3 | 111 |
| | 2.4.4 | The Cone Theorem for the Space of Divisors | 113 |
| 2.5 | Types of Contraction Morphisms and the Minimal Model Program | | 116 |
| | 2.5.1 | Classification of Contraction Morphisms | 116 |
| | 2.5.2 | Flips | 118 |
| | 2.5.3 | Decrease of Canonical Divisors | 123 |
| | 2.5.4 | The Existence and the Termination of Flips | 124 |
| | 2.5.5 | Minimal Models and Canonical Models | 126 |
| | 2.5.6 | The Minimal Model Program | 128 |
| 2.6 | Minimal Model Program with Scaling | | 130 |
| 2.7 | The Existence of Rational Curves | | 132 |
| | 2.7.1 | Deformation of Morphisms | 133 |
| | 2.7.2 | The Bend-and-Break Method | 134 |
| 2.8 | The Lengths of Extremal Rays | | 139 |
| 2.9 | The Divisorial Zariski Decomposition | | 141 |
| 2.10 | Polyhedral Decompositions of a Cone of Divisors | | 147 |
| | 2.10.1 | Rationality of Sections of Nef Cones | 148 |
| | 2.10.2 | Polyhedral Decomposition according to Canonical Models | 149 |
| | 2.10.3 | Polyhedral Decomposition according to Minimal Models | 152 |
| | 2.10.4 | Applications of Polyhedral Decompositions | 155 |
| 2.11 | Multiplier Ideal Sheaves | | 160 |
| | 2.11.1 | Multiplier Ideal Sheaves | 160 |
| | 2.11.2 | Adjoint Ideal Sheaves | 164 |

# Contents

vii

| | | |
|---|---|---|
| 2.12 | Extension Theorems | 168 |
| | 2.12.1 Extension Theorems I | 168 |
| | 2.12.2 Extension Theorems II | 175 |
| **3** | **The Finite Generation Theorem** | **179** |
| 3.1 | Setting of the Inductive Proof | 179 |
| 3.2 | PL Flips | 183 |
| | 3.2.1 Restriction of Canonical Rings to Divisors | 185 |
| | 3.2.2 The Existence of PL Flips | 189 |
| 3.3 | The Special Termination | 192 |
| 3.4 | The Existence and Finiteness of Minimal Models | 197 |
| 3.5 | The Nonvanishing Theorem | 204 |
| 3.6 | Summary | 210 |
| 3.7 | Algebraic Fiber Spaces | 214 |
| | 3.7.1 Algebraic Fiber Spaces and Toroidal Geometry | 214 |
| | 3.7.2 The Weak Semistable Reduction Theorem and the Semipositivity Theorem | 217 |
| 3.8 | The Finite Generation Theorem | 220 |
| 3.9 | Generalizations of the Minimal Model Theory | 222 |
| | 3.9.1 The Case with a Group Action | 223 |
| | 3.9.2 The Case when the Base Field Is Not Algebraically Closed | 224 |
| 3.10 | Remaining Problems | 225 |
| | 3.10.1 The Abundance Conjecture | 225 |
| | 3.10.2 The Case of Numerical Kodaira Dimension 0 | 227 |
| | 3.10.3 Generalization to Positive Characteristics | 228 |
| 3.11 | Related Topics | 229 |
| | 3.11.1 Boundedness Results | 229 |
| | 3.11.2 Minimal Log Discrepancies | 231 |
| | 3.11.3 The Sarkisov Program | 233 |
| | 3.11.4 Rationally Connected Varieties | 234 |
| | 3.11.5 The Category of Smooth Algebraic Varieties | 235 |

| | |
|---|---|
| *References* | 237 |
| *Index* | 245 |

# Preface

In this book, we explain the proof of the theorem on finite generation of canonical rings of algebraic varieties. The proof is based on the minimal model theory. The main goal of this book is to go through the progress on the minimal model theory in the past 30 years.

The finite generation of canonical rings is one of the main goals of the minimal model theory. Even though it has been proved now, the minimal model theory is not yet complete. There are open problems such as the existence of minimal models of an arbitrary algebraic variety, the abundance conjecture, and the generalization to positive characteristics.

The minimal model theory was developed by many mathematicians from different countries. Although the theory explained in this book works only for algebraic varieties defined over a base field of characteristic 0, we apply various methods in a wider range which contains not only algebraic arguments over characteristic 0, but also methods over positive characteristics and analytic methods. The breakthrough was accomplished by international collaborations combining algebraic, geometric, and analytic methods.

The content of this book is based on my lectures at the University of Tokyo in 2005, 2007, 2010, 2011, the Felix Klein lecture at Bonn University in 2006, series lectures in Current Developments in Mathematics at Harvard University in 2007, and intensive lectures at Osaka University in 2007.

I would like to thank Professor Takeshi Saito of the University of Tokyo for the encouragement to write this book. I would also like to express my deep gratitude to the anonymous referees who carefully read the manuscript and made many suggestions.

x                                   *Preface*

In addition, I would like to thank Professor Burt Totaro for recommending the translation of this book. I would also like to thank Professor Chen Jiang for numerous clarifications, corrections, and modifications as well as an excellent translation. In particular, the proof of Lemma 2.1.8, the modifications of the condition (1) of Theorem 2.12.1 and (3) of the BCHM condition, Remark 3.1.4, and Theorem 3.4.6 are all due to him.

# Introduction

First, we will explain the overall plan of this book. Proofs and accurate definitions will not be given here, but will appear in subsequent chapters with more precise details.

Let $X$ be a smooth projective complex algebraic variety, or in other words, a closed subvariety in a complex projective space. Denote $\dim X = n$ and take a positive integer $m$. A regular $m$-canonical differential form on $X$ is defined locally by $h(x)(dx_1 \wedge \cdots \wedge dx_n)^{\otimes m}$ where $x_1, \ldots, x_n$ are local coordinates of $X$ and $h$ is a regular function on $X$. The set of regular $m$-canonical differential forms on $X$ is a finite-dimensional $\mathbf{C}$-linear space denoted by $H^0(X, mK_X)$. Here $K_X$ is the canonical divisor. For example, when $m = 0$, this space is just $\mathbf{C}$, and when $m = 1$, this space consists of all regular canonical differential forms.

For two positive integers $m, m'$, we can define a multiplication map

$$H^0(X, mK_X) \otimes H^0(X, m'K_X) \to H^0(X, (m + m')K_X),$$

inducing a graded ring

$$R(X, K_X) = \bigoplus_{m=0}^{\infty} H^0(X, mK_X)$$

over the complex number field, which is called the *canonical ring* of $X$.

Two algebraic varieties $X, Y$ are said to be *birationally equivalent* if there are non-empty Zariski open subsets $U \subset X$, $V \subset Y$ such that there is an isomorphism $U \cong V$. In this case $Y$ is called a *birational model* of $X$, and the canonical ring is a *birational invariant*, that is, $R(X, K_X) \cong R(Y, K_Y)$. Birational invariants reflect intrinsic properties of algebraic varieties.

The main theorem of this book is the following proved by Birkar–Cascini–Hacon–M$^c$Kernan ([16]):

# 2 Introduction

**Theorem 1** (*Finite generation of canonical rings*) *For any smooth projective complex algebraic variety* $X$, *the canonical ring* $R(X, K_X)$ *is a finitely generated graded* **C**-*algebra.*

The proof uses the *minimal model program* (*MMP*). The main part of this book is devoted to the foundation of the MMP.

If the transcendental degree of the canonical ring is $n + 1$, then $X$ is said to be of *general type*. In this case one can show that there exists a "minimal model" $X'$ birationally equivalent to $X$ with good properties. An MMP is a sequence of operations constructing $X'$ starting from $X$. In general $X'$ has singularities, but the singularities are mild so that the birational invariance $R(X, K_X) \cong R(X', K_{X'})$ still holds as the smooth case. The finite generation of canonical rings of minimal models is a consequence of the "*basepoint-free theorem.*"

When $X$ is not of general type, by applying the "*semipositivity theorem*" of algebraic fiber spaces, one can reduce the problem to the case of "log general type," and then derive the finite generation by the "log version" of the MMP.

The MMP is a process of changing birational models one after another. During this process, algebraic varieties with singularities naturally appear. However, those singularities are special kinds of normal singularities. The singularities in the MMP are very interesting research objects for their own sake. With the development of higher dimensional algebraic geometry, it is gradually becoming more common to consider algebraic varieties with singularities.

Proofs in the minimal model theory often use induction on integral invariants such as dimensions and Picard numbers. In order for this to work well, it is necessary to enhance the category of objects we consider. Here we extend to the *log version* and the *relative version*.

In the log version, instead of a single algebraic variety $X$, we consider a couple $(X, B)$ consisting of $X$ and an **R**-divisor $B$ on $X$. For historical reasons, this is called a *log pair* and $B$ is called a *boundary divisor*. Here an **R**-*divisor* $B = \sum b_j B_j$ is a formal **R**-linear combination of subvarieties $B_j$ of codimension 1 with real coefficients $b_j$. It is called a **Q**-*divisor* if $b_j$ are rational numbers. Instead of the canonical divisor $K_X$, the *log canonical divisor* $K_X + B$ plays the main role.

Conditions on singularities are imposed onto the log pair $(X, B)$. In this book, we mainly consider the "KLT condition" (Kawamata log terminal condition) and the "DLT condition" (divisorially log terminal condition). For example, when $X$ is smooth and the support $\sum B_j$ of $B$ is a "normal crossing

# Introduction

divisor," these conditions correspond to inequalities $0 < b_j < 1$ and $0 < b_j \leq 1$, respectively.

In the relative version, all objects are considered over a base variety. Instead of a single algebraic variety $X$, we consider a morphism $f: X \to S$ to a base variety.

In summary, we are going to consider a log pair $(X, B)$ with the KLT or DLT condition and a projective morphism $f: X \to S$ to another algebraic variety. Sometimes we use $f: (X, B) \to S$ for short to keep in mind the log version and the relative version at the same time.

The *log canonical ring* is defined as

$$R(X/S, K_X + B) = \bigoplus_{m=0}^{\infty} f_*(\mathcal{O}_X(\llcorner m(K_X + B) \lrcorner)).$$

Here the symbol $\llcorner \lrcorner$ means *round down*, that is, to replace each coefficient by the nearest integer from below and $f_*$ is the direct image of sheaves. $R(X/S, K_X + B)$ is a graded $\mathcal{O}_S$-algebra.

The log and relative version of the finite generation of canonical rings is as follows:

**Theorem 2** *Let $f: (X, B) \to S$ be a projective morphism from a KLT pair defined over the complex number field, where $B$ is a **Q**-divisor. Then the log canonical ring $R(X/S, K_X + B)$ is a finitely generated graded $\mathcal{O}_S$-algebra.*

In Chapter 1, we will give basic definitions used in this book. The main idea is to associate a variety with a divisor called boundary and to consider them as a pair. Such "logarithmization" makes it possible to introduce many new methods. Log pairs are allowed to have mild singularities. Usually in algebraic geometry, nonsingular varieties are the central objects, but singularities of pairs are indispensable and play important roles in the minimal model theory. We will also explain two big theorems in characteristic 0 (the Hironaka desingularization theorem and the Kodaira vanishing theorem), both of which are main tools of this book. In particular, it is known that the vanishing theorem fails when the characteristic is not 0, so most results of this book are in characteristic 0. Then we will describe the classification theory of low-dimensional algebraic varieties. The goal of this part is to provide examples, and it is logically independent.

In Chapter 2, we will explain the outline of the minimal model theory. There are two main theorems: the basepoint-free theorem and the cone theorem. Using these theorems we formulate the MMP. The minimality of a log pair $(X, B)$ is tested by the "numerical property" of the log canonical divisor. If the pair is not minimal, then there exists an "extremal ray" by the "cone theorem," and it induces a "contraction morphism" by the "basepoint-free theorem." There are three types of contraction morphisms: "divisorial contractions," "small contractions," and "Mori fiber spaces." For small contractions we need to consider another birational map called "flip" on the "opposite side." We will also explain new contents such as an effective version of the basepoint-free theorem and the MMP with scaling. In addition, we describe an important extension theorem developed from the theory of multiplier ideal sheaves.

In Chapter 3, we will give the proof of the finite generation of canonical rings, which is the main topic of this book. To this end, we show the existence of minimal models for varieties of general type. The "existence of flips" is proved as a special case of the finite generation of canonical rings. Furthermore, the "termination of flips" is proved under the assumption of "general type," which finishes the proof of the finite generation theorem in the general type case. In the end, we apply the semipositivity theorem of Hodge bundles, which is also a result to be held only in characteristic 0.

The content of Chapter 2 is basically the same as [76]. This book is a sequel of [76] and [67]. [76] summarized the results of the minimal model theory in its early stages, and has been cited in much literature. In there, the minimal model theory was already described in the log version and the relative version, which is consistent with the direction of the development afterward. We are proud to have played a certain role in producing a standard literature on the minimal model theory. At that stage, the basepoint-free theorem and the cone theorem were proved and the existence of minimal models was reduced to two conjectures on flips. In subsequent developments, the existence of flips was proved, along with the termination of flips in some special but important cases. The purpose of Chapter 3 is to explain those developments.

**Remark 3** (1) In the proof, it is necessary to consider not only $\mathbf{Q}$-divisors, but also $\mathbf{R}$-divisors. However, the finite generation theorem only holds when $B$ is a $\mathbf{Q}$-divisor.

(2) Although in our discussion we assumed that the base field is the complex number field $\mathbf{C}$, all proofs work for algebraically closed fields in characteristic 0. Moreover, the results can be extended to algebraically nonclosed fields after necessary modifications. On the other hand, it is expected that the same conclusions (theorems in the minimal model theory and the finite generation of canonical rings) still hold true in positive characteristics, but the arguments in this book fail for two reasons. First, the desingularization theorem will be used in many places, which is still an open problem in positive characteristics; second, the vanishing theorem is a key tool in the

*Introduction* 5

proofs, which has counterexamples in positive characteristics. Therefore, there is almost no progress in positive characteristics. (Added in 2023: This claim held only at the time of the publication of the Japanese version.)

(3) In this book, all results are stated in the log and relative version. If this seems annoying to you, just take the boundary $B$ to be 0, take $S$ to be a point Spec $k$, and replace the direct image sheaf $f_*F$ by the space of global sections $H^0(X, F)$, but the point of the proof will not change at all. However, as the proofs in the MMP are inductive, it is indispensable to state the log and relative version. Also, when dealing with algebraic varieties of nongeneral type, even if we start from an ordinary algebraic variety without boundary, log pairs naturally appear from the structure of algebraic fiber spaces.

(4) The finite generation of canonical rings is one of the main goals of the MMP in the beginning. Even though it is proved now, the existence of minimal models still remains open in the general case.

As prerequisites, we hope the reader has some familiarity with algebraic varieties. For this, it is sufficient to have standard knowledge from the textbook of Hartshorne ([44]). In particular, the theory of cohomologies of coherent sheaves is a basic tool; the concept of linear systems of divisors and the correspondence of Cartier divisors and invertible sheaves on a normal algebraic variety are important, which will be explained in Chapter 1; also it is better to have knowledge of algebraic surface theory as in [44, Chapter V]; but it is not necessary to understand every detail in [44] because, other than Section 2.7, this book does not deal with general schemes but only deals with irreducible reduced separated schemes of finite type over an algebraically closed field (i.e. algebraic varieties). The Kodaira vanishing theorem and the Hironaka desingularization theorem are important theorems cited in this book (the statements of the theorems will be given). These are indispensable tools for the discussions in this book, but it is not necessary to understand the proofs.

# 1

# Algebraic Varieties with Boundaries

A boundary of an algebraic variety is a divisor with real coefficients. In this chapter, we introduce basic concepts of algebraic varieties with boundaries. Using the language of numerical geometry, we define cones of curves and divisors. According to the Hironaka desingularization theorem, it is possible to use birational morphisms to make algebraic varieties smooth and divisors normal crossing. We focus on log canonical divisors of algebraic varieties with boundaries, and define concepts of KLT (Kawamata log terminal) pairs and DLT (divisorially log terminal) pairs. The Kodaira vanishing theorem for smooth projective varieties can be extended to KLT or DLT pairs by constructing covering spaces using the covering trick. We also discuss the classification of algebraic varieties and singularities in lower dimensions.

## 1.1 Q-divisors and R-divisors

The linear equivalence class of a divisor determines a coherent sheaf which is called a divisorial sheaf. Algebraic geometry often deals with coherent sheaves, but this book focuses on the language of divisors. It is like dealing with differential forms themselves instead of cohomology classes of differential forms in differential geometry.

Fix a base field $k$. An *algebraic variety* $X$ is an irreducible reduced separated scheme of finite type over $k$.

An algebraic variety $X$ is attached with the structure sheaf $\mathcal{O}_X$ and a local ring $\mathcal{O}_{X,P}$ at each point $P$. If the local ring $\mathcal{O}_{X,P}$ is a regular local ring, then $X$ is said to be *nonsingular* at $P$. In this book, we mostly work over characteristic 0, so we will use the word *smooth* instead of nonsingular which sounds better.

## 1.1 Q-divisors and R-divisors

$\dots = n$, $X$ is smooth if and only if for every closed point $P$ on $X$, maximal ideal $\mathfrak{m}_P$ of the local ring is generated by $n$ elements $x_1, \dots, x_n$. Such $x_1, \dots, x_n$ is called a *regular system of parameters* or *local coordinates*. When $k = \mathbf{C}$, this is equivalent to saying that the set of closed points of $X$ forms a complex manifold.

The set of all smooth points $\text{Reg}(X)$ of an algebraic variety $X$ is a non-empty open subset of $X$, and its complement set $\text{Sing}(X) = X \setminus \text{Reg}(X)$, which is a proper closed subset of $X$, is called the *singular locus* of $X$.

An algebraic variety $X$ is said to be *normal* if the local ring at every point is an integrally closed domain. Since normal local rings of dimension 1 are regular, the singular locus of a normal algebraic variety is a closed subset of codimension at least 2. That is, it is the closure of several points with codimensions at least 2.

Every algebraic variety $X$ can be easily modified into a normal one: There is a unique finite morphism $f : X^\nu \to X$ from a normal algebraic variety which is isomorphic over $\text{Reg}(X)$. This is called the *normalization* of $X$.

Normality can be checked by Serre's criterion ([94]):

**Theorem 1.1.1** *An algebraic variety $X$ is normal if and only if the following two conditions are satisfied:*

*(1) ($R_1$) Its singular locus is a closed subset of codimension at least 2.*

*(2) ($S_2$) For any open subset $U$ and any closed subset $Z$ of codimension at least 2, the restriction map $\Gamma(U, \mathcal{O}_X) \to \Gamma(U \setminus Z, \mathcal{O}_X)$ is bijective.*

From now on, we will always assume that $X$ is a normal algebraic variety. A *prime divisor* on $X$ is a closed subvariety of codimension 1. A *divisor* is a formal finite sum of prime divisors $D = \sum d_i D_i$. Unless otherwise stated, the coefficients $d_i$ are integers and $D_i$ are distinct prime divisors. In other words, divisors are elements in the free Abelian group $Z^1(X)$ generated by all prime divisors on $X$.

$D$ is said to be *effective* if all coefficients $d_i$ are nonnegative. For two divisors $D, D'$, we write the inequality $D \geq D'$ if $D - D'$ is effective. $D$ is said to be *reduced* if $d_i = 1$ for all $i$.

Let $D$ be a prime divisor on a normal algebraic variety $X$ and let $P$ be the generic point of $D$, then the local ring $\mathcal{O}_{X,P}$ is a *discrete valuation ring* with the function field $k(X)$ as its quotient field.

For a rational function $h \in k(X)^*$, its divisor $\text{div}(h)$ is defined as

$$\text{div}(h) = \sum v_D(h) D.$$

Here the sum runs over all prime divisors $D$, and $v_D$ is the discrete valuation of the local ring at the generic point of $D$. It is known that the right-hand side

8    *1 Algebraic Varieties with Boundaries*

is a finite sum. Any divisor defined by a nonzero rational function is called a *principal divisor*.

For a divisor $D$, the corresponding *divisorial sheaf* $\mathcal{O}_X(D)$ is defined as the following: for any open subset $U$ of $X$,

$$\Gamma(U, \mathcal{O}_X(D)) = \{h \in k(X)^* \mid \mathrm{div}(h)|_U + D|_U \geq 0\} \cup \{0\}.$$

Also we define

$$H^0(X, D) = H^0(X, \mathcal{O}_X(D)).$$

If a nonzero global section $s$ of $\mathcal{O}_X(D)$ corresponds to a rational function $h$, we define the divisor of $s$ by

$$\mathrm{div}(s) = \mathrm{div}(h) + D,$$

which is effective. Generally we can also define the divisor $\mathrm{div}(s)$ of a rational section $s$ of $\mathcal{O}_X(D)$ by the corresponding rational function $h$ as the above equation, but in this case $\mathrm{div}(s)$ is not necessarily effective. For example, if we take $s_1$ to be the rational section corresponding to the rational function $h = 1$, then the corresponding divisor is just $D$.

There is an isomorphism $(\mathcal{O}_X(D))_\eta \cong \mathcal{O}_{X,\eta}$ on the generic point $\eta$ of $X$. Moreover, by taking the dual, we have

$$\mathcal{O}_X(D)^* := \mathrm{Hom}(\mathcal{O}_X(D), \mathcal{O}_X) \cong \mathcal{O}_X(-D),$$

hence the divisorial sheaf $\mathcal{O}_X(D)$ is a *reflexive sheaf of rank 1*. Here a *reflexive sheaf* is a coherent sheaf which is isomorphic to its double dual: $F^{**} \cong F$.

A divisor $D$ is called a *Cartier divisor* if its divisorial sheaf $\mathcal{O}_X(D)$ is invertible. In other words, this is to say that, in a neighborhood of each point $P$, this divisor is a principal divisor defined by some rational function depending on $P$. To distinguish from Cartier divisors, we call this divisor a *Weil divisor* or an *integral divisor*. Denote by $\mathrm{Div}(X)$ the set of all Cartier divisors. There is an inclusion $\mathrm{Div}(X) \subset Z^1(X)$, and they coincide when $X$ is smooth.

Two divisors $D, D'$ on an algebraic variety $X$ are said to be *linearly equivalent*, and denoted by $D \sim D'$, if $D - D'$ is a principal divisor. Note that $D \sim D'$ if and only if $\mathcal{O}_X(D) \cong \mathcal{O}_X(D')$. In other words, divisorial sheaves can be viewed as linear equivalence classes of divisors. Here $D, D'$ are not necessarily Cartier divisors.

The relative version is as follows. Given a *morphism* $f: X \to S$ between algebraic varieties, two divisors $D, D'$ on $X$ are said to be *relatively linearly equivalent* over $S$, and denoted by $D \sim_S D'$, if there exists an open covering $\{S_i\}$ of $S$ such that $D|_{S_i} \sim D'|_{S_i}$ after restriction over each $S_i$. Here we remark that in some other references, $D, D'$ are defined to be relatively

ists a Cartier divisor $B$ on $S$ such that
definitions are not the same and the
er. But under certain conditions, for
with connected geometric fibers, it is
icide.
braic variety $X$ is called a *normal*
t $P$ there is a regular system of
$_P$ and an integer $1 \leq r \leq n$ such
rm $z_1 \cdots z_r = 0$ locally around $P$.
component of $B$ is smooth. Also, the union of
ucible components of $B$ is again a normal crossing divisor.

For an algebraic variety $X$ and a closed subset $B$, the set of points at which $X$ is smooth and $B$ is a normal crossing divisor is an open subset of $X$, which is denoted by $\mathrm{Reg}(X, B)$. The complement set $\mathrm{Sing}(X, B) = X \setminus \mathrm{Reg}(X, B)$ is called the *singular locus* of $(X, B)$.

**Remark 1.1.2** A normal crossing divisor defined above is also called a *simple normal crossing divisor* in many references.

If $X$ is a complex algebraic manifold and $z_1, \ldots, z_n$ are regular local coordinates on the complex manifold associated to $X$, then a normal crossing divisor $B$ satisfying the same condition as above is not necessarily a simple normal crossing divisor in the algebraic setting. In fact, irreducible components of $B$ may have self-intersection. So, we use the term "simple" in the algebraic setting in order to distinguish with the analytic setting.

For example, in the affine plane $\mathbf{C}^2$ with coordinates $x, y$, the closed subset defined by the equation $x^2 + y^2 + y^3 = 0$ is irreducible but has self-intersection at the point $(0,0)$, therefore it is a normal crossing divisor on the complex manifold, but not a simple normal crossing divisor.

One feature of this book is to consider divisors which don't necessarily have integral coefficients. If the coefficients $d_i$ in $D = \sum d_i D_i$ are rational numbers (respectively, real numbers), then $D$ is called a *Q-divisor* (respectively, an *R-divisor*). Note that a Q-divisor is also an R-divisor. Those are elements in $Z^1(X) \otimes \mathbf{Q}$ or $Z^1(X) \otimes \mathbf{R}$, respectively, and these vector spaces are usually denoted by $Z^1(X)_\mathbf{Q}$ and $Z^1(X)_\mathbf{R}$. We will see soon that the range of discussions is expanded widely by considering Q-divisors and R-divisors.

Let $D = \sum d_i D_i$ be an R-divisor on $X$, where $D_i$ are distinct prime divisors. $D$ is said to be *effective* if all coefficients $d_i$ are nonnegative. For two R-divisors $D, D'$, we write the inequality $D \geq D'$ if $D - D'$ is effective. $D$ is said to be *reduced* if $d_i = 1$ for all $i$. The *support* of $D$ is the union

10          *1 Algebraic Varieties with*

of all $D_i$ with $d_i \neq 0$, and is denoted by $\mathrm{Supp}(D)$. Set
and $D^- = \sum_{d_i < 0}(-d_i)D_i$, then $D^+$ and $D^-$ are effecti
no common irreducible component and the equality $D = D$
$D^+, D^-$ are called the *positive part* and the *negative part* of $D$, r
For two **R**-divisors $D = \sum_i d_i D_i$ and $D' = \sum_i d'_i D_i$, define th
mum to be $\max\{D, D'\} = \sum_i \max\{d_i, d'_i\}D_i$. For example, $D^+ = \max$
$D^- = \max\{-D, 0\}$. Similarly we can define $\min\{D, D'\} = \sum_i \min\{d_i, d$
The *round up* (respectively, *round down*) of an **R**-divisor is defined via th
round up (respectively, round down) of coefficients:

$$\ulcorner D \urcorner = \sum \ulcorner d_i \urcorner D_i, \quad \llcorner D \lrcorner = \sum \llcorner d_i \lrcorner D_i.$$

A **Q**-divisor (respectively, an **R**-divisor) is said to be a **Q**-*Cartier divisor* (respectively, an **R**-*Cartier divisor*) if it is an element of $\mathrm{Div}(X) \otimes \mathbf{Q}$ (respectively, $\mathrm{Div}(X) \otimes \mathbf{R}$). Note that if a **Q**-divisor is an **R**-Cartier divisor, then it is automatically a **Q**-Cartier divisor. For a **Q**-Cartier divisor $D$, there exists a positive integer $m$ such that $mD$ is a Cartier divisor. However, in general there might not be a nonzero multiple to make an **R**-Cartier divisor Cartier. $X$ is said to be *factorial* (respectively, **Q**-*factorial*), if all Weil divisors on $X$ are Cartier divisors (respectively, **Q**-Cartier divisors).

Two **R**-divisors $D, D'$ are said to be **R**-*linearly equivalent*, denoted by $D \sim_{\mathbf{R}} D'$, if $D - D'$ can be written as an **R**-linear combination of principal divisors. The relative version and **Q**-linear equivalence can be defined similarly.

**Remark 1.1.3** Considering **R**-divisors is now essential to the development of the minimal model theory. This book can be viewed as a revised version of [76], in which only **Q**-divisors are treated. Later in [61], **R**-divisors already played a central role. The *divisorial Zariski decomposition* (which is called the sectional decomposition in [61]) is defined via limits of **Q**-divisors, so **R**-divisors appear naturally. Moreover, it is proved in [60] that the existence of Zariski decomposition (in a good sense, not only in codimension 1) into **R**-divisors implies the finite generation of canonical rings.

**Example 1.1.4** We give examples for a **Q**-Cartier Weil divisor which is not Cartier and a Weil divisor which is not **Q**-Cartier.

(1) Let $X$ be the hypersurface defined by the equation $xy = z^2$ in the 3-dimensional affine space $\mathbf{A}^3$ with coordinates $x, y, z$, which is an algebraic surface with an ordinary double point at the origin $(0, 0, 0)$. The line $D$ defined by the equation $x = z = 0$ is a prime divisor on $X$. At lease two

equations are needed to define $D$ in $X$, so $D$ is not a Cartier divisor. On the other hand, $\mathrm{div}(x) = 2D$ on $X$, so $D$ is **Q**-Cartier.

(2) Let $X$ be the hypersurface defined by the equation $xy = zw$ in $\mathbf{A}^4$ with coordinates $x, y, z, w$, which is a 3-fold with an ordinary double point at the origin $(0, 0, 0, 0)$. The 2-dimensional linear subspace $D_1$ defined by the equation $x = z = 0$ is a prime divisor on $X$, which is not a **Q**-Cartier divisor (see Example 1.2.4 for the reason). It is the same for $D_2$ defined by $x = w = 0$. However, the sum $D_1 + D_2 = \mathrm{div}(x)$ is a Cartier divisor. See Example 2.5.4(2) for related discussions.

## 1.2 Rational Maps and Birational Maps

A *rational map* $f : X \dashrightarrow Y$ between algebraic varieties is a morphism $f : U \to Y$ from a non-empty open subset $U$ of $X$. Since $f$ might not be defined on the whole $X$, such a map is denoted by a dashed arrow in this book. If there is another non-empty open subset $U'$ and a morphism $f' : U' \to Y$ which coincides with $f$ on $U \cap U'$, then we consider $f = f'$ as the same rational map. The *domain of definition* of a rational map $f$ is defined to be the largest $U$ such that there is a morphism $f : U \to Y$ representing $f$.

The *graph* $\Gamma_f$ of a rational map $f : X \dashrightarrow Y$ is defined to be the closure of the graph $\Gamma \subset U \times Y$ of the morphism $f : U \to Y$ in $X \times Y$.

A rational map $f : X \dashrightarrow Y$ is said to be a *birational map* if there exist non-empty open subsets $U, V$ on $X, Y$ such that $f$ induces an isomorphism $U \cong V$. In this situation, the inverse map $f^{-1} : Y \dashrightarrow X$ is also a birational map.

A morphism $f : X \to Y$ is said to be a *birational morphism* if it is a birational map. If $U$ is the largest open subset of $X$ on which $f$ induces an isomorphism $U \cong V$, then $\mathrm{Exc}(f) = X \setminus U$ is called the *exceptional set* of $f$. In this situation, $V$ is the domain of definition of $f^{-1}$. A prime divisor contained in the exceptional set is called an *exceptional divisor* over $Y$ or an $f$-*exceptional divisor*. Generally, a divisor whose support is contained in the exceptional set is also called an exceptional divisor over $Y$ or an $f$-exceptional divisor.

$X$ and $Y$ are said to be *birationally equivalent* if there exists a birational map $f : X \dashrightarrow Y$. In this case, we also say that one is a *birational model* to the other.

For a morphism $f : Y \to X$ and a closed subset $D$ of $X$, the inverse image $f^{-1}(D)$ is a closed subset of $Y$. In this book, $f^{-1}(D)$ only means the set-theoretic inverse image, and we do not consider its scheme structure. However, for a divisor we can define its direct image and inverse image as the following.

First, we define the *inverse image* or *pullback* of a Cartier divisor. For a morphism $f: Y \to X$ and an invertible sheaf $L$ on $X$, we can always define the pullback $f^*L$ which is an invertible sheaf on $Y$. On the other hand, for a Cartier divisor $D$ on $X$, we can define its pullback only if the image $f(Y)$ is not contained in the support of $D$. In this situation, the pullback $f^*D$ is defined by pulling back the local equation of $D$. If $D$ is given by a rational section $s$ of the invertible sheaf $\mathcal{O}_X(D)$ by $\mathrm{div}(s) = D$, then the pullback $f^*D$ is given by the rational section $f^*s$ of the invertible sheaf $f^*\mathcal{O}_X(D)$.

For an **R**-Cartier divisor $D$, if we write it as an **R**-linear combination of Cartier divisors $D = \sum d_i D_i$, then we can define the pullback by $f^*D = \sum d_i f^*D_i$. Here $D_i$ are Cartier divisors, not prime divisors. In other words, the pullback of **R**-Cartier divisors can be defined by extending the coefficients of the pullback map $f^*: \mathrm{Div}(X) \to \mathrm{Div}(Y)$ of Cartier divisors. Note that this definition does not depend on the expression of $D$. The pullback $f^*D$ is also called the *total transform* of $D$.

On the other hand, we cannot define the pullback for a general divisor which is not an **R**-Cartier divisor. However, if the morphism $f: Y \to X$ is a birational map, we can define another form of "pullback" (the strict transform by the inverse map $f^{-1}$) as the following.

Let $f: X \dashrightarrow Y$ be a birational map and let $D$ be a prime divisor on $X$. For the domain of definition $U$, if $D \cap U \neq \emptyset$, then the image $(f|_U)(D \cap U)$ is a locally closed subvariety of $Y$. If its closure is a prime divisor on $Y$, then we denote the closure by $f_*D$; if $D \cap U = \emptyset$ or the image $(f|_U)(D \cap U)$ has codimension at least 2, then we set $f_*D = 0$. Here $f_*D$ is called the *strict transform* or *birational transform* of $D$. Generally for **R**-divisors, the definition can be extended by linearity $f_*(\sum d_i D_i) = \sum d_i f_*(D_i)$ and we consider the linear map $f_*: Z^1(X)_\mathbf{R} \to Z^1(Y)_\mathbf{R}$ by extending the coefficients.

**Example 1.2.1** For a birational projective morphism $f: Y \to X$ and any prime divisor $D$ on $X$, the strict transform $f_*^{-1}D$ on $Y$ is again a prime divisor, which is not 0.

In fact, the inverse map $f^{-1}$ is well defined at the generic point of $D$, and there is no prime divisor contracted by $f^{-1}$, hence the strict transform is a prime divisor.

**Remark 1.2.2** A birational map $f: X \dashrightarrow Y$ between normal algebraic varieties induces an isomorphism between function fields $k(X) \cong k(Y)$. For a prime divisor $D$ on $X$ whose strict transform $f_*D$ is nonzero, this isomorphism identifies the local rings at generic points of $D$ and $f_*D$.

## 1.2 Rational Maps and Birational Maps

When we consider birationally equivalent algebraic varieties as a whole, we identify the divisors defining the same discrete valuation ring, which is equivalent to identifying prime divisors connected by strict transforms.

Similarly, for a subvariety $Z$ of higher codimension, we can define the strict transform $f_* Z$ in a similar way: If $Z \cap U \neq \emptyset$ and $f|_U$ induces an isomorphism at the generic point of $Z$, then we define $f_* Z$ to be the closure of $(f|_U)(Z \cap U)$ in $Y$. We refer to Section 1.4 for the definition in general case.

A birational map $f : X \dashrightarrow Y$ is said to be *surjective in codimension* 1 if the map $f_* : Z^1(X) \to Z^1(Y)$ is surjective, that is, for any prime divisor $E \subset Y$ there is a prime divisor $D$ on $X$ such that $E = f_* D$. Moreover, it is said to be *isomorphic in codimension* 1 if $f_* : Z^1(X) \to Z^1(Y)$ is bijective. The minimal model theory mainly deals with the phenomenon in codimension 1, so these maps play important roles.

**Example 1.2.3** A classical example of birational maps is a *blowup*. In this book, blowing up along a smooth *center* is important. A blowup is obtained by gluing the following local construction.

(1) Define the rational map $f : X = \mathbf{A}^n \dashrightarrow Y = \mathbf{P}^{r-1}$ from the $n$-dimensional affine space to a projective space by $f(x_1, \ldots, x_n) = [x_1 : \ldots : x_r]$. Let $Z$ be the linear subspace of $X$ defined by $x_1 = \cdots = x_r = 0$, then the domain of definition of $f$ is $U = X \setminus Z$.

The graph $X' \subset X \times Y$ of $f$ is defined by $x_i y_j = x_j y_i$ $(1 \leq i, j \leq r)$, where $y_1, \ldots, y_r$ are the homogeneous coordinates of $Y$. The first projection $p : X' \to X$ is the blowup along center $Z$. $E = p^{-1}(Z)$ is the exceptional set of the birational morphism $p$, which is a prime divisor. $p$ induces an isomorphism $X' \setminus E \to X \setminus Z$. Moreover, $E \cong Z \times \mathbf{P}^{r-1}$.

In this case, $p$ is surjective in codimension 1, but $p^{-1}$ is not.

(2) Let $X_1$ be a subvariety of $X$ which is not contained in $Z$. The strict transform $X_1' = p_*^{-1}(X_1)$ of $X_1$ is the closure of $p^{-1}(X_1 \setminus Z)$. In this case, $p_1 = p|_{X_1'} : X_1' \to X_1$ is the blowup of $X_1$ along center $Z \cap X_1$. In particular, the case $Z \subset X_1$ is important. If $X_1 \subset Z$, we can think $X_1' = \emptyset$, in other words, the variety disappears after the blowing up.

If $X_1 \not\subset Z$, $p_1$ is a birational morphism. However, the exceptional set $\mathrm{Exc}(p_1)$ does not necessarily coincide with $E \cap X_1'$. For example, consider $n = 4, r = 2, X_1 \subset \mathbf{A}^4$ is the subvariety defined by the equation $x_1 x_3 + x_2 x_4 = 0$. This is the situation in Example 1.1.4(2). In this case, $Z \subset X_1$, the exceptional set $C$ of $p_1 : X_1' \to X_1$ is isomorphic to $\mathbf{P}^1$, and $p_1(C)$ is the origin. Hence $p_1$ is isomorphic in codimension 1, and so is $p_1^{-1}$.

14  *1 Algebraic Varieties with Boundaries*

**Example 1.2.4** Consider the situations in Example 1.1.4.

(1) For a **Q**-Cartier Weil divisor which is not Cartier, the pullback might not be a Weil divisor but only a **Q**-divisor.

The blowup $f\colon X' \to X$ of $X$ along the origin $Z = (0,0,0)$ as the center gives a resolution of singularity. The exceptional set $C \subset X'$ is isomorphic to $\mathbf{P}^1$. We have $f^*D = f_*^{-1}D + \frac{1}{2}C$.

The projection formula $(f^*D \cdot C) = (D \cdot f_*C)$ stated later (before Proposition 1.4.3) can be confirmed by the following facts: $(f_*^{-1}D \cdot C) = 1$, $(C^2) = -2$, and $f_*C = 0$.

(2) Non-**Q**-Cartier divisors cannot be pulled back according to the projection formula.

Consider the blowup $p_1\colon X_1' \to X_1$ at the end of Example 1.2.3(2). We change the notation by $f\colon X' \to X$. Then $X'$ is smooth. As the exceptional set $C \subset X'$ is isomorphic to $\mathbf{P}^1$ which is only 1-dimensional, $p_1$ is isomorphic in codimension 1.

If the pullbacks $f^*D_1, f^*D_2$ of $D_1, D_2$ exist, they would have to coincide with the strict transforms $f_*^{-1}D_1, f_*^{-1}D_2$ since there are no other prime divisors in the supports of $f^{-1}(D_1), f^{-1}(D_2)$. However, intersecting with $C$, we have $(f_*^{-1}D_1 \cdot C) = -1$ and $(f_*^{-1}D_2 \cdot C) = 1$. Since $f_*C = 0$, this violates the projection formula $(f^*D \cdot C) = (D \cdot f_*C)$ which holds for pullbacks.

A coherent sheaf $F$ on an algebraic variety $X$ is said to be *generated by global sections* if the natural homomorphism $H^0(X, F) \otimes \mathcal{O}_X \to F$ is surjective.

For a Cartier divisor $D$, its *complete linear system* is defined by $|D| = \{D' \mid D \sim D' \geq 0\}$, and its *base locus* is defined by $\mathrm{Bs}\,|D| = \bigcap_{D' \in |D|} \mathrm{Supp}(D')$. When $\mathrm{Bs}\,|D| = \emptyset$, $|D|$ is said to be *free*, which is equivalent to that the corresponding coherent sheaf $\mathcal{O}_X(D)$ is generated by global sections.

$D$ is also said to be free if $|D|$ is free, and $D$ is said to be *semi-ample* if there exists a positive integer $m$ such that $mD$ is free.

More generally, a finite-dimensional linear subspace $V \subset H^0(X, D)$ corresponds to a (not necessarily complete) *linear system* $\Lambda = \{\mathrm{div}(s) \mid s \in V \setminus \{0\}\}$. As an algebraic variety, $\Lambda$ is isomorphic to the projective space $\mathbf{P}(V^*) := (V \setminus \{0\})/k^*$. The base locus of $\Lambda$ is defined similarly by $\mathrm{Bs}\,\Lambda = \bigcap_{D' \in \Lambda} \mathrm{Supp}(D')$, and $\Lambda$ is said to be free if $\mathrm{Bs}\,\Lambda$ is empty, which is equivalent to that the natural homomorphism $V \otimes \mathcal{O}_X \to \mathcal{O}_X(D)$ is surjective.

The *fixed part* of a linear system $\Lambda$ is the effective divisor $F = \min_{D' \in \Lambda} D'$. In other words, $F$ is the maximal divisor such that $F \leq D'$ for all $D' \in \Lambda$.

## 1.2 Rational Maps and Birational Maps

In this case, the image of the natural injection $H^0(X, D - F) \to H^0(X, D)$ contains $V$. Being viewed as a subspace of $H^0(X, D - F)$, $V$ corresponds to the linear system $\Lambda' = \{D' - F \mid D' \in \Lambda\}$, which is called the *movable part* of $\Lambda$. We write $\Lambda = \Lambda' + F$. By definition, the support of $F$ coincides with the codimension 1 components of Bs $\Lambda$.

A non-empty linear system $\Lambda$ induces a rational map $\Phi_\Lambda : X \dashrightarrow \mathbf{P}(V) := (V^* \setminus \{0\})/k^*$ to its dual projective space. The domain of definition of $\Phi_\Lambda$ contains $U = X \setminus \text{Bs}\,\Lambda$; for $P \in U$, $\Phi_\Lambda(P)$ is the point in $\mathbf{P}(V)$ corresponding to the hyperplane $\{s \in V \mid s(P) = 0\}$ of $V$. In other words, if we take a basis $s_1, s_2, \ldots, s_m \in V$, then we can define $\Phi_\Lambda(P) = [s_1(P): s_2(P): \cdots : s_m(P)] \in \mathbf{P}(V)$. Note that here $s_i(P)$ is not a well-defined value, but $[s_1(P): s_2(P): \cdots : s_m(P)]$ is a well-defined point as long as $P \in U$. In particular, when $\Lambda$ is free, $\Phi_\Lambda$ is a morphism. The rational map given by the movable part of a linear system coincides with the rational map given by the original linear system.

In general, for a morphism $f : Y \to X$ and a linear system $\Lambda$ on $X$, the *pullback* is defined by $f^*\Lambda = \{f^*D' \mid D' \in \Lambda\}$. If there is a morphism to a projective space, a free linear system can be obtained by pulling back the linear system consisting of all hyperplanes.

The base locus of a linear system can be removed in the following sense:

**Proposition 1.2.5** *Let $\Lambda$ be a linear system of Cartier divisors on a normal algebraic variety $X$. Then there exists a birational projective morphism $f : Y \to X$ from a normal algebraic variety $Y$ such that the pullback has the form $f^*\Lambda = \Lambda_1 + F$, where $F$ is the fixed part of $f^*\Lambda$ and the linear system $\Lambda_1$ is free.*

*Proof* Let $V \subset H^0(X, D)$ be the linear subspace corresponding to $\Lambda$. The image of the natural map $V \otimes \mathcal{O}_X \to \mathcal{O}_X(D)$ can be written as $I\mathcal{O}_X(D)$, where $I$ is an ideal sheaf on $X$. Take $f$ to be the normalization of the blowup along $I$, then the inverse image ideal sheaf $I\mathcal{O}_Y$ is an invertible sheaf on $Y$. Then $I\mathcal{O}_Y(f^*D)$ is the image of the natural map $V \otimes \mathcal{O}_Y \to \mathcal{O}_Y(f^*D)$, so it can be written as the form $\mathcal{O}_Y(f^*D - F)$ for some effective divisor $F$. Since the natural map $V \otimes \mathcal{O}_Y \to \mathcal{O}_Y(f^*D - F)$ is surjective, the linear system $\Lambda_1 = f^*\Lambda - F$ is free and $F$ is the fixed part of $f^*\Lambda$. $\qquad\square$

For an $\mathbf{R}$-divisor $D$ on a normal proper algebraic variety $X$, the set of global sections $H^0(X, \llcorner D \lrcorner)$ is a finite-dimensional $k$-linear space. Considering all positive integer multiples $mD$ of $D$ and taking a direct sum, we define the *section ring* of $D$ by

$$R(X, D) = \bigoplus_{m=0}^{\infty} H^0(X, \llcorner m D \lrcorner).$$

Here $m$ runs over all nonnegative integers. It admits a graded $k$-algebra structure defined by

$$H^0(X, \llcorner m D \lrcorner) \otimes H^0(X, \llcorner m' D \lrcorner) \to H^0(X, \llcorner (m + m') D \lrcorner)$$

since

$$\llcorner m D \lrcorner + \llcorner m' D \lrcorner \leq \llcorner (m + m') D \lrcorner.$$

The *Iitaka–Kodaira dimension* $\kappa(X, D)$ of an **R**-divisor can be defined by the transcendental degree of the section ring:

$$\kappa(X, D) = \begin{cases} \text{tr.deg}_k \, R(X, D) - 1 & \text{if } R(X, D) \neq k; \\ -\infty & \text{otherwise.} \end{cases}$$

The Iitaka–Kodaira dimension takes value among $-\infty, 0, 1, \ldots, n = \dim X$. In particular, when it takes the maximal value, that is, when $\kappa(X, D) = \dim X$, $D$ is said to be *big*. For example, ample divisors are big.

If $R(X, D) = k$, that is, $H^0(X, \llcorner m D \lrcorner) = 0$ for any $m > 0$, then $\kappa(X, D)$ is defined to be $-\infty$ instead of $-1$. The reason is the following lemma:

**Lemma 1.2.6** ([47, Theorem 10.2], [116, Theorem II.3.7]) *There exist positive real numbers* $c_1, c_2$ *such that for any sufficiently large and sufficiently divisible integer m,*

$$c_1 m^{\kappa(X,D)} \leq \dim H^0(X, \llcorner m D \lrcorner) \leq c_2 m^{\kappa(X,D)}.$$

**Remark 1.2.7** The canonical ring is the section ring of the canonical divisor, which is proved to be finitely generated for smooth projective varieties ([16]), and one of the main goals of this book is to explain the proof. However, in general, the section ring $R(X, D)$ of a divisor $D$ is not necessarily finitely generated. There exist examples such that the *anti-canonical ring* (i.e. the section ring of the *anti-canonical divisor* $-K_X$) of a 2-dimensional variety is not finitely generated ([125], see also Example 2.4.8). Also, the anti-canonical ring $R(X, -K_X)$ is not a birational invariant.

The relative version is as follows. Let $f \colon X \to S$ be a proper morphism from a normal algebraic variety. The *relative global sections* of a coherent sheaf $F$ on $X$ are given by the direct image sheaf $f_* F$. $F$ is said to be *generated by relative global sections* if the natural homomorphism $f^* f_* F \to F$ is surjective.

A Cartier divisor $D$ on $X$ is said to be *relatively free* if the corresponding coherent sheaf $\mathcal{O}_X(D)$ is generated by relative global sections. $D$ is said to be *relatively semi-ample* if there exists a positive integer $m$ such that the multiple $mD$ is relatively free.

For an **R**-divisor $D$ on $X$, the direct image sheaf $f_*(\mathcal{O}_X(\llcorner D \lrcorner))$ is a coherent $\mathcal{O}_S$-module. Considering all positive integer multiples $mD$ of $D$ and taking a direct sum, we define the *relative section ring* of $D$ by

$$R(X/S, D) = \bigoplus_{m=0}^{\infty} f_*(\mathcal{O}_X(\llcorner mD \lrcorner)),$$

which is a graded $\mathcal{O}_S$-algebra.

The *relative Iitaka–Kodaira dimension* is defined by the Iitaka–Kodaira dimension of the generic fiber. Here we always assume that $f$ is surjective with irreducible geometric generic fiber, and define

$$\kappa(X/S, D) = \kappa(X_{\bar{\eta}}, D|_{X_{\bar{\eta}}}).$$

Here $X_\eta$ is the *generic fiber* which is the fiber of $f$ over the generic point $\eta$ of $S$ and $X_{\bar{\eta}}$ is the *geometric generic fiber* which is the base change of $X_\eta$ to the algebraic closure of $k(S)$. $D$ is said to be *relatively big* or $f$-*big* if $\kappa(X/S, D) = \dim X_{\bar{\eta}}$. In Section 1.5.1, we will give an equivalent definition for (relative) bigness using Kodaira's lemma (Corollary 1.5.10).

## 1.3 Canonical Divisors

A normal algebraic variety $X$ is automatically associated with a Weil divisor $K_X$ which is called the canonical divisor. $K_X$ is the key player of this book. The canonical ring is the section ring of the canonical divisor. The minimal model program (MMP) is a sequence of operations that "minimizes" the canonical divisor.

As $X$ is normal, the singular locus $\mathrm{Sing}(X)$ is a closed subset of $X$ of codimension at least 2. Since the complement set $U = X \setminus \mathrm{Sing}(X)$ is smooth, the sheaf of differentials $\Omega^1_{X/k}$ is a locally free sheaf of rank $n = \dim X$ over $U$. The determinant $\omega_U = \det(\Omega^1_{X/k}|_U)$ is an invertible sheaf on $U$. Taking a nonzero rational section $\theta_U$ of $\omega_U$, we get a canonical divisor $K_U = \mathrm{div}(\theta_U)$ of $U$. Since $X \setminus U$ contains no prime divisors of $X$, the restriction map of divisors $Z^1(X) \to Z^1(U)$ is bijective. Denote by $K_X \in Z^1(X)$ the corresponding divisor of $K_U \in Z^1(U)$, which is called the *canonical divisor* of $X$.

**Remark 1.3.1** (1) By construction, $K_X$ depends on the choice of $\theta_U$. However, traditionally our discussions proceed as if the canonical divisor is a fixed one. Nevertheless, in this book, all discussions are independent of the choice of $\theta_U$.

On the other hand, the corresponding divisorial sheaf $\omega_X = \mathcal{O}_X(K_X)$ is uniquely determined. It is called the *canonical sheaf*. The canonical sheaf $\omega_X$ is a natural subject. However, since we consider the pair with an **R**-divisor called the "boundary divisor," the canonical divisor is easier to handle.

(2) In this book, the following situation appears frequently: Let $f : Y \to X$ be a birational morphism between normal algebraic varieties and let $B$ be an **R**-divisor on $X$ such that $K_X + B$ is **R**-Cartier. Consider the pullback $f^*(K_X + B)$. By using the isomorphism between function fields $f^* : k(X) \to k(Y)$, we can take the same rational differential form $\theta$ which defines $K_X$ and $K_Y$ (in particular, $K_X = f_* K_Y$), then the **R**-divisor $C$ can be defined by the equation $f^*(K_X + B) = K_Y + C$. Here $C$ is uniquely determined as the sum of the strict transform $f_*^{-1} B$ and an **R**-divisor supported on the exceptional set of $f$.

We will discuss general boundary divisors later, here we first consider the case when $X$ is a smooth algebraic variety and $B = \sum B_i$ is a normal crossing divisor. Denote $n = \dim X$. The sheaf of differentials $\Omega_X^1(\log B)$ with at most *logarithmic poles* along $B$ is naturally defined as a locally free sheaf of rank $n$ with the following property. For any closed point $P \in X$, choose a regular system of parameters $x_1, \ldots, x_n$ of the local ring $\mathcal{O}_{X,P}$ such that the local equation of $B$ is $x_1 \cdots x_r = 0$ for some integer $r$. In this case, the stalk $\Omega_X^1(\log B)_P$ is a free $\mathcal{O}_{X,P}$-module with basis $dx_1/x_1, \ldots, dx_r/x_r, dx_{r+1}, \ldots, dx_n$.

The determinant $\Omega_X^n(\log B)$ of $\Omega_X^1(\log B)$ is isomorphic to $\mathcal{O}_X(K_X + B)$. Therefore, $K_X + B$ is called the logarithmic canonical divisor or just *log canonical divisor*. This is the origin of the terminology "log."

In general, a log canonical divisor $K_X + B$ is a sum of the canonical divisor and an effective **R**-divisor. Usually certain conditions on singularities will be imposed on the pair $(X, B)$, which will be discussed in Sections 1.10 and 1.11. The *log canonical ring* is defined to be $R(X, K_X + B)$, and the *log Kodaira dimension* is defined to be $\kappa(X, K_X + B)$.

Let $X$ be a smooth projective variety. $R(X) = R(X, K_X)$ is the *canonical ring* of $X$. $P_m(X) = \dim H^0(X, mK_X)$ is called the *m-genus*, which is an important birational invariant having been studied for a long time. Its growth order $\kappa(X, K_X)$ is called the *Kodaira dimension*, sometimes simply denoted by $\kappa(X)$. $X$ is said to be of *general type* if $K_X$ is big.

## 1.3 Canonical Divisors

When working with induction on dimensions, one key is the adjunction formula.

Let $D$ be a smooth prime divisor on a smooth algebraic variety $X$. Then the log canonical divisor and the canonical divisor of the prime divisor satisfy the following *adjunction formula*:

$$(K_X + D)|_D = K_D.$$

In this formula, $K_X|_D$ and $D|_D$ have no natural meaning, but the adjunction itself is given by the map

$$\mathrm{Res}_D : \Omega_X^n(\log D) \to \Omega_D^{n-1},$$

which is induced by the residue map

$$\mathrm{Res}_D : \Omega_X^1(\log D) \to \mathcal{O}_D.$$

The residue map is a natural map which is independent of the choice of coordinates. Therefore, the adjunction formula is also a natural formula. Note that this adjunction formula still holds if $D$ is normal and $D \cap \mathrm{Sing}(X)$ has codimension at least 2 in $D$, since we can first apply the above adjunction formula to $D \setminus \mathrm{Sing}(X) \subset X \setminus \mathrm{Sing}(X)$, then extend it to $D$ by the normality.

When $D$ is not a prime divisor but a normal crossing divisor, if we take an irreducible component $D_1$ of $D$ and write $E = (D - D_1)|_{D_1}$, then we have the adjunction formula

$$(K_X + D)|_{D_1} = K_{D_1} + E.$$

Here the restriction $E$ is well defined since the intersection of $D - D_1$ and $D_1$ is of codimension 1 on $D_1$.

More generally, we can consider the adjunction formula as a relation between canonical divisors of relevant varieties. For example, consider a surjective finite morphism $f : Y \to X$ between smooth algebraic varieties, which is a ramified cover whose ramification locus is a smooth prime divisor $D$ on $X$ with ramification index $m$. The set-theoretic inverse image $E = f^{-1}(D)$ is a smooth prime divisor on $Y$ and $f^*D = mE$. In this case, the *ramification formula* or the adjunction formula with respect to the ramification is the following:

$$K_Y = f^*K_X + (m - 1)E.$$

If written as

$$K_Y = f^*\left(K_X + \frac{m - 1}{m}D\right),$$

20                    1 Algebraic Varieties with Boundaries

then it looks like the adjunction formula for subvarieties. The latter formula is
the origin of considering log canonical divisors with boundary divisors with
rational coefficients. Also, if you write

$$K_Y + E = f^*(K_X + D),$$

you will find that "ramification is killed by log setting."

As another example of the adjunction formula, consider the blowup of an
$n$-dimensional smooth algebraic variety $X$ along an $r$-codimensional smooth
subvariety $Z$. The blowup $f : Y \to X$ is a birational morphism with
exceptional set $E$, which is a prime divisor isomorphic to a $\mathbf{P}^{r-1}$-bundle over
$Z$. The changing of canonical divisors is given by

$$K_Y = f^*K_X + (r-1)E.$$

As shown in the following example, if $X$ is a singular normal algebraic
variety and $D$ is a prime divisor on $X$ intersecting $\mathrm{Sing}(X)$ such that
$D \cap \mathrm{Sing}(X)$ contains an irreducible component of codimension 1 on $D$, then
the singularities contribute to the adjunction formula. This phenomenon is
called the *subadjunction formula*, which is very important.

**Example 1.3.2** Let $X$ be the quadric surface defined by the equation
$xy + z^2 = 0$ in the projective space $\mathbf{P}^3$ with homogeneous coordinates
$x, y, z, w$. $X$ has a singularity at $[0 : 0 : 0 : 1]$. Let $H$ be a hyperplane
section of $X$, then $K_X \sim -2H$.

The projective line $L$ defined by $x = z = 0$ is a prime divisor on $X$. We have
$\mathrm{div}(x) = 2L$ on $X$, hence $L \sim_{\mathbf{Q}} \frac{1}{2}H$. Therefore, $(K_X + L)|_L \sim_{\mathbf{Q}} -\frac{3}{2}H|_L$
since $L|_L \sim_{\mathbf{Q}} \frac{1}{2}H|_L$. On the other hand, $K_L \sim -2H|_L$. Therefore, we have
the subadjunction formula $(K_X + L)|_L = K_L + \frac{1}{2}H|_L$ (see Remark 1.11.14).

## 1.4 Intersection Numbers and Numerical Geometry

Problems in algebraic geometry are equivalent to solving systems of polynomial equations, which are highly nonlinear. *Numerical geometry* attempts
to linearize those problems using intersection numbers. In the following two
sections, we explain basic definitions in numerical geometry. In Chapter 2,
we will explain the basepoint-free theorem and the cone theorem which are
important theorems in numerical geometry. The explanation here is according
to Kleiman ([80]). We refer to the original paper for the proof of the ampleness
criterion.

## 1.4 Intersection Numbers and Numerical Geometry

All definitions here will be for a proper morphism $f : X \to S$ between algebraic varieties over a field $k$. In the case $S = \operatorname{Spec} k$, the definitions are for a proper algebraic variety $X$. We use words "relative" or "over $S$" for all definitions in this section. In the case $S = \operatorname{Spec} k$, those words can be removed. For simplicity, one can just consider $S = \operatorname{Spec} k$ and ignore the word "relative," the context will be almost the same. However, it is indispensable to consider the relative version in applications.

In the following definition, $k$ is an arbitrary field, and $X$ is of finite type over $k$, not necessarily irreducible or reduced. However, in this book when considering Cartier divisors, $X$ is always assumed to be a normal algebraic variety.

A closed subvariety $Z$ on $X$ is called a *relative subvariety* over $S$ if $f(Z)$ is a closed point of $S$. In particular, if $\dim Z = 1$, it is called a *relative curve* over $S$. Denote $\dim Z = t$ and take $t$ invertible sheaves $L_1, \ldots, L_t$ on $X$. Then the *intersection number* $(L_1 \cdots L_t \cdot Z)$ is defined as the coefficient of the following polynomial ([80, p. 296])

$$\chi(Z, L_1^{\otimes m_1} \otimes \cdots \otimes L_t^{\otimes m_t} \otimes \mathcal{O}_Z) = (L_1 \cdots L_t \cdot Z) m_1 \cdots m_t + (\text{other terms}).$$

Here $m_1, \ldots, m_t$ are variables with integer values and

$$\chi(Z, \bullet) = \sum (-1)^p \dim_k H^p(Z, \bullet)$$

is the *Euler–Poincaré characteristic*. Here $X$ itself is not necessarily proper, but $Z$ is proper as $f(Z)$ is a point, hence the cohomology groups are finite-dimensional.

The intersection number $(L_1 \cdots L_t \cdot Z)$ takes integer value and it is a symmetric $t$-linear form with respect to $L_1, \ldots, L_t$ ([80, p. 296]). That is, it is independent of the order of $L_i$ and

$$((L_1^{\otimes n_1} \otimes L_1'^{\otimes n_1'}) \cdots L_t \cdot Z) = n_1(L_1 \cdots L_t \cdot Z) + n_1'(L_1' \cdots L_t \cdot Z).$$

For Cartier divisors $D_1, \ldots, D_t$, define

$$(D_1 \cdots D_t \cdot Z) = (\mathcal{O}_X(D_1) \cdots \mathcal{O}_X(D_t) \cdot Z).$$

In particular, when $\dim Z = 1$, taking $\nu : Z^\nu \to Z$ to be the normalization where $Z^\nu$ is a smooth projective curve, then by the Riemann–Roch theorem,

$$(D_1 \cdot Z) = \deg_{Z^\nu}(\nu^*(\mathcal{O}_X(D_1)|_Z)).$$

When $Z = X$, we simply write $(D_1 \cdots D_t) = (D_1 \cdots D_t \cdot X)$. If, moreover, all $D_i$ are the same $D$, then write $(D_1 \cdots D_t) = (D^t)$.

By multi-linearity, the definition of $(D_1 \cdots D_t \cdot Z)$ can be extended to the case when $D_i$ are **R**-Cartier divisors, which takes value in real numbers.

**Remark 1.4.1** (1) Here we use Euler–Poincaré characteristic to give a simple definition for intersection numbers, but the correct geometric definition of intersection numbers is by adding up local intersection numbers to get the global intersection number. This is how the number of "intersection points" is defined originally. Using the geometric definition, for effective **R**-Cartier divisors $D_i$ and a $t$-dimensional relative subvariety $Z$, if the intersection $\bigcap_{i=1}^{t} \text{Supp}(D_i) \cap Z$ is non-empty and of dimension 0, then the intersection number is positive, and if the intersection is empty, then the intersection number is 0. These two definitions of intersection numbers coincide.

(2) By using the definition of intersection numbers of divisorial sheaves, we can define the *self-intersection number* of a divisor, which seems to be a weird name. For example, for an effective Cartier divisor $D$ on an $n$-dimensional algebraic variety, the self-intersection number $(D^n)$ can be either positive or nonpositive.

(3) In this book, a *curve* is an irreducible reduced projective variety of dimension 1. The intersection number considered in this book is mainly the intersection number of a Cartier divisor with a curve.

Among all curves, *rational curve* plays a very important role in the minimal model theory (see Sections 2.7 and 2.8). A rational curve is a curve whose normalization is isomorphic to $\mathbf{P}^1$. In general a rational curve might have singularities and not necessarily be isomorphic to $\mathbf{P}^1$ itself.

**Example 1.4.2** The intersection number of a divisor and a curve can be defined if this divisor is a **Q**-Cartier divisor. However, the intersection number is not necessarily an integer if the divisor is not Cartier. In general it cannot be defined if the divisor is not **Q**-Cartier.

Consider $X$, as in Example 1.1.4 or 1.2.4, and let $\bar{X}$ be its compactification in the projective space $\mathbf{P}^3$ or $\mathbf{P}^4$.

(1) $\bar{X}$ is defined by the equation $xy = z^2$ in $\mathbf{P}^3$ with homogeneous coordinates $u, x, y, z$. The compactification $\bar{D}$ of $D$ is a prime divisor defined by $x = z = 0$. In this case, $(\bar{D}^2) = \frac{1}{2}$.

In fact, take a plane $\bar{H}$, then $\bar{H}|_{\bar{X}} \sim \text{div}(x) = 2\bar{D}$ and $(\bar{H} \cdot \bar{D}) = 1$.

(2) $\bar{X}$ is defined by the equation $xy = zw$ in $\mathbf{P}^4$ with homogeneous coordinates $u, x, y, z, w$. The compactifications $\bar{D}_1, \bar{D}_2$ of $D_1, D_2$ are prime divisors defined by $x = z = 0$, $x = w = 0$. Take the curve $C$ defined by $y = z = w = 0$. $\bar{D}_1 + \bar{D}_2$ is a Cartier divisor and $((\bar{D}_1 + \bar{D}_2) \cdot C) = 1$. The blowup $f_1 \colon Y_1 \to \bar{X}$ is isomorphic in codimension 1. If intersection numbers $(\bar{D}_i \cdot C)$ $(i = 1, 2)$ could be defined, by the projection formula stated later (before Proposition 1.4.3), $(\bar{D}_i \cdot C) = (f_{1*}^{-1} \bar{D}_i \cdot f_{1*}^{-1} C)$.

## 1.4 Intersection Numbers and Numerical Geometry 23

The right-hand side can be calculated to be $1, 0$ for $i = 1, 2$. This is absurd since the relations between $\bar{D}_1, \bar{D}_2$ and $C$ are symmetric.

Two invertible sheaves $L, L'$ are said to be *relatively numerically equivalent*, denoted by $L \equiv_S L'$, if $(L \cdot C) = (L' \cdot C)$ for any relative curve $C$. When the base is clear, we just write $L \equiv L'$. The Abelian group consisting of isomorphism classes of all invertible sheaves is denoted by $\mathrm{Pic}(X)$ and the subgroup consisting of all invertible sheaves relatively numerically equivalent to $\mathcal{O}_X$ is denoted by $\mathrm{Pic}^{\tau}(X/S)$. The quotient group $\mathrm{Pic}(X)/\mathrm{Pic}^{\tau}(X/S)$ is a finitely generated Abelian group ([80, p. 323]), which is called the *relative Neron–Severi group*, and is denoted by $\mathrm{NS}(X/S)$. $\rho(X/S) = \mathrm{rank}\,\mathrm{NS}(X/S)$ is called the *relative Picard number*. When $S = \mathrm{Spec}\,k$, it is just called the *Picard number* and is denoted by $\rho(X)$.

If $L_1 \equiv_S \mathcal{O}_X$, then the equality $(L_1 \cdot L_2 \cdots L_t \cdot Z) = 0$ holds for arbitrary $L_2, \ldots, L_t, Z$ ([80, p. 304]). Also, for any coherent sheaf $F$ on a relative subvariety $Z$, $\chi(Z, F) = \chi(Z, F \otimes L_1)$ holds ([80, p. 311]).

Two **R**-Cartier divisors $D, D'$ are said to be *relatively numerically equivalent*, denoted by $D \equiv_S D'$ or $D \equiv D'$, if $(D \cdot C) = (D' \cdot C)$ for any relative curve $C$. The numerical equivalence class of $D$ is denoted by $[D]$. The set of all numerical equivalence classes of **R**-Cartier divisors coincides with $\mathrm{NS}(X/S) \otimes \mathbf{R}$, which is a $\rho(X/S)$-dimensional real vector space and is denoted by $N^1(X/S)$.

If $X$ is a smooth complete complex manifold, $D \equiv D'$ is equivalent to having the same cohomology class $[D] = [D'] \in H^2(X, \mathbf{R})$.

Fix an integer $t$, a finite formal linear sum of $t$-dimensional relative subvarieties $Z = \sum a_j Z_j$ is called a *relative $t$-cycle*. The coefficients $a_i$ can be integers, rational numbers, or real numbers depending on the situation. By linearity, intersection numbers can be defined for relative $t$-cycles. In this book, we only consider the case $t = 1$ or $\dim X - 1$.

Relative 1-cycles $C, C'$ are said to be *numerically equivalent*, denoted by $C \equiv_S C'$, if $(D \cdot C) = (D \cdot C')$ for any Cartier divisor $D$. The set $N_1(X/S)$ of all numerical equivalence classes of relative 1-cycles with real coefficients is a finite-dimensional real vector space. $N_1(X/S)$ and $N^1(X/S)$ are dual linear spaces to each other.

Let $g \colon Y \to X$ be a proper morphism from another algebraic variety. For a relative subvariety $Z$ on $Y$ over $S$, the *direct image* $g_* Z$ as an algebraic cycle is defined as the following: if $\dim g(Z) = \dim Z$, then $g_* Z = [k(Z) : k(g(Z))] g(Z)$; if $\dim g(Z) < \dim Z$, then $g_* Z = 0$. Here $g(Z)$ is the set-theoretic image of $Z$, and $[k(Z) : k(g(Z))]$ is the extension degree of function fields. If $g$ is a birational morphism, then $g_* Z$ coincides with the strict transform defined before. Also, for a relative $t$-cycle $Z = \sum a_j Z_j$, its direct image can be defined as $g_* Z = \sum a_j g_* Z_j$ by linearity.

For a relative $t$-cycle $Z$ and invertible sheaves $L_1, \ldots, L_t$ on $X$, the *projection formula*

$$(g^* L_1 \cdots g^* L_t \cdot Z) = (L_1 \cdots L_t \cdot g_* Z)$$

holds ([80, p. 299]). In this book, we often use this formula for $t = 1$ in which case

$$(g^* L \cdot C) = (L \cdot g_* C).$$

**Proposition 1.4.3** ([80, p. 304]) *Let $f \colon X \to S$ and $g \colon Y \to X$ be two proper morphisms. Consider the pullback $g^* L$ of an invertible sheaf $L$ on $X$.*

*(1) If $L \equiv_S 0$, then $g^* L \equiv_S 0$. Therefore, $g$ induces a natural linear map $g^* \colon N^1(X/S) \to N^1(Y/S)$.*

*(2) When $g$ is surjective, conversely, if $g^* L \equiv_S 0$, then $L \equiv_S 0$, hence the pullback map $g^*$ is injective.*

*Proof* (1) For any relative curve $C'$ on $Y$,

$$\begin{aligned} (g^* L \cdot C') &= (L \cdot g_* C') \\ &= \begin{cases} [k(C') : k(g(C'))](L \cdot g(C')) & \text{if } \dim g(C') = 1, \\ 0 & \text{if } \dim g(C') = 0, \end{cases} \end{aligned}$$

which implies the assertion.

(2) If $g$ is surjective, for any relative curve $C$ on $X$, there exists a relative curve $C'$ on $Y$ such that $C = g(C')$, which proves the assertion. $\square$

On the other hand, let $h \colon S \to T$ be a proper morphism, then the identity map on $\mathrm{Div}(X)$ induces a surjective linear map $(1/h)^* \colon N^1(X/T) \to N^1(X/S)$. By taking the dual, $(1/h)_* \colon N_1(X/S) \to N_1(X/T)$ is injective.

For proper morphisms $g \colon Y \to X$ and $f \colon X \to S$, the composition of $g^* \colon N^1(X/S) \to N^1(Y/S)$ and $(1/f)^* \colon N^1(Y/S) \to N^1(Y/X)$ is 0.

## 1.5 Cones of Curves and Cones of Divisors

Cones and polytopes in finite-dimensional vector spaces play important roles in this book. In Chapter 2, morphisms from algebraic varieties can be constructed by using faces of convex cones (the cone theorem). Also, in Chapter 3, a sequence of rational maps can be analyzed from the behavior of a sequence of polytopes.

## 1.5 Cones of Curves and Cones of Divisors

### 1.5.1 Pseudo-Effective Cones and Nef Cones

We will define the closed convex cone generated by numerical equivalence classes of curves in the real vector space $N_1(X/S)$ and the closed convex cones generated by numerical equivalence classes of effective divisors and nef divisors in the dual space $N^1(X/S)$.

A subset $\mathscr{C}$ in a finite-dimensional vector space $V$ is called a *convex cone* if for any $a, a' \in \mathscr{C}$ and $r > 0$, $a + a' \in \mathscr{C}$ and $ra \in \mathscr{C}$ hold. It is called a *closed convex cone* if moreover it is a closed subset.

For an element $u \in V^*$ in the dual space, define $\mathscr{C}_{u \geq 0} = \{v \in \mathscr{C} \mid (u \cdot v) \geq 0\}$. $\mathscr{C}_{u=0}$ and $\mathscr{C}_{u<0}$ can be defined similarly. The *dual closed convex cone* of a closed convex cone $\mathscr{C}$ is defined by

$$\mathscr{C}^* = \bigcap_{v \in \mathscr{C}} V^*_{v \geq 0} = \{u \in V^* \mid (u \cdot v) \geq 0 \text{ for any } v \in \mathscr{C}\}.$$

As $\mathscr{C}$ is a closed convex cone, $v \in \mathscr{C}$ is equivalent to $(u \cdot v) \geq 0$ for all $u \in \mathscr{C}^*$. That is, $\mathscr{C} = \mathscr{C}^{**}$.

For a morphism $f : X \to S$, an invertible sheaf $L$ on $X$ is said to be *relatively ample*, or *ample over S*, or *f-ample*, if there exists an open covering $\{S_i\}$ of $S$, positive integers $m, N$, and *locally closed* immersions $g_i : X_i = f^{-1}(S_i) \to \mathbf{P}^N \times S_i$ such that $L^{\otimes m}|_{X_i} \cong g_i^* p_1^* \mathcal{O}_{\mathbf{P}^N}(1)$, where $p_1 : \mathbf{P}^N \times S_i \to \mathbf{P}^N$ is the first projection. Here the left-hand side is the $m$th tensor power of $L$, and the right-hand side is the pullback of the invertible sheaf corresponding to a hyperplane section by the first projection and $g_i$. A Cartier divisor $D$ is said to be *relatively ample* if its divisorial sheaf $\mathcal{O}_X(D)$ is relatively ample. A morphism admitting a relatively ample invertible sheaf is said to be *quasi-projective*. In particular, if all immersions $g_i$ are closed immersions, then the morphism is said to be *projective*.

Here we recall the following useful fact. Let $f : X \to S$ and $g : Y \to X$ be two projective morphisms, let $A$ be an $f$-ample Cartier divisor on $X$, and let $B$ be a $g$-ample Cartier divisor on $Y$. Then $ng^*A + B$ is ample over $S$ for sufficiently large $n$ ([44, II.7.10]).

In the following, $X$ is assumed to be normal and the morphism $f : X \to S$ is assumed to be projective.

In general, the convex cone consisting of numerical equivalence classes of all effective **R**-Cartier divisors is neither closed nor open. This is because there might be infinitely many prime divisors showing up when considering a limit of effective divisors in $N^1(X/S)$. The closure of this cone is denoted by $\overline{\text{Eff}}(X/S)$, which is called the *relative pseudo-effective cone*. An **R**-Cartier divisor $D$ is said to be *relatively pseudo-effective* if its numerical equivalence class $[D]$ is contained in $\overline{\text{Eff}}(X/S)$.

The set of interior points of the closed convex cone $\overline{\text{Eff}}(X/S)$ is called the *relative big cone* and is denoted by $\text{Big}(X/S)$. Recall that in Section 1.2, we introduced the definition of an **R**-Cartier divisor $D$ being *relatively big* or $f$-*big*. By Kodaira's lemma later (Corollary 1.5.8), it can be shown that an **R**-Cartier divisor $D$ is relatively big if and only if its numerical equivalence class $[D]$ is contained in $\text{Big}(X/S)$.

An **R**-Cartier divisor $D$ is said to be *relatively nef* or $f$-*nef* if $(D \cdot C) \geq 0$ for any relative curve $C$. This is also called *relatively numerically effective*. "Nef" is an abbreviation, but it is commonly used now. The set of numerical equivalence classes of all nef **R**-Cartier divisors is a closed convex cone in $N^1(X/S)$, which is denoted by $\overline{\text{Amp}}(X/S)$ and called the *relative nef cone*.

The set of interior points of the relative nef cone is called the *relative ample cone* and is denoted by $\text{Amp}(X/S)$. An **R**-Cartier divisor $D$ is said to be *relatively ample* or $f$-*ample* if its numerical equivalence class is contained in $\text{Amp}(X/S)$. This definition will be justified by Kleiman's theorem discussed later (Theorem 1.5.4): For a Cartier divisor $D$, being $f$-ample in this sense is equivalent to being $f$-ample in the original sense.

By definition, the sum of a relatively ample **R**-Cartier divisor and a relatively nef **R**-Cartier divisor is again a relatively ample **R**-Cartier divisor.

In the dual space $N_1(X/S)$, the cone of relative curves is the convex cone generated by numerical equivalence classes of all relative curves, which is in general neither open nor closed. Its closure is called the *closed cone of relative curves*, which is denoted by $\overline{\text{NE}}(X/S)$. By definition, the latter is the dual closed convex cone of the relative nef cone and the relative ample cone:

$$\overline{\text{Amp}}(X/S) = \{u \in N^1(X/S) \mid (u \cdot v) \geq 0 \text{ for all } v \in \overline{\text{NE}}(X/S)\} \text{ and}$$

$$\text{Amp}(X/S) = \{u \in N^1(X/S) \mid (u \cdot v) > 0 \text{ for all } v \in \overline{\text{NE}}(X/S) \setminus \{0\}\}.$$

**Remark 1.5.1** The cones $\overline{\text{Amp}}(X/S)$ and $\overline{\text{NE}}(X/S)$ considered here contain interior points, but contain no linear subspaces. This is because $f : X \to S$ is assumed to be a projective morphism. For example, $\overline{\text{NE}}(X/S)$ contains no lines since the intersection number of a relatively ample divisor with a nonzero element in $\overline{\text{NE}}(X/S)$ is always positive by Theorem 1.5.4. A relatively ample divisor is also called a *polarization* as it gives the positive direction.

The structures of relative nef cones and closed cones of relative curves are important themes of this book.

## 1.5 Cones of Curves and Cones of Divisors

**Proposition 1.5.2** ([80, p. 337]) *Let $f: X \to S$ and $g: Y \to X$ be two projective morphisms, and let $L$ be an invertible sheaf on $X$.*

*(1) If $L$ is $f$-nef, then the pullback $g^*L$ is $(f \circ g)$-nef.*
*(2) If $g$ is surjective and $g^*L$ is $(f \circ g)$-nef, then $L$ is $f$-nef.*
*(3) If $g$ is surjective, then*

$$g^*\overline{\mathrm{Amp}}(X/S) = \overline{\mathrm{Amp}}(Y/S) \cap g^*N^1(X/S).$$

*(4) Assume that $g$ is surjective. If moreover $g$ is a* finite *morphism, then*

$$g^*\mathrm{Amp}(X/S) = \mathrm{Amp}(Y/S) \cap g^*N^1(X/S),$$

*otherwise*

$$g^*\overline{\mathrm{Amp}}(X/S) = \partial\overline{\mathrm{Amp}}(Y/S) \cap g^*N^1(X/S).$$

*Here $\partial$ is the boundary of the closed convex cone.*

*Proof* The proof of (1) and (2) is similar to that of Proposition 1.4.3. (3) follows from (2).

(4) When $g$ is a finite morphism, the pullback of a relatively ample invertible sheaf is again a relatively ample invertible sheaf, hence the former assertion follows. On the other hand, when $g$ is not a finite morphism, the pullback of a relatively ample invertible sheaf is never a relatively ample invertible sheaf, hence the latter assertion follows from (3). $\qquad\square$

It was shown that a nonfinite morphism gives a face of the relative nef cone. Conversely, sometimes it is possible to construct a nonfinite morphism from a face of the relative nef cone; this is the contraction theorem in the minimal model theory.

**Example 1.5.3** (1) Let $X$ be a smooth projective *complex algebraic surface* and let $C$ be a curve on $X$ with negative self-intersection $(C^2) < 0$.

For any curve $C'$ different from $C$, the intersection number is always nonnegative: $(C \cdot C') \geq 0$. Denote by $\mathscr{C}' \subset N_1(X)$ the closed convex cone generated by the numerical equivalence classes of all curves $C'$ different from $C$, then the closed cone of curves $\overline{\mathrm{NE}}(X)$ is generated by $\mathscr{C}'$ and $[C]$.

Since $(C \cdot C') \geq 0$ for all $C' \in \mathscr{C}'$, $[C] \notin \mathscr{C}'$. Therefore, one can see that $[C]$ generates an *extremal ray* of $\overline{\mathrm{NE}}(X)$. Taking the dual, we get a face $F = \overline{\mathrm{Amp}}(X)_{C=0}$ of $\overline{\mathrm{Amp}}(X)$.

According to a result of Grauert ([33]), there exists a compact complex analytic surface $Y$ with only normal singularities and a birational morphism $f: X \to Y$ between complex analytic surfaces such that $C$ is contracted to a point. That is, $f(C)$ is a point and there is an isomorphism

$f: X \setminus C \to Y \setminus f(C)$. However, $Y$ is in general not an algebraic variety. But according to a result of Artin ([6]), if $C \cong \mathbf{P}^1$, then $Y$ is a projective algebraic surface and $f$ becomes a birational morphism between algebraic varieties.

In this sense, it may or may not be possible to construct a morphism from a face of the nef cone.

(2) Let $X$ be an *Abelian variety*, that is, a smooth projective algebraic variety with an algebraic group structure. In this case, any prime divisor $D$ on $X$ is nef, and

$$\mathrm{Amp}(X) = \{v \in N^1(X) \mid (v^n) > 0\}^0.$$

Here $n = \dim X$ and $^0$ on the right-hand side means one of the connected components.

### 1.5.2 Kleiman's Criterion and Kodaira's Lemma

In this subsection, we introduce Kleiman's ampleness criterion. We also prove Kodaira's lemma, which characterizes big divisors.

**Theorem 1.5.4** (*Kleiman's criterion* [80]) *For a projective morphism $f: X \to S$ between algebraic varieties, a Cartier divisor $D$ on $X$ is relatively ample if and only if its numerical equivalence class is contained in the relative ample cone* $\mathrm{Amp}(X/S)$.

**Remark 1.5.5** Kleiman's criterion is a paraphrase of Nakai's criterion for projectivity and ampleness using the language of cones of divisors. In Kleiman's criterion as well as Nakai's criterion, $X$ is not necessarily assumed to be irreducible or reduced. It is also not necessarily assumed to be projective, and whether a proper scheme is projective can be determined by whether $\mathrm{Amp}(X)$ is not empty.

As ampleness is an algebro-geometric property which is nonlinear, we can say that it is linearized by Kleiman's criterion using conditions in numerical geometry. This is a typical example of numerical geometry.

An invertible sheaf $L$ on a projective algebraic variety $X$ induces a functional $h_L$ on the dual space $N_1(X)$. By Kleiman's criterion, $L$ is ample if and only if $h_L$ is positive on the closed cone of curves $\overline{\mathrm{NE}}(X)$.

This condition is strictly stronger than the condition that $h_L(C) = (L \cdot C) > 0$ for any curve $C$. We explain this by the following example:

**Example 1.5.6** (Mumford's example) Let $\Gamma$ be a smooth projective complex algebraic curve of genus at least 2 and let $F$ be a locally free sheaf on $\Gamma$ of rank 2 and of degree 0. The last condition means that $\bigwedge^2 F \equiv \mathcal{O}_\Gamma$. Assume that $F$

## 1.5 Cones of Curves and Cones of Divisors 29

is *stable*, that is, $\deg(M) < 0$ for any invertible subsheaf $M$ of $F$. Such $F$ can be constructed by using unitary representations of the fundamental group $\pi_1(\Gamma)$. In this case, for any surjective morphism $f: C \to \Gamma$ from a smooth projective curve, $f^*F$ is also stable.

Let $X = \mathbf{P}(F)$ be the corresponding $\mathbf{P}^1$-bundle over $\Gamma$ and $L = \mathcal{O}_{\mathbf{P}(F)}(1)$. Let $C_0$ be a curve on $X$. If it is not a fiber of $f$, take $f: C \to \Gamma$ to be the composition of the normalization $g: C \to C_0$ and the projection $C_0 \to \Gamma$. In this case, $g^*L$ is an invertible sheaf which is a quotient of $f^*F$, hence its degree is positive. If $C_0$ is a fiber of $f$, then $(L \cdot C_0) = 1$. That is, the inequality $(L \cdot C_0) > 0$ holds for any curve $C_0$ on $X$. On the other hand, $(L^2) = 0$ since $\deg(F) = 0$, which means that $L$ is not ample.

The following Kodaira's lemma gives a characterization of big divisors.

**Theorem 1.5.7** (*Kodaira's lemma*) *(1) A Cartier divisor $D$ on a normal projective algebraic variety $X$ is big if and only if there exists a positive integer $m$, an ample Cartier divisor $A$, and an effective Cartier divisor $E$ such that $mD = A + E$.*

*(2) For a surjective projective morphism $f: X \to S$ from a normal algebraic variety to a quasi-projective algebraic variety, a Cartier divisor $D$ on $X$ is relatively big if and only if there exists a positive integer $m$, a relatively ample Cartier divisor $A$, and an effective Cartier divisor $E$ such that $mD = A + E$.*

In other words, big divisors are divisors bigger than ample divisors.

*Proof* (1) As ample divisors are big, the condition is sufficient.

Conversely, assume that $D$ is big. Denote $n = \dim X$. Take a very ample Cartier divisor $A$ and a general element in its complete linear system $Y \in |A|$. Consider the following exact sequence:

$$0 \to \mathcal{O}_X(mD - Y) \to \mathcal{O}_X(mD) \to \mathcal{O}_Y(mD|_Y) \to 0.$$

Look at the first part of the corresponding long exact sequence

$$0 \to H^0(X, mD - Y) \to H^0(X, mD) \to H^0(Y, mD|_Y),$$

as $\dim Y = n-1$, the dimension of the last term is bounded by $cm^{n-1}$ for some constant $c$. But by the bigness, the central term goes much larger, so the first term is not 0 for sufficiently large $m$. Hence there exists an effective divisor $E$ with linear equivalence $mD - Y \sim E$. In this case, $mD - E \sim Y$ is ample and the proof is completed.

(2) As the restriction of relatively ample (respectively, effective) divisors on the generic fiber are ample (respectively, effective), the condition is sufficient.

30    *1 Algebraic Varieties with Boundaries*

Conversely, assume that $D$ is relatively big. By the argument of (1), for a relatively ample Cartier divisor $A$, there exists a sufficiently large $m$ such that the direct image sheaf $f_*(\mathcal{O}_X(mD - A)) \neq 0$. Take a sufficiently ample Cartier divisor $B$ on $S$ such that

$$H^0(X, mD - A + f^*B) = H^0(S, f_*(\mathcal{O}_X(mD - A)) \otimes \mathcal{O}_S(B)) \neq 0.$$

Then there exists an effective Cartier divisor $E$ with linear equivalence $mD - A + f^*B \sim E$. In this case, $mD - E \sim A - f^*B$ is relatively ample and the proof is completed. $\qquad\square$

As a corollary, together with Kleiman's criterion, the definition of relative big cones is justified:

**Corollary 1.5.8** *For a surjective projective morphism $f : X \to S$ from a normal algebraic variety to a quasi-projective algebraic variety, a Cartier divisor $D$ on $X$ is relatively big if and only if the numerical equivalence class $[D]$ is contained in the relative big cone $\mathrm{Big}(X/S)$.*

*Proof* By Kleiman's criterion and Kodaira's lemma, $D$ is relatively big if and only if $[D]$ is an interior point of the closed convex cone generated by effective divisors. $\qquad\square$

**Corollary 1.5.9** $\overline{\mathrm{Amp}}(X/S) \subset \overline{\mathrm{Eff}}(X/S)$.

*Proof* As ample divisors are big, we have an inclusion $\mathrm{Amp}(X/S) \subset \mathrm{Big}(X/S)$ between open cones. The conclusion follows by taking closures. $\qquad\square$

Kodaira's lemma can be generalized as the following:

**Corollary 1.5.10** *(1) An **R**-Cartier divisor $D$ on a normal projective algebraic variety $X$ is big if and only if there exists an ample **R**-Cartier divisor $A$, and an effective **R**-Cartier divisor $E$ such that $D = A + E$.*
*(2) For a surjective projective morphism $f : X \to S$ from a normal algebraic variety to a quasi-projective algebraic variety, an **R**-Cartier divisor $D$ on $X$ is relatively big if and only if there exists a relatively ample **R**-Cartier divisor $A$ and an effective **R**-Cartier divisor $E$ such that $D = A + E$.*

*Proof* (1) Assume that $D = A + E$. Then there exists an ample **Q**-Cartier divisor $A'$ and an effective **R**-Cartier divisor $E'$ such that we can write $A = A' + E'$, hence $D$ is big.

### 1.5 Cones of Curves and Cones of Divisors

Conversely, assume that $D$ is big. By the proof of Kodaira's lemma, for sufficiently large $m$, there exists an ample Cartier divisor $A$ and an effective divisor $E$ such that $\llcorner mD\lrcorner = A + E$. Since $mD - \llcorner mD\lrcorner$ is effective, the assertion is proved.

(2) It is similarly deduced from the relative version of Kodaira's lemma. $\square$

**Proposition 1.5.11** *Let* $f : Y \to X$ *be a birational morphism between normal projective algebraic varieties and let* $D$ *be an* **R**-*Cartier divisor on* $X$. *Then* $D$ *is big if and only if the pullback* $f^*D$ *is big.*

*Proof* For a rational function $h \in k(X) \cong k(Y)$, $\mathrm{div}_X(h) + \llcorner mD\lrcorner \geq 0$ is equivalent to $\mathrm{div}_X(h) + mD \geq 0$. Here the subscript $X$ means taking the corresponding divisor on $X$. The latter is equivalent to $\mathrm{div}_Y(h) + mf^*D \geq 0$, which is then equivalent to $\mathrm{div}_Y(h) + \llcorner mf^*D\lrcorner \geq 0$. Therefore, the natural homomorphism $H^0(X, \llcorner mD\lrcorner) \to H^0(Y, \llcorner mf^*D\lrcorner)$ is bijective, and the assertion is concluded. $\square$

**Theorem 1.5.12** ([92, Theorem 2.2.16]) *Let* $X$ *be an* $n$-*dimensional projective algebraic variety and let* $D$ *be a nef* **R**-*Cartier divisor. Then* $D$ *is big if and only if* $(D^n) > 0$.

*Proof* If $D$ is big, then we can write $D = A + E$ for some ample **Q**-divisor $A$ and some effective **R**-divisor $E$. In this case, since $D$ and $A$ are nef,

$$(D^n) = (D^{n-1} \cdot A) + (D^{n-1} \cdot E) \geq (D^{n-1} \cdot A)$$
$$= (D^{n-2} \cdot A^2) + (D^{n-2} \cdot A \cdot E) \geq \cdots \geq (A^n) > 0.$$

Here we use the fact that if $D_1, \ldots, D_n$ are **R**-divisors on $X$ such that $D_1, \ldots, D_{n-1}$ are nef and $D_n$ is either effective or nef, then $(D_1 \cdots D_n) \geq 0$.

Conversely, to show that $D$ is big provided that $D$ is nef and $(D^n) > 0$, we will show the following slightly generalized statement: If for two nef **R**-Cartier divisors $L, M$ we have $(L^n) > n(L^{n-1} \cdot M)$, then $L - M$ is big. The theorem follows by taking $M = 0$.

First, we assume that $L, M$ are ample **Q**-Cartier divisors. We may assume that they are both very ample by taking a common multiple. Taking $m$ general elements $M_i \in |M|$ $(1 \leq i \leq m)$, by the exact sequence

$$0 \to \mathcal{O}_X(m(L - M)) \to \mathcal{O}_X(mL) \to \bigoplus_i \mathcal{O}_{M_i}(mL),$$

the Riemann–Roch theorem, and the Serre vanishing theorem, when $m \to \infty$, we have

$$\dim H^0(X, m(L-M))$$

$$\geq \dim H^0(X, mL) - \sum_{i=1}^{m} \dim H^0(M_i, mL|_{M_i})$$

$$= \frac{(L^n)}{n!}m^n - \sum_{i=1}^{m} \frac{(L^{n-1} \cdot M_i)}{(n-1)!}m^{n-1} + O(m^{n-1})$$

$$= \frac{(L^n) - n(L^{n-1} \cdot M)}{n!}m^n + O(m^{n-1}).$$

Here note that for each $M_i$, the dimension of $H^0(M_i, mL|_{M_i})$ is independent of the choice of $M_i$, and it can be estimated by $O(m^{n-2})$. Therefore, $L - M$ is big.

Then we consider the general case. We may take two sufficiently small ample **R**-Cartier divisors $H, H'$ such that $H' - H$ is big and $L + H, M + H'$ are ample **Q**-Cartier divisors. Here $H, H'$ can be taken sufficiently small in the sense that $((L + H)^n) > n((L + H)^{n-1} \cdot (M + H'))$ holds. Then we already showed that $L + H - M - H'$ is big, which implies that $L - M$ is big. $\qquad \square$

We can investigate how cones of divisors change under birational maps:

**Lemma 1.5.13** *Let $\alpha \colon X \dashrightarrow X'$ be a birational map between normal **Q**-factorial varieties which is isomorphic in codimension 1 and let $f \colon X \to S$ and $f' \colon X' \to S$ be projective morphisms with $f = f' \circ \alpha$.*

*(1) $\alpha$ induces an isomorphism $\alpha_* \colon N^1(X/S) \to N^1(X'/S)$ between real linear spaces.*

*(2) $\alpha_*(\overline{\mathrm{Eff}}(X/S)) = \overline{\mathrm{Eff}}(X'/S)$.*

*(3) If $\alpha$ is not an isomorphism, then $\alpha_*(\mathrm{Amp}(X/S)) \cap \mathrm{Amp}(X'/S) = \emptyset$.*

*Proof* (1) Since $\alpha$ is isomorphic in codimension 1, there is a 1–1 correspondence between prime divisors on $X$ and $X'$. Hence $Z^1(X) \cong Z^1(X')$. Take a divisor $D$ on $X$ and take its strict transform $D' = \alpha_* D$. Applying the desingularization theorem discussed in Section 1.6, there exists a smooth algebraic variety $W$ and birational projective morphisms $g \colon W \to X$ and $g' \colon W \to X'$ such that we can write $g^* D = (g')^* D' + E$, where $g_* E = 0$ and $g'_* E = 0$. Assume that $D \equiv_S 0$, then $g^* D \equiv_S 0$.

In the following we will show that $D' \equiv_S 0$. We may assume that $E \neq 0$ otherwise it is obvious. Write $E = E^+ - E^-$ into the positive part and the negative part. If $E^+ \neq 0$, then by the negativity lemma (Lemma 1.6.3), there exists a curve $C$ contracted by $g'$ such that $(E^+ \cdot C) < 0$ and $(E^- \cdot C) \geq 0$. On the other hand, $((g')^* D' \cdot C) = (D' \cdot g'_* C) = 0$ and $(g^* D \cdot C) = 0$, a contradiction. We can get a contradiction similarly if $E^- \neq 0$.

## 1.6 The Hironaka Desingularization Theorem

(2) follows from (1) as the strict transform of an effective divisor is again effective.

(3) As the intersection is an open cone, if the intersection is non-empty, then there exists a relatively ample divisor $D$ on $X$ such that $\alpha_* D$ is a relatively ample divisor on $X'$. Since $\alpha$ is isomorphic in codimension 1, for any integer $m$, $\alpha_* : f_* \mathcal{O}_X(mD) \to f'_* \mathcal{O}_{X'}(mD')$ is an isomorphism. Therefore,

$$X = \mathrm{Proj}_S \left( \bigoplus_{m=0}^{\infty} f_* \mathcal{O}_X(mD) \right) \cong \mathrm{Proj}_S \left( \bigoplus_{m=0}^{\infty} f'_* \mathcal{O}_{X'}(mD') \right) = X'$$

and $\alpha$ is an isomorphism. $\qquad\square$

## 1.6 The Hironaka Desingularization Theorem

The desingularization theorem was established by Hironaka for algebraic varieties in characteristic 0. Although it is expected that the same theorem holds for positive characteristics and mixed characteristics, it is only proved in dimension 2 and for positive characteristics in dimension 3, while it remains open in the general case. Together with the Kodaira vanishing theorem, they are very important theorems in characteristic 0. Here we introduce the desingularization theorem ([45]) without proof.

**Theorem 1.6.1** (*Hironaka desingularization theorem*) *(1) For any algebraic variety $X$ defined over a field of characteristic 0, there exists a smooth algebraic variety $Y$ and a birational projective morphism $f : Y \to X$.*

*(2) For any algebraic variety $X$ defined over a field of characteristic 0 and a proper closed subset $B$ of $X$, there exists a smooth algebraic variety $Y$, a normal crossing divisor $C$ on $Y$, and a birational projective morphism $f : Y \to X$ with the following properties:*

    *(a) If $B$ is non-empty, then the set-theoretic inverse image $f^{-1}(B)$ is a union of several irreducible components of $C$.*

    *(b) The exceptional set $\mathrm{Exc}(f)$ is a union of several irreducible components of $C$.*

*For each statement, we can assume further the following properties hold:*

*(1') $f$ is isomorphic over the smooth locus $\mathrm{Reg}(X) = X \setminus \mathrm{Sing}(X)$, and the exceptional set $\mathrm{Exc}(f)$ coincides with the set-theoretic inverse image of the singular locus $f^{-1}(\mathrm{Sing}(X))$.*

*(2') $f$ is isomorphic over $\mathrm{Reg}(X, B)$ and the exceptional set $\mathrm{Exc}(f)$ coincides with the set-theoretic inverse image $f^{-1}(\mathrm{Sing}(X, B))$.*

A birational morphism with the property in (1) is called a *resolution of singularities* of the algebraic variety $X$. A birational morphism with the property in (2) is called a *log resolution* of the pair $(X, B)$. For the definition of normal crossing divisors please refer to Section 1.1.

**Remark 1.6.2** (1) If replacing the two conditions for the log resolution by the condition that $f^{-1}(B) \cup \text{Exc}(f)$ is a normal crossing divisor, we call it a *log resolution in weak sense*. This is called a log resolution in some literature. On the other hand, if we assume furthermore that $\text{Exc}(f)$ is the support of an $f$-ample divisor in condition (b), we call it a *log resolution in strong sense*. In this case, the $f$-ample divisor supported on $\text{Exc}(f)$ has negative coefficients according to Lemma 1.6.3 below.

(2) Hironaka's desingularization can be obtained by blowing up along smooth centers finitely many times. Since there exists a relatively ample divisor supported on the exceptional divisor with negative coefficient for a blowup along a smooth center, Hironaka's desingularization obtained in this way is a log resolution in strong sense.

By Theorem 1.6.4, starting from any log resolution, one can construct a log resolution in strong sense by further taking blowups along the exceptional set.

(3) In the latter part of the above theorem, a normal crossing divisor is in the sense of the Zariski topology, which is a "simple normal crossing divisor." It does not hold for normal crossing divisors in the complex analytic sense. For example, take the divisor $B$ defined by the equation $x^2 + y^2 z = 0$ in $X = \mathbf{C}^3$. The singular locus of $B$ is the line defined by $x = y = 0$ and $B$ is a normal crossing divisor in the complex analytic sense if $z \neq 0$. However, the origin $P = (0, 0, 0)$ has the so-called *pinch point* singularity, no blowup which is isomorphic outside $P$ can make $B$ a normal crossing divisor.

(4) The above theorem is proved in Hironaka's original paper ([45]), but it has been shown that there exists a more precise "canonical resolution" in subsequent developments. The canonical resolution admits strong functoriality such that any local isomorphism (isomorphism between two open subsets) of the pair $(X, B)$ lifts to a local isomorphism of $(Y, C)$. However, the canonical resolution is not unique, it is only shown that there exists a universal choice ([11, 46, 140, 142]).

**Lemma 1.6.3** (Negativity lemma) *Let $f: X \to Y$ be a birational projective morphism between normal algebraic varieties and let $D$ be an $\mathbf{R}$-Cartier divisor on $X$ supported in the exceptional set $\text{Exc}(f)$.*

## 1.6 The Hironaka Desingularization Theorem

*(1) If $D$ is nonzero and effective, then there exists a curve $C$ which is contracted by $f$ and passes through a general point of an irreducible component of $D$ such that $(D \cdot C) < 0$.*

*(2) If $D$ is $f$-nef and nonzero, then the coefficients of $D$ are all negative. Furthermore, the support of $D$ coincides with the set-theoretic inverse image $f^{-1}(f(\mathrm{Supp}(D)))$.*

*(3) If $D$ is $f$-ample, then the support of $D$ coincides with $\mathrm{Exc}(f)$.*

*Proof* We may assume that $Y$ is affine. Consider $0 \le i \le \dim f(\mathrm{Supp}(D))$ and $j = \dim X - 2 - i$. Take $Y_i$ by cutting $Y$ by general hyperplane sections $i$ times and take $X_{ij}$ by cutting $f^{-1}(Y_i)$ by general hyperplane sections $j$ times. Since $i + j = \dim X - 2$, $X_{ij}$ is a normal algebraic surface. Let $Y_{ij}$ be the normalization of $f(X_{ij})$, then $f$ induces a birational projective morphism $f_{ij}: X_{ij} \to Y_{ij}$. Note that $D_{ij} = D|_{X_{ij}}$ is an **R**-Cartier divisor supported in the exceptional set $\mathrm{Exc}(f_{ij})$.

(1) Since $D$ is nonzero and effective, so is $D_{ij}$ for some $i, j$. By the Hodge index theorem, applying Corollary 1.13.2 to $\pi: \tilde{X}_{ij} \to Y_{ij}$ and $\pi^* D_{ij}$, where $\tilde{X}_{ij}$ is a resolution of $X_{ij}$, we get $(D_{ij})^2 < 0$. In particular, there exists an irreducible component $C$ of $D_{ij}$ such that $(D_{ij} \cdot C) < 0$. View $C$ as a curve in $X$, we have $(D \cdot C) < 0$. Note that by construction, $C$ comes from cutting an irreducible component of $D$ by hyperplane sections, so such $C$ passes through a general point of an irreducible component of $D$.

(2) We may write $D_{ij} = D_{ij}^+ - D_{ij}^-$ in terms of its positive and negative parts. Since $D_{ij}$ is $f_{ij}$-nef, $(D_{ij}^+)^2 \ge (D_{ij}^+ \cdot D_{ij}) \ge 0$. By the Hodge index theorem (Corollary 1.13.2), $D_{ij}^+ = 0$. Hence the coefficients of $D_{ij}$ are negative. As $i, j$ varies, any coefficient of $D$ appears as the coefficient of some $D_{ij}$. So, the coefficients of $D$ are all negative. If the support of $D$ does not coincide with $f^{-1}(f(\mathrm{Supp}(D)))$, then there is a curve $C$ intersecting $\mathrm{Supp}(D)$ properly such that $f(C)$ is a point. Then $(D \cdot C) < 0$, a contradiction.

(3) By (2), all coefficients of $D$ are negative. If the support of $D$ does not coincide with $\mathrm{Exc}(f)$, then there is a curve $C$ not contained in $\mathrm{Supp}(D)$ such that $f(C)$ is a point. Then $(D \cdot C) \le 0$, a contradiction. $\square$

Let $X$ be a smooth algebraic variety and let $B$ be a normal crossing divisor on $X$. A smooth subvariety $Z$ is called a *permissible center* with respect to the pair $(X, B)$ if the following is satisfied: For the local ring $\mathcal{O}_{X,P}$ at every point $P \in X$, there exists a regular system of parameters $z_1, \dots, z_n$ and integers $r, s, t$ such that the equations of $B, Z$ are $z_1 \cdots z_r = 0$, $z_s = \cdots = z_t = 0$, respectively. Here, $0 \le r \le n$ and $0 \le s \le t \le n$, but there is no specific relation between $r$ and $s, t$.

36         *1 Algebraic Varieties with Boundaries*

The blowup $f : Y \to X$ along a permissible center $Z$ with respect to $(X, B)$ is called a *permissible blowup*. In this case, the exceptional set $E$ is a smooth prime divisor on $Y$ and coincides with the set-theoretic inverse image $f^{-1}(Z)$. The sum $C = f_*^{-1}B + E$ with the strict transform is a normal crossing divisor on $Y$. We have $K_Y = f^*K_X + (t - s)E$ and $f^*B = f_*^{-1}B + \max\{r - s + 1, 0\}E$.

The desingularization theorem also contains the following statement:

**Theorem 1.6.4** ([45]) *Let $X$ be a smooth algebraic variety defined over a field of characteristic 0, let $B$ be a normal crossing divisor on $X$, and let $f : Y \to X$ be a proper birational morphism from another smooth algebraic variety $Y$. Then there exists a sequence of blowups $f_i : X_i \to X_{i-1}$ ($i = 1, \ldots, n$) and a birational morphism $g : X_n \to Y$ with the following properties:*

*(1) $X = X_0$ and $f \circ g = f_1 \circ \cdots \circ f_n$.*

*(2) $f_i$ is a permissible blowup with respect to $(X_{i-1}, B_{i-1})$. Here $B = B_0$ and the normal crossing divisor $B_i$ on $X_i$ is defined inductively by $B_i = f_{i*}^{-1}B_{i-1} + \mathrm{Exc}(f_i)$.*

## 1.7 The Kodaira Vanishing Theorem

The Kodaira vanishing theorem holds only in characteristic 0. There are counterexamples in positive characteristics ([118]). The vanishing theorem and its generalizations are indispensable tools for the minimal model. Here we introduce the Kodaira vanishing theorem ([82]) without proof.

**Theorem 1.7.1** (*Kodaira vanishing theorem*) *Let $X$ be a smooth projective complex algebraic variety and let $D$ be an ample divisor on $X$. Then for any positive integer $p > 0$, $H^p(X, K_X + D) = 0$. Here $K_X$ is the canonical divisor of $X$.*

The Kodaira vanishing theorem is a theorem in complex differential geometry established for a compact complex manifold $X$. Let $L$ be a holomorphic line bundle on a compact complex manifold $X$. $L$ is always endowed with a $C^\infty$ Hermitian metric $h$. The curvature of the corresponding connection of $h$ determines a $C^\infty$ $(1, 1)$-form on $X$. In this case, the following assertion holds by the *Kodaira embedding theorem*:

**Theorem 1.7.2** ([83]) *Let $X$ be a compact complex manifold and let $L$ be a line bundle with a Hermitian metric $h$. If the curvature $\sqrt{-1}\Theta$ is positive definite everywhere, then $X$ has a projective complex algebraic variety structure and $L$ is the line bundle corresponding to an ample divisor.*

## 1.7 The Kodaira Vanishing Theorem

We have the following implications:

| Algebraic geometry | $\Rightarrow$ | Complex differential geometry | $\Rightarrow$ | Numerical geometry |
|---|---|---|---|---|
| Ample divisors | $\Rightarrow$ | Line bundles with positive curvatures | $\Rightarrow$ | Numerically positive divisors |

The feature of the Kodaira vanishing theorem is that the canonical divisor appears in the statement and it provides a more accurate vanishing compared to the Serre vanishing theorem below. This paves the way for geometric applications. To be applied in higher dimensional algebraic geometry, the Kodaira vanishing theorem is greatly generalized and used in many directions, as will be discussed in Section 1.9.

**Remark 1.7.3** The Kodaira vanishing theorem is originally proved for algebraic varieties defined over complex numbers, but it holds also for algebraic varieties defined over any field in characteristic 0, since a field in characteristic 0 which is finitely generated over the prime field $\mathbf{Q}$ can be always embedded into $\mathbf{C}$.

**Theorem 1.7.4** (*Serre vanishing theorem* [126], [44, III.5.2]) *Let X be a projective scheme over a field k, let L be an ample sheaf on X, and let F be a coherent sheaf on X. Then there exists a positive integer $m_0$ such that for any integer $m \geq m_0$, the following assertions hold:*

*(1) $F \otimes L^{\otimes m}$ is generated by global sections.*
*(2) For any positive integer $p > 0$, $H^p(X, F \otimes L^{\otimes m}) = 0$.*

The Serre vanishing theorem holds without conditions on characteristics of the field $k$ and singularities of $X$. It has much more applicability than the Kodaira vanishing theorem, but it is weaker.

The log version of the Kodaira vanishing theorem can be proved by the adjunction formula ([117]):

**Corollary 1.7.5** *Let X be a smooth projective algebraic variety defined over a field of characteristic 0, let B be a normal crossing divisor on X, and let D be an ample divisor on X. Then for any positive integer $p > 0$, $H^p(X, K_X + B + D) = 0$.*

*Proof* We do induction on the dimension $n$ of $X$ and the number $r$ of prime divisors of $B$. If $r = 0$, this is just the Kodaira vanishing theorem. If $r > 0$, take a prime divisor $B_1$ of $B$, denote $B' = B - B_1$ and $C = B'|_{B_1}$. By the adjunction formula, we get an exact sequence

$$0 \to \mathcal{O}_X(K_X + B' + D) \to \mathcal{O}_X(K_X + B + D) \to \mathcal{O}_{B_1}(K_{B_1} + C + D|_{B_1}) \to 0.$$

38    *1 Algebraic Varieties with Boundaries*

By the inductive hypothesis, for any positive integer $p > 0$, $H^p(X, K_X + B' + D) = H^p(B_1, K_{B_1} + C + D|_{B_1}) = 0$. This concludes the assertion.    □


## 1.8 The Covering Trick

The *covering trick* is a classical method to construct new algebraic varieties from a given one by using *cyclic coverings*. However, in this method, the new algebraic variety may have singularities even if the given algebraic variety is smooth. Therefore, we describe how to construct a covering without creating new singularities.

First, we describe the construction of cyclic coverings. Let $X$ be an algebraic variety over an algebraically closed field $k$, let $h$ be a nonzero rational function on $X$, and let $m$ be a positive integer coprime to the characteristic of $k$. When $k = \mathbf{C}$, $m$ can be taken arbitrarily. Consider the function field extension $K = k(X)[h^{1/m}]$, take $Y$ to be the normalization of $X$ in $K$ with the natural morphism $f : Y \to X$. The extension $Y/X$ is a Galois extension with a cyclic Galois group, and the extension degree $m' = [k(Y) : k(X)]$ is a divisor of $m$.

$Y$ can be constructed as the following. Assume that $X$ is covered by affine open subsets $U_i = \mathrm{Spec}(A_i)$. The fractional field of $A_i$ is the function field $k(X)$. Take $B_i$ to be the normalization of $A_i$ in $K$, then $Y$ is obtained by gluing affine varieties $\mathrm{Spec}(B_i)$.

**Example 1.8.1** Let $X$ be a smooth complex algebraic variety, let $D$ be a divisor on $X$, and let $s$ be a global section of $\mathcal{O}_X(mD)$. The zero divisor $\mathrm{div}(s)$ of $s$ and the divisor $\mathrm{div}(h)$ of the rational function $h$ corresponding to $s$ is related by

$$\mathrm{div}(s) = \mathrm{div}(h) + mD.$$

Here $\mathrm{div}(s)$ is an effective divisor but $\mathrm{div}(h)$ is not necessarily effective and might have poles along $D$ in general.

Assume that $B = \mathrm{div}(s)$ is reduced and is a smooth subvariety of $X$. Consider $Y$ to be the cyclic covering of $X$ induced by $h$. In this case, $Y$ is smooth and $f : Y \to X$ is a finite morphism branched along $B$. Here $D$ is not contained in the branch locus. Indeed, for any point $P$ in $B$, take a regular system of parameters $z_1, \ldots, z_n$ such that $B = \mathrm{div}(z_1)$, then the regular system of parameters of any point $Q$ over $P$ can be taken as $z_1^{1/m}, z_2, \ldots, z_n$.

One should be careful that if $B = \mathrm{div}(s)$ has singularities, then $Y$ has singularities correspondingly. When the support of $B$ is a normal crossing divisor, $Y$ has at worst *toric singularities*, which is easier to handle. This will be discussed in Section 3.7.

## 1.8 The Covering Trick

We can produce a more useful covering by considering the *Kummer covering*, a generalization of cyclic covering.

**Theorem 1.8.2** ([52]) *Let $X$ be a smooth projective algebraic variety defined over an algebraically closed field of characteristic 0 and let $B$ be a normal crossing divisor on $X$. Fix a positive integer $m_i$ for each irreducible component $B_i$ of $B$. Then there exists a smooth projective algebraic variety $Y$ and a finite morphism $f : Y \to X$ with the following properties:*

*(1) The set-theoretic inverse image $C = f^{-1}(B)$ is a normal crossing divisor.*
*(2) For each $i$, there exists a reduced divisor $C_i$ such that the pullback of $B_i$ as a divisor can be written as $f^* B_i = m_i C_i$. Here a reduced divisor is a divisor with all coefficients equal to 1.*
*(3) $f$ is a Galois covering and the Galois group $G$ is an Abelian group.*

One feature of this covering is that it is a finite morphism branched along a normal crossing divisor such that the covering space is again smooth. Note that the branch locus of $f$ is a normal crossing divisor containing $B$, but they do not coincide in general. Moreover, since $X$ is smooth, $f$ is a *flat morphism*.

*Proof* Denote $n = \dim X$. Take a very ample divisor $A$ such that $m_i A - B_i$ is very ample for all $i$. For each $i$, take $n$ general global sections $s_{ij}$ ($j = 1, \ldots, n$) in $H^0(X, m_i A - B_i)$. We may assume that for each $i, j$, $M_{ij} = \mathrm{div}(s_{ij})$ is smooth and $\sum_{i,j} M_{ij} + \sum_i B_i$ is a normal crossing divisor.

Take the rational function $h_{ij}$ corresponding to $s_{ij}$ and take $f_{ij} : Y_{ij} \to X$ to be the normalization of $X$ in $k(X)[h_{ij}^{1/m_i}]$. It is easy to see that the branch locus is $M_{ij} + B_i$ and the ramification index is $m_i$.

Take $f : Y \to X$ to be the normalization of the fiber product of all $f_{ij} : Y_{ij} \to X$ over $X$. In other words, $Y$ is just the normalization of $X$ in the field $k(X)[h_{ij}^{1/m_i}]_{ij}$. We will check that this $Y$ satisfies the required properties.

For any point $P$ in $X$, denote by $B_{i_l}$ ($l = 1, \ldots, r$) and $M_{j_q k_q}$ ($q = 1, \ldots, s$) the irreducible components of $\sum_{i,j} M_{ij} + \sum_i B_i$ containing $P$. Note that $r + s \leq \dim X = n$.

If $r = 0$, that is, $P$ is not contained in the support of $B$, then by construction, $Y_{ij}$ is smooth over a neighborhood of $P$, and there is nothing to prove. So we may assume that $r \geq 1$.

By the numbers of $M_{ij}$, for each $i_l$, there exists at least one $p_l$ such that $M_{i_l p_l}$ does not contain $P$. Denote $\bar{h}_{j_q k_q} = h_{j_q k_q} / h_{i_l p_l}$ if $j_q = i_l$; otherwise $\bar{h}_{j_q k_q} = h_{j_q k_q} h_A^{m_{j_q}}$, where $h_A$ is a local equation of the divisor $A$. In this case,

$$h_{i_1 p_1} h_A^{m_{i_1}}, \ldots, h_{i_r p_r} h_A^{m_{i_r}}, \bar{h}_{j_1 k_1}, \ldots, \bar{h}_{j_s k_s}$$

40      *1 Algebraic Varieties with Boundaries*

is a part of a regular system of parameters of $\mathcal{O}_{X,P}$. Indeed, in a neighborhood of $P$, these functions are exactly the defining equations of

$$B_{i_1}, \ldots, B_{i_r}, M_{j_1 k_1}, \ldots, M_{j_s k_s},$$

which form a normal crossing divisor in a neighborhood of $P$. The localization $Y \times_X \operatorname{Spec} \mathcal{O}_{X,P}$ is étale over the normalization of $\operatorname{Spec} \mathcal{O}_{X,P}$ in

$$k(X)\left[h_{i_1 p_1}^{1/m_{i_1}}, \ldots, h_{i_r p_r}^{1/m_{i_r}}, h_{j_1 k_1}^{1/m_{j_1}}, \ldots, h_{j_s k_s}^{1/m_{j_s}}\right]$$
$$= k(X)\left[h_{i_1 p_1}^{1/m_{i_1}} h_A, \ldots, h_{i_r p_r}^{1/m_{i_r}} h_A, \bar{h}_{j_1 k_1}^{1/m_{j_1}}, \ldots, \bar{h}_{j_s k_s}^{1/m_{j_s}}\right].$$

Therefore, $Y$ is smooth. The properties on $C$ and $C_i$ can be checked similarly. $\qquad\square$

The covering in the above theorem preserves smoothness by adding branch locus artificially. The covering below is a natural construction for a **Q**-Cartier Weil divisor which is not Cartier:

**Proposition 1.8.3** *Let $X$ be a normal algebraic variety defined over an algebraically closed field of characteristic $0$ and let $D$ be a divisor on $X$. Assume that for some positive integer $r$, $rD$ is Cartier and moreover $\mathcal{O}_X(rD) \cong \mathcal{O}_X$. Take $r$ to be the minimal one, then there exists a Galois finite morphism $f : Y \to X$ from a normal algebraic variety whose Galois group is the cyclic group of order $r$ such that $f$ is étale in codimension $1$ and $f^*D$ is a Cartier divisor on $Y$.*

*Proof* Fix an everywhere nonzero global section $s$ of $\mathcal{O}_X(rD)$. The corresponding rational function $h$ satisfies $\operatorname{div}_X(h) = -rD$. Take $Y$ to be the normalization of $X$ in the function field extension $L = k(X)[h^{1/r}]$. $L$ is a field as $r$ is minimal. Then $-f^*(D) = \operatorname{div}_Y(h^{1/r})$ is Cartier. It is easy to see that $f$ is étale over the locally free locus of $\mathcal{O}_X(D)$, and in particular, $f$ is étale over $X \setminus \operatorname{Sing}(X)$. $\qquad\square$

Such $f : Y \to X$ is called the *index $1$ cover* of the divisor $D$. In particular, if $D = K_X$, it is called the *canonical cover*.

**Remark 1.8.4** (1) This covering is not unique, it depends on the choice of $s$. Take another global section $s'$, there is a nowhere $0$ function $u$ such that $s' = us$. The normalization of $X$ in $k(X)[u^{1/r}]$ gives an étale covering $X' \to X$, and the base change to $X'$ gives an isomorphism $Y \times_X X' \cong Y' \times_X X'$. Here $Y'$ is the cyclic covering obtained by $s'$. Therefore, this covering is unique up to étale base changes.

*1.9 Generalizations of the Kodaira Vanishing Theorem* 41

(2) Fix a point $P \in X$ and take $r_P$ to be the minimal positive integer such that $r_P D$ is Cartier in a neighborhood of $P$, then $f^{-1}(P)$ consists of $r/r_P$ points by construction. In particular, $f$ is étale over the points where $D$ is Cartier.

## 1.9 Generalizations of the Kodaira Vanishing Theorem

According to [76], we generalize the Kodaira vanishing theorem to different directions in order to apply it to higher dimensional algebraic geometry. The generalized vanishing theorems will be used as the key point of proofs in each part of this book.

In this section, we always assume that the base field is of characteristic 0.

First, we extend the Kodaira vanishing theorem to **R**-divisors:

**Theorem 1.9.1** *Let $X$ be a smooth projective algebraic variety and let $D$ be an ample **R**-divisor on $X$ such that the support of $\lceil D \rceil - D$ is a normal crossing divisor. Then for any positive integer $p > 0$, $H^p(X, K_X + \lceil D \rceil) = 0$.*

Here we prove the following equivalent theorem:

**Theorem 1.9.2** *Let $X$ be a smooth projective algebraic variety, let $B$ be an **R**-divisor on $X$ with coefficients in $(0, 1)$ and supported on a normal crossing divisor, and let $D$ be an integral divisor on $X$. Assume that $D - (K_X + B)$ is an ample **R**-divisor. Then for any positive integer $p > 0$, $H^p(X, D) = 0$.*

*Proof* Write $B = \sum b_i B_i$. Here $B_i$ are prime divisors and $\sum B_i$ is a normal crossing divisor. As ampleness is an open condition, for each $i$ we can take a fraction $n_i/m_i$ $(0 < n_i < m_i)$ sufficiently close to $b_i$ such that $D - (K_X + \sum(n_i/m_i)B_i)$ is an ample **Q**-divisor. In the following we may assume that $B = \sum(n_i/m_i)B_i$.

Take the covering $f: Y \to X$ as in Theorem 1.8.2 for irreducible components $B_i$ of $B$ with positive integers $m_i$. By construction, $f^*B$ is a divisor with integral coefficients. The Galois group $G$ acts on the invertible sheaf $\mathcal{O}_Y(K_Y - f^*(K_X + B))$ equivariantly in the following way. The action of $G$ on the tangent sheaf $T_Y$ induces the action on the canonical sheaf $\mathcal{O}_Y(K_Y)$, and the action of $G$ on $\mathcal{O}_Y(-f^*(K_X + B))$ is induced from that on $\mathcal{O}_Y$ as $-f^*(K_X + B)$ is a $G$-invariant divisor. Since $f$ is flat, the direct image sheaf $f_*\mathcal{O}_Y(K_Y - f^*(K_X + B))$ is a locally free sheaf with a $G$-action and the $G$-invariant part $L = (f_*\mathcal{O}_Y(K_Y - f^*(K_X + B)))^G$ is an invertible sheaf. Here since $G$ is Abelian, $f_*\mathcal{O}_Y$ decomposes to a direct sum of invertible sheaves corresponding the $G$-eigenspaces, and hence $L$ is invertible.

$L$ can be written as the form of a divisorial sheaf $\mathcal{O}_X(E)$. In order to determine $E$, we only need to look at the generic points of the branched divisor of $f$. First, any prime divisor not contained in $B$ is not an irreducible component of $E$. Indeed, for any finite Galois covering $g: W \to Z$ between smooth varieties with Galois group $G$, we have a natural isomorphism $(g_*\omega_W)^G \cong \omega_Z$, which means that over $U = X \setminus \mathrm{Supp}(B)$, $L|_U = (f_*\mathcal{O}_Y(K_Y - f^*(K_X + B)))^G|_U = (f_*\mathcal{O}_Y(K_Y - f^*(K_X)))^G|_U \simeq \mathcal{O}_U$.

For the generic point $P$ of $B_i$, set $x_1$ to be the regular parameter of the discrete valuation ring $\mathcal{O}_{X,P}$. Then for a point $Q$ on $Y$ over $P$, $y_1 = f^*x_1^{1/m_i}$ is a regular parameter and the invertible sheaf $\mathcal{O}_Y(K_Y - f^*(K_X + B))$ is generated by the section $y_1^{-(m_i-1)+n_i}$. Since $0 < n_i < m_i$, $G$-invariant sections are generated by 1. Therefore, it turns out that $E = 0$. In summary, $L = (f_*\mathcal{O}_Y(K_Y - f^*(K_X + B)))^G = \mathcal{O}_X$.

As the pullback of an ample divisor by a finite morphism is ample, the pullback $f^*(D-(K_X+B))$ is again ample. By the Kodaira vanishing theorem, for any positive integer $p > 0$, $H^p(Y, K_Y + f^*(D - (K_X + B))) = 0$. As $f$ is finite, there is no higher direct image, hence $H^p(X, f_*\mathcal{O}_Y(K_Y + f^*(D - (K_X+B)))) = 0$. As the $G$-invariant part is a direct summand, $H^p(X, D) = 0$. $\qquad\square$

Next, we prove the relative version of the vanishing theorem:

**Theorem 1.9.3** *Let $X$ be a smooth algebraic variety, let $B$ be an* **R***-divisor on $X$ with coefficients in $(0, 1)$ and supported on a normal crossing divisor, let $D$ be an integral divisor on $X$, and let $f: X \to S$ be a projective morphism to another algebraic variety. Assume that $D - (K_X + B)$ is a relatively ample* **R***-divisor. Then for any positive integer $p > 0$,*

$$R^p f_*(\mathcal{O}_X(D)) = 0.$$

We will prove the following equivalent theorem:

**Theorem 1.9.4** *Let $X$ be a smooth algebraic variety, let $f: X \to S$ be a projective morphism to another algebraic variety and let $D$ be a relatively ample* **R***-divisor on $X$ such that the support of $\lceil D \rceil - D$ is a normal crossing divisor. Then for any positive integer $p > 0$,*

$$R^p f_*(\mathcal{O}_X(K_X + \lceil D \rceil)) = 0.$$

*Proof* As the assertion is local on $S$, we may assume that $S$ is affine. Replacing the integral part of $D$ by a linearly equivalent one while keeping $\lceil D \rceil - D$ unchanged, we may assume that the support of $D$ is a normal crossing divisor.

## 1.9 Generalizations of the Kodaira Vanishing Theorem

However, $D$ is not necessarily effective. We may assume that $D$ is a $\mathbf{Q}$-divisor as ampleness is an open condition.

Shrinking $S$ if necessary, we can find a sufficiently large integer $m$ such that $mD$ is an integral divisor and there exists a closed immersion $g: X \to \mathbf{P}^N \times S$ such that $\mathcal{O}_X(mD) \cong g^* p_1^* \mathcal{O}_{\mathbf{P}^N}(1)$, where $p_1$ is the first projection.

Next, take a projective algebraic variety $\bar{S}$ to be the compactification of $S$, and take $\bar{X}$ to be the normalization of the closure of $X$ in $\mathbf{P}^N \times \bar{S}$. The projective morphism $\bar{f}: \bar{X} \to \bar{S}$ and the finite morphism $\bar{g}: \bar{X} \to \mathbf{P}^N \times \bar{S}$ are naturally induced.

Here $\bar{X}$ is possibly singular and the extension of $D$ is a $\mathbf{Q}$-Cartier divisor $\bar{D}$ defined by $\mathcal{O}_{\bar{X}}(m\bar{D}) \cong \bar{g}^* p_1^* \mathcal{O}_{\mathbf{P}^N}(1)$. Since $\bar{D}$ is relatively ample over $\bar{S}$, we can choose an ample Cartier divisor $A_1$ on $\bar{S}$ such that $\bar{D} + \bar{f}^* A_1$ is ample. As $S$ is affine, we may assume that the support of $A_1$ is contained in $\bar{S} \setminus S$.

Take $h: Y \to \bar{X}$ to be a log resolution of the pair $(\bar{X}, \bar{D} + \bar{f}^* A_1)$ in strong sense. As $X$ is smooth and the support of $D$ is a normal crossing divisor, $h$ can be assumed to be the identity over $X$. We may choose a $\mathbf{Q}$-Cartier divisor $A_2$ supported in the exceptional set of $h$ such that $\bar{D}' = h^* \bar{D} + h^* \bar{f}^* A_1 + A_2$ is ample. By construction, the support of $\bar{D}'$ is a normal crossing divisor, and by Theorem 1.9.1, for any positive integer $p$, $H^p(Y, K_Y + \lceil \bar{D}' \rceil) = 0$. Note that the support of $h^* \bar{f}^* A_1 + A_2$ is contained in $Y \setminus X$.

Consider the following spectral sequence:

$$E_2^{p,q} = H^p(\bar{S}, R^q(\bar{f} \circ h)_*(\mathcal{O}_Y(K_Y + \lceil \bar{D}' \rceil))) \Rightarrow H^{p+q}(Y, K_Y + \lceil \bar{D}' \rceil).$$

For any positive integer $m_1$, replacing $A_1$ by $m_1 A_1$, the above argument still works. When $m_1$ is sufficiently large, by the Serre vanishing theorem, for any positive integer $p$ and any integer $q$,

$$H^p(\bar{S}, R^q(\bar{f} \circ h)_*(\mathcal{O}_Y(K_Y + \lceil \bar{D}' \rceil))) = 0.$$

Also the coherent sheaf $R^q(\bar{f} \circ h)_*(\mathcal{O}_Y(K_Y + \lceil \bar{D}' \rceil))$ on $\bar{S}$ is generated by global sections.

By the spectral sequence, when $q > 0$, $H^0(\bar{S}, R^q(\bar{f} \circ h)_*(\mathcal{O}_Y(K_Y + \lceil \bar{D}' \rceil))) = 0$. Therefore, $R^q(\bar{f} \circ h)_*(\mathcal{O}_Y(K_Y + \lceil \bar{D}' \rceil)) = 0$. We conclude the theorem by restricting on $S$. $\qquad \square$

The next lemma shows that the conditions in the definitions of KLT and LC (log canonical) defined in Sections 1.10 and 1.11 are birational properties:

**Lemma 1.9.5** *Let $f: Y \to X$ be a proper birational morphism between smooth algebraic varieties and let $B, C$ be $\mathbf{R}$-divisors on $X, Y$ supported on normal crossing divisors such that $f^*(K_X + B) = K_Y + C$. Then the* coefficients of $B$ are all contained in the open interval $(-\infty, 1)$ if and only if so are the coefficients of $C$.

The same also holds for the condition that the coefficients are contained in the half-open interval $(-\infty, 1]$. Moreover, in this case, assume that the irreducible components of $B$ with coefficients exactly 1 are disjoint, then the coefficients of $C - f_*^{-1} B$ are all contained in the open interval $(-\infty, 1)$.

*Proof* As $B = f_*C$, if the coefficients of $C$ are all contained in the open interval $(-\infty, 1)$, then the coefficients of $B$ are all contained in the open interval $(-\infty, 1)$.

Conversely, assume that the coefficients of $B$ are all contained in the open interval $(-\infty, 1)$. First, we consider the case that $f$ is a permissible blowup with respect to the pair $(X, B)$. Set $B = \sum b_i B_i$. Suppose that the center $Z$ of the blowup is of codimension $r$ and contained in $B_1, \ldots, B_s$. Note that $r \geq s$. The coefficient $e$ of the exceptional divisor $E$ of $f$ in $C$ is given by

$$ e = \sum_{j=1}^{s} b_j + 1 - r. $$

As $b_j < 1$, we have $e < 1$. Since the coefficients of other prime divisors of $C$ coincide with those of $B$, the coefficients of $C$ are all contained in the open interval $(-\infty, 1)$.

The general case can be reduced to the above case by applying Theorem 1.6.4. The later part can be proved similarly. $\square$

We can also prove the following lemma which will be used in Section 1.11:

**Lemma 1.9.6** *Fix an $n$-dimensional pair $(X, B)$ of a normal algebraic variety and an effective $\mathbf{R}$-divisor such that $K_X + B$ is $\mathbf{R}$-Cartier and let $P$ be a point on $X$. Take effective Cartier divisors $D_1, \ldots, D_n$ passing through $P$ such that $P$ is an irreducible component of $\bigcap D_i$. Then there exists a log resolution $f : Y \to (X, B + \sum D_i)$ such that if we write $K_Y + C = f^*(K_X + B + \sum D_i)$, then there exists an irreducible component $C_1$ of $C$ with coefficient at least 1 and $f(C_1) = \{P\}$.*

*Proof* We may assume that $X$ is affine by shrinking $X$. Write $D_i = \mathrm{div}(h_i)$, where $h_i$ are regular functions on $X$. Define the morphism $h : X \to Z = \mathbf{A}^n$ by $h = (h_1, \ldots, h_n)$. By assumption, $h$ is quasi-finite in a neighborhood of $P$. Take $E_1, \ldots, E_n$ to be coordinate hyperplanes of $Z$, and $h^* E_i = D_i$ by construction. Take $g : Z' \to Z$ to be the blowup at the origin and take $F$ to be the exceptional divisor, we get $g^*(K_Z + \sum E_i) = K_{Z'} + F + \sum g_*^{-1} E_i$. As differential forms on $Z$ with poles along $\sum E_i$ can be pulled back by $h$,

## 1.9 Generalizations of the Kodaira Vanishing Theorem

$h^*(K_Z + \sum E_i) \leq K_X + B + \sum D_i$. By taking a log resolution $f: Y \to (X, B + \sum D_i)$ which factors through $X \times_Z Z'$, we may assume that the exceptional set contains a prime divisor $C_1$ mapping onto $F$, and this satisfies the requirements. $\qquad\square$

Using the relative version of the vanishing theorem, it is easy to show the following generalization:

**Theorem 1.9.7** ([76, Theorem 1.2.3]) *Let $X$ be a smooth algebraic variety, let $f: X \to S$ be a projective morphism to another algebraic variety, and let $D$ be a relatively nef and relatively big $\mathbf{R}$-divisor on $X$ such that the support of $\ulcorner D \urcorner - D$ is a normal crossing divisor. Then for any positive integer $p > 0$,*

$$R^p f_*(\mathcal{O}_X(K_X + \ulcorner D \urcorner)) = 0.$$

*Proof* Since the assertion is local on $S$, we may assume that $S$ is affine. By Kodaira's lemma, we can write $D = A + E$ for some relatively ample $\mathbf{R}$-Cartier divisor $A$ and some effective $\mathbf{R}$-Cartier divisor $E$. If $0 < \epsilon < 1$, then $D - \epsilon E = (1 - \epsilon)D + \epsilon A$ is relatively ample.

Take $g: Y \to X$ to be a log resolution of $(X, D + E)$ in strong sense, and take $h: Y \to S$ to be the composition with $f$. We can choose a sufficiently small effective $\mathbf{R}$-divisor $A'$ supported on the exceptional set of $g$ such that $-A'$ is $g$-ample and $D' = g^*(D - \epsilon E) - A'$ is $h$-ample. By Theorem 1.9.4, for any positive integer $p$,

$$R^p h_*(\mathcal{O}_Y(K_Y + \ulcorner D' \urcorner)) = R^p g_*(\mathcal{O}_Y(K_Y + \ulcorner D' \urcorner)) = 0.$$

By the spectral sequence

$$E_2^{p,q} = R^p f_*(R^q g_*(\mathcal{O}_Y(K_Y + \ulcorner D' \urcorner))) \Rightarrow R^{p+q} h_*(\mathcal{O}_Y(K_Y + \ulcorner D' \urcorner)),$$

$R^p f_*(g_*(\mathcal{O}_Y(K_Y + \ulcorner D' \urcorner))) = 0$ holds for $p > 0$.

Take $\epsilon$ and $A'$ to be sufficiently small, then $\ulcorner D' \urcorner = \ulcorner g^* D \urcorner$. Take $B = \ulcorner D \urcorner - D$ and $g^*(K_X + B) = K_Y + C$, by Lemma 1.9.5, the coefficients of $C$ are less than 1. Therefore, by

$$g^*(K_X + \ulcorner D \urcorner) = g^*(K_X + B + D) = K_Y + C + g^* D \leq K_Y + \ulcorner g^* D \urcorner$$

(here note that $C + g^* D$ is an integral divisor) and $g_*(K_Y + \ulcorner g^* D \urcorner) = K_X + \ulcorner D \urcorner$, we have

$$g_*(\mathcal{O}_Y(K_Y + \ulcorner D' \urcorner)) = \mathcal{O}_X(K_X + \ulcorner D \urcorner),$$

which proves the theorem. $\qquad\square$

46          *1 Algebraic Varieties with Boundaries*

Higher dimensional algebraic geometry became greatly developed since the following result was proved:

**Corollary 1.9.8** (*Kawamata–Viehweg vanishing theorem* [53, 139]) *Let $X$ be a smooth projective algebraic variety and let $D$ be a nef and big $\mathbf{R}$-divisor on $X$ such that the support of $\lceil D \rceil - D$ is a normal crossing divisor. Then for any positive integer $p > 0$,*

$$H^p(X, K_X + \lceil D \rceil) = 0.$$

## 1.10 KLT Singularities for Pairs

We can define various singularities for a pair $(X, B)$, where $X$ is a normal algebraic variety and $B$ is an $\mathbf{R}$-divisor on $X$. $B$ is called the *boundary* of the pair for historical reasons. These singularities appear naturally in the minimal model theory. Vanishing theorems can be also generalized to these singularities. The characteristic of the base field is always assumed to be 0 if not specified.

First, we define the KLT condition. This is a very natural condition corresponding to the $L^2$-condition in complex analysis. It does not depend on the choice of log resolutions. Furthermore, it is easy to handle since it satisfies the so-called "open condition" in the sense that it is stable under *perturbation* of the divisors. The KLT condition defines a category in which the minimal model theory works most naturally and easily.

For simplicity, sometimes we denote a pair $(X, B)$ and a morphism $f : X \to S$ together by a morphism $f : (X, B) \to S$.

**Definition 1.10.1** A pair $(X, B)$ is *KLT* if it satisfies the following conditions:

(1) $K_X + B$ is $\mathbf{R}$-Cartier.
(2) The coefficients of $B$ are contained in the open interval $(0, 1)$.
(3) There exists a log resolution $f : Y \to (X, B)$ such that if we write $f^*(K_X + B) = K_Y + C$, then the coefficients $c_j$ of $C = \sum c_j C_j$ are contained in $(-\infty, 1)$. Here $C_j$ are distinct prime divisors.

Condition (1) is necessary in order to define the $\mathbf{R}$-divisor $C$ in condition (3). The support of $C$ is contained in the union of the set-theoretic inverse image of the support of $B$ and the exceptional set of $f$, which is a normal crossing divisor. The coefficients $c_j$ of $C$ play an important role in higher dimensional algebraic geometry. Further, $-c_j$ is called the *discrepancy coefficient* and $1 - c_j$ is called the *log discrepancy coefficient*.

## 1.10 KLT Singularities for Pairs

Historically, KLT singularity is just called *log terminal singularity* in [54].

Condition (3) in the definition of KLT does not depend on the choice of log resolutions:

**Proposition 1.10.2** *Assume that $(X, B)$ satisfies conditions (1) and (2) in Definition 1.10.1 and there exists a log resolution $f : Y \to (X, B)$ in weak sense satisfying condition (3). Then $(X, B)$ is KLT. Moreover, for any log resolution $f' : Y' \to (X, B)$ in weak sense, condition (3) in Definition 1.10.1 holds.*

*Proof* For two log resolutions $f_1 : Y_1 \to X$, $f_2 : Y_2 \to X$, there exists a third log resolution $f_3 : Y_3 \to X$ dominating them. That is, there exist morphisms $g_i : Y_3 \to Y_i$ $(i = 1, 2)$ such that $f_3 = f_i \circ g_i$. Therefore, the assertion follows from Lemma 1.9.5. $\qquad\qquad\square$

The following proposition is obvious:

**Proposition 1.10.3** *(1) A pair $(X, B)$ is KLT if and only if there exists an open covering $\{X_i\}$ of $X$ such that the pairs $(X_i, B|_{X_i})$ are all KLT.*

*(2) Let $(X, B)$ be a KLT pair and let $B'$ be another effective $\mathbf{R}$-divisor such that $B \geq B'$ and $B - B'$ is $\mathbf{R}$-Cartier, then $(X, B')$ is again KLT.*

*(3) When $X$ is a normal complex analytic variety, we can define the KLT condition similarly by using complex analytic resolution of singularities. When $X$ is a complex algebraic variety, for a pair $(X, B)$, the algebraic KLT condition and the analytic KLT condition are equivalent.*

**Remark 1.10.4** Take regular functions $h_1, \ldots, h_r$ on the polydisk $X = \Delta^n = \{(z_1, \ldots, z_n) \in \mathbf{C}^n \mid |z_i| < 1\}$ and write the corresponding divisors by $B_i = \mathrm{div}(h_i)$. Take real numbers $b_i \in (0, 1)$. Then $(X, B = \sum b_i B_i)$ is KLT if and only if the function $h = \prod |h_i|^{-b_i}$ is locally $L^2$ everywhere.

Indeed, the $L^2$-condition on integrability can be studied via resolutions of singularities. When the support of $B$ is a normal crossing divisor, the absolute value of a regular function with poles along $B$ satisfies the $L^2$-condition if and only if the coefficients of $B$ are in $(-\infty, 1)$, which is exactly the KLT condition.

We introduce quotient singularities as an important example of KLT pairs.

An algebraic variety $X$ is said to have *quotient singularities* if it is a quotient variety of a smooth algebraic variety in an étale neighborhood of each point $P$. That is, there exists a neighborhood $U$ of $P$, an étale morphism $g : V \to U$ such that $P \in g(V)$, and a smooth algebraic variety $\tilde{V}$ with a finite group action $G$ such that $V \cong \tilde{V}/G$.

48    *1 Algebraic Varieties with Boundaries*

**Example 1.10.5** Fix a positive integer $r$ and integers $a_1, \ldots, a_n$. Define the action of the cyclic group $G = \mathbf{Z}/(r)$ on the affine space $\tilde{X} = \mathbf{A}^n$ by $z_i \mapsto \zeta^{a_i} z_i$. Here $(z_1, \ldots, z_n)$ are coordinates of $\tilde{X}$ and $\zeta$ is a primitive $r$th root of 1. Then the quotient space $X = \tilde{X}/G$ has only quotient singularities. The image $P_0$ of the origin might or might not be an isolated singularity, depending on the values of $a_i$. $X$ is said to have a cyclic quotient singularity of *type* $\frac{1}{r}(a_1, \ldots, a_n)$ at $P_0$.

**Proposition 1.10.6** *For an algebraic variety $X$ with quotient singularities, the pair $(X, 0)$ with divisor $B = 0$ is KLT.*

*Proof* As discrepancy coefficients remain unchanged under étale morphisms, we may assume that $X$ is a global quotient variety. That is, there is a smooth algebraic variety $\tilde{X}$ and a finite group $G$ such that $X = \tilde{X}/G$. It is not hard to see that $K_X$ is $\mathbf{Q}$-Cartier, indeed, $|G| \cdot K_X$ is Cartier.

Take a log resolution $f : Y \to X$ and write $f^* K_X = K_Y + C$. Take $\tilde{Y}$ to be the normalization of $Y$ in the function field $k(\tilde{X})$ and take $\tilde{f} : \tilde{Y} \to \tilde{X}$ and $\pi_Y : \tilde{Y} \to Y$ to be the induced morphisms, write $\tilde{f}^* K_{\tilde{X}} = K_{\tilde{Y}} + \tilde{C}$. Take a prime divisor $E$ on $Y$ contained in the exceptional set of $f$ and take a prime divisor $\tilde{E}$ on $\tilde{Y}$ such that $\pi_Y(\tilde{E}) = E$. Denote the coefficients of $E, \tilde{E}$ in $C, \tilde{C}$ by $c, \tilde{c}$, respectively, denote the ramification index of $\tilde{E}$ with respect to $\pi_Y$ by $e$, then we have

$$ce = \tilde{c} + e - 1.$$

Here $\tilde{c} \leq 0$ as $\tilde{X}$ is smooth, hence $c < 1$.    $\square$

A KLT pair admits the following special log resolution. We call it a *very log resolution* in this book.

**Proposition 1.10.7** *Let $(X, B)$ be a KLT pair consisting of a normal algebraic variety and an $\mathbf{R}$-divisor. Then there exists a log resolution $f : Y \to (X, B)$ such that if we write $f^*(K_X + B) = K_Y + C$, then the support of the $\mathbf{R}$-divisor $C' = \max\{C, 0\}$ is a disjoint union of smooth prime divisors.*

*Proof* Fix a log resolution $f_0 : Y_0 \to (X, B)$ and write $f_0^*(K_X + B) = K_{Y_0} + C_0$. Choose two prime divisors in $C_0$ and blowup along their intersection, we get $g_1 : Y_1 \to Y_0$. The composition with $f_0$ gives a new log resolution $f_1 : Y_1 \to X$. We will show that a very log resolution can be constructed by repeating this operation.

Write $C_0 = \sum c_{0j} C_{0j}$. Take a positive number $n$ such that the inequality $c_{0j} \leq 1 - \frac{1}{n}$ holds for all $j$. $n$ will be fixed in the following process.

## 1.10 KLT Singularities for Pairs

For any log resolution $f: Y \to (X, B)$, write $f^*(K_X + B) = K_Y + C$ and $C = \sum c_j C_j$. Note that $c_j \le 1 - \frac{1}{n}$ for all $j$ by the proof of Lemma 1.9.5. We define a sequence of integers $r(f) = (r_3(f), \dots, r_{2n}(f))$ by the formula

$$r_i(f) = \# \left\{ (j_1, j_2) \mid j_1 < j_2, C_{j_1} \cap C_{j_2} \ne \emptyset, 2 - \frac{i}{n} < c_{j_1} + c_{j_2} \le 2 - \frac{i-1}{n} \right\}.$$

For two sequences $(r_3, \dots, r_{2n})$ and $(r_3', \dots, r_{2n}')$, we consider the lexicographical order. As $r_i \ge 0$, the set of sequences $(r_3, \dots, r_{2n})$ satisfies the *DCC* (short for *descending chain condition*). That is, there is no infinite strictly decreasing chain.

For a given $f$, take the minimal $i$ such that $r_i(f) \ne 0$ and take a pair $(j_1, j_2)$ realizing it. That is, $j_1 < j_2, C_{j_1} \cap C_{j_2} \ne \emptyset$, and $2 - \frac{i}{n} < c_{j_1} + c_{j_2} \le 2 - \frac{i-1}{n}$. Take $g: Y' \to Y$ to be the blowup along center $Z = C_{j_1} \cap C_{j_2}$, denote $f' = f \circ g$, and write $(f')^*(K_X + B) = K_{Y'} + C'$. The coefficient $e$ of the exceptional divisor $E = \mathrm{Exc}(g)$ in $C'$ satisfies $1 - \frac{i}{n} < e \le 1 - \frac{i-1}{n}$.

The construction of $Y'$ kills the intersection of $C_{j_1}$ and $C_{j_2}$, and produces the intersections of $E$ with the strict transforms of $C_{j_1}, C_{j_2}$, and $C_j$ which intersect with $C_{j_1} \cap C_{j_2}$. Note that $e + c_j \le 2 - \frac{i}{n}$ as $c_j \le 1 - \frac{1}{n}$. So these new intersections do not contribute to $r_k(f')$ for $k \le i$. Therefore, $r_k(f') = r_k(f) = 0$ for $k < i$ and $r_i(f') = r_i(f) - 1$, which means that $r(f') < r(f)$. Since there is no infinite strictly decreasing chain for the sequence $r(f)$, eventually we can get a log resolution $f$ such that $r_i(f) = 0$ for all $i$. This concludes the proof. $\square$

Note that the log resolution in the above proposition is obtained by blowing up repeatedly, it does not satisfy condition (2') in Theorem 1.6.1. Also, the proposition cannot be extended to DLT pairs.

We can generalize the vanishing theorem to KLT pairs:

**Theorem 1.10.8** ([76, 1.2.5]) *Let $X$ be a normal algebraic variety, let $f: X \to S$ be a projective morphism, let $B$ be an $\mathbf{R}$-divisor on $X$, and let $D$ be a $\mathbf{Q}$-Cartier integral divisor on $X$. Assume that $(X, B)$ is KLT and $D - (K_X + B)$ is relatively nef and relatively big. Then for any positive integer $p$, $R^p f_*(\mathcal{O}_X(D)) = 0$.*

*Proof* Take a log resolution $g: Y \to (X, B)$, denote $h = f \circ g$, and write $g^*(K_X + B) = K_Y + C$. Note that $g^*D - (K_Y + C)$ is $h$-nef and $h$-big. Here note that the coefficients of $g^*D$ are not necessarily integers.

By Theorem 1.9.7, for any positive integer $p$, $R^p g_*(\mathcal{O}_Y(\lceil g^*D - C \rceil)) = R^p h_*(\mathcal{O}_Y(\lceil g^*D - C \rceil)) = 0$. Hence $R^p f_*(g_*(\mathcal{O}_Y(\lceil g^*D - C \rceil))) = 0$.

For a rational function $r \in k(X) \cong k(Y)$, if $\mathrm{div}_X(r) + D \ge 0$, then $\mathrm{div}_Y(r) + g^*D \ge 0$. In this case, $\mathrm{div}_Y(r) + \lfloor g^*D \rfloor \ge 0$ and then $\mathrm{div}_Y(r) +$

$\ulcorner g^*D - C \urcorner \geq 0$ since the coefficients of $C$ are contained in the open interval $(-\infty, 1)$. This shows that the natural inclusion

$$g_*(\mathcal{O}_Y(\ulcorner g^*D - C \urcorner)) \subset g_*(\mathcal{O}_Y(\ulcorner g^*D \urcorner)) \simeq \mathcal{O}_X(D)$$

is indeed an isomorphism and the proof is finished. $\qquad\qquad\square$

**Remark 1.10.9** In a KLT pair $(X, B)$, $X$ has only *rational singularities*, and hence is *Cohen–Macaulay* ([76, 1.3.6]).

This asserts that KLT is a "good" singularity. On the other hand, LC to be introduced in Section 1.11 is not "good" in this sense. This fact will not be used in this book.

Consider a pair $(X, B)$ consisting of a normal algebraic variety and an effective **R**-divisor such that $K_X + B$ is **R**-Cartier. In Chapter 2, we will introduce the multiplier ideal sheaf in order to measure how far this pair is from being KLT.

The set of points $P \in X$ in whose neighborhood the pair $(X, B)$ is not KLT is a closed subset of $X$. It is called the *non-KLT locus* of the pair $(X, B)$. The cosupport of the multiplier ideal sheaf coincides with the non-KLT locus. Also, the vanishing theorem can be generalized using multiplier ideal sheaves (see Section 2.11).

# 1.11 LC, DLT, and PLT Singularities for Pairs

The KLT condition is easy to handle since it is an open condition with respect to changes of coefficients of divisors. However, in the minimal model theory, since it is necessary to consider the limits of divisors, it is necessary to consider the closed condition called the LC condition. Among LC pairs, we call by $\overline{\text{KLT}}$ pairs the pairs obtained by increasing boundaries of KLT pairs. The property of general LC pairs is not so good, but for $\overline{\text{KLT}}$ pairs it is possible to have similar discussions as for KLT pairs. Besides, there are conditions called DLT and PLT (purely log terminal) between KLT and LC, which are a little complicated but very useful. In this book, we develop the minimal model theory mainly for DLT pairs. The characteristic of the base field is always assumed to be 0 if not specified.

## 1.11.1 Various Singularities

**Definition 1.11.1** A pair $(X, B)$ is *LC* if it satisfies the following conditions:

(1) $K_X + B$ is **R**-Cartier.

## 1.11 LC, DLT, and PLT Singularities for Pairs

(2) The coefficients of $B$ are contained in the half-open interval $(0, 1]$.
(3) There exists a log resolution $f: Y \to (X, B)$ such that if we write $f^*(K_X + B) = K_Y + C$, then the coefficients $c_j$ of $C = \sum c_j C_j$ are contained in the half-open interval $(-\infty, 1]$. Here $C_j$ are distinct prime divisors.

When $(X, B)$ is an LC pair, $(X, B)$ is said to have *log canonical singularities*. Same as Proposition 1.10.2, condition (3) above does not depend on the choice of log resolutions. Also, the same assertion as in Proposition 1.10.3 holds for LC pairs.

**Example 1.11.2** The property of singularities of LC pairs is not always good.
Let $Z$ be a smooth projective $n$-dimensional algebraic variety such that $K_Z \sim 0$, that is, $\omega_Z \cong \mathcal{O}_Z$. Take an ample invertible sheaf $L$ and take the total space $Y = \mathrm{Spec}_Z(\bigoplus_{m=0}^{\infty} L^{\otimes m})$ of the dual sheaf $L^*$. $Y$ admits an $\mathbf{A}^1$-bundle structure over $Z$. Denote $X = \mathrm{Spec}(\bigoplus_{m=0}^{\infty} H^0(Z, L^{\otimes m}))$, there is a natural birational morphism $f: Y \to X$ which contracts the 0-section $E$ of $Y \to Z$ to a point $P = f(E)$.
By the adjunction formula $(K_Y + E)|_E \sim K_E \sim 0$, we have $K_Y + E \sim 0$ and $K_X \sim 0$, which implies that $f^*K_X \sim K_Y + E$. Hence $(X, 0)$ is LC.
The higher direct images of $\mathcal{O}_Y$ are supported on the singular point $P$ of $X$:

$$R^p f_* \mathcal{O}_Y \cong \bigoplus_{m=0}^{\infty} H^p(Z, L^{\otimes m}) \supset H^p(Z, \mathcal{O}_Z).$$

For $p = n$, $H^n(Z, \mathcal{O}_Z) \neq 0$, hence $X$ is not a *rational singularity*. Moreover, if $Z$ is an Abelian variety, then for $0 < p \leq n$, the right-hand side is not 0, and $X$ is not *Cohen–Macaulay*.

As the property of singularities of LC pairs is not always good, we consider intermediate conditions:

**Definition 1.11.3** A pair $(X, B)$ is *DLT* if it satisfies the following conditions:

(1) $K_X + B$ is $\mathbf{R}$-Cartier.
(2) The coefficients of $B$ are contained in the half-open interval $(0, 1]$.
(3) There exists a log resolution $f: Y \to (X, B)$ such that if we write $f^*(K_X + B) = K_Y + C$, then the coefficients $c_j$ of $C = \sum c_j C_j$ are contained in the open interval $(-\infty, 1)$ for those $C_j$ contained in the exceptional set of $f$.

A pair $(X, B)$ is *PLT* if it satisfies the above conditions (1) and (2) and the following condition (3'):

52     *1 Algebraic Varieties with Boundaries*

(3') For any log resolution $f: Y \to (X, B)$, if we write $f^*(K_X + B) = K_Y + C$, then the coefficients $c_j$ of $C = \sum c_j C_j$ are contained in the open interval $(-\infty, 1)$ for those $C_j$ contained in the exceptional set of $f$.

**Remark 1.11.4** (1) In [76], a condition called *WLT* (short for *weak log terminal*) is considered. The definition of WLT is by assuming further that the log resolution in condition (3) of the definition of DLT is in strong sense. By using similar argument as in Proposition 1.10.2, it can be shown that DLT and WLT are indeed equivalent ([136]). In this book, we will just use DLT rather than WLT.

(2) For a log resolution $f: Y \to X$ of $(X, B)$, when considering the relation $f^*(K_X + B) = K_Y + C$, sometimes we just write "a morphism $f: (Y, C) \to (X, B)$."

**Example 1.11.5** (1) Take the affine space $X = \mathbf{A}^n$ and coordinate hyperplanes $B_1, \ldots, B_n$, denote $B = \sum b_i B_i$. Then $(X, B)$ is KLT (respectively, PLT, DLT) if and only if $0 \le b_i < 1$ for all $i$ (respectively, $0 \le b_i \le 1$ for all $i$ and $b_i < 1$ except for at most one $i$, $0 \le b_i \le 1$ for all $i$). Furthermore, DLT and LC coincide.

(2) Let $X = \mathbf{A}^2/\mathbf{Z}_2$ be the quotient of the 2-dimensional affine space $\mathbf{A}^2$ with coordinates $x, y$ by the order 2 cyclic group $\mathbf{Z}_2$ action $(x, y) \mapsto (-x, -y)$. That is, it is a cyclic quotient singularity of type $\frac{1}{2}(1, 1)$. This singularity is the same as the ordinary double point in Example 1.1.4(1). Denote the image of the coordinate axes in $X$ by $B_1, B_2$ and take $B = \sum b_i B_i$. Then $(X, B)$ is KLT (respectively, PLT, LC) if and only if $0 \le b_i < 1$ for all $i$ (respectively, $0 \le b_{i_1} \le 1$ for one $i_1$ and $0 \le b_{i_2} < 1$ for the other $i_2$, $0 \le b_i \le 1$ for all $i$). Furthermore, PLT and DLT coincide.

Indeed, the blowup $f: Y \to X$ of $X$ along the image of the origin $(0, 0)$ is a log resolution. The exceptional set $E$ is isomorphic to $\mathbf{P}^1$, $f^* B_i = f_*^{-1} B_i + \frac{1}{2} E$, and $f^* K_X = K_Y$. So the assertion can be checked easily.

(3) Take $X = \mathbf{A}^2$ to be the 2-dimensional affine space with coordinates $x, y$ and a prime divisor $D = \operatorname{div}(x^2 + y^3)$. We determine the necessary and sufficient condition for the pair $(X, dD)$ to be KLT or LC for a real number $d$ (see Figure 1.1).

We can construct a log resolution of $(X, dD)$ in the following way. First, take the blowup $f_1: Y_1 \to X$ along the origin $P_0 = (0, 0)$, the exceptional set $E_1$ is a prime divisor isomorphic to $\mathbf{P}^1$. The strict transform $D_1 = f_{1*}^{-1} D$ is smooth, $E_1$ and $D_1$ intersect at one point $P_1$.

## 1.11 LC, DLT, and PLT Singularities for Pairs

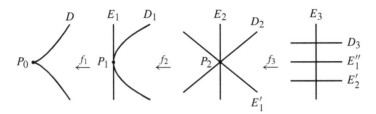

Figure 1.1 A log resolution of $(X, D)$.

Take the blowup $f_2 \colon Y_2 \to Y_1$ along $P_1$, the exceptional set $E_2$ is a prime divisor isomorphic to $\mathbf{P}^1$. Three smooth prime divisors $E_2$, $D_2 = f_{2*}^{-1} D_1$, and $E_1' = f_{2*}^{-1} E_1$ intersect at one point $P_2$.

Take the blowup $f_3 \colon Y = Y_3 \to Y_2$ along $P_2$, the exceptional set $E_3$ is a prime divisor isomorphic to $\mathbf{P}^1$. The union of four prime divisors $E_3$, $D_3 = f_{3*}^{-1} D_2$, $E_1'' = f_{3*}^{-1} E_1'$, and $E_2' = f_{3*}^{-1} E_2$ is a normal crossing divisor.

The composition $f \colon Y \to X$ is a log resolution of $(X, dD)$. We have $K_Y = f^* K_X + E_1'' + 2 E_2' + 4 E_3$ and $f^* D = D_3 + 2 E_1'' + 3 E_2' + 6 E_3$. Therefore, the pair $(X, dD)$ is KLT (respectively, LC) if and only if $0 \le d < 5/6$ (respectively, $0 \le d \le 5/6$).

(4) Consider the example in Examples 1.1.4(2) or 1.2.4(2). In addition to the prime divisors $D_1, D_2$, consider prime divisors $D_3, D_4$ defined by $y = z = 0$ or $y = w = 0$. Note that $D_3 + D_4$ and $K_X$ are Cartier divisors. Take $B = \sum_{i=1}^{4} D_i$ and consider the pair $(X, B)$. Take the resolution of singularities $f \colon X' \to X$ as in Example 1.2.4(2), then $B' = \sum_{i=1}^{4} f_*^{-1} D_i$ is a normal crossing divisor. As $f$ is isomorphic in codimension 1, $f^*(K_X + B) = K_{X'} + B'$.

The pair $(X, B)$ is LC but not DLT. Here, as the exceptional set of $f$ is not a normal crossing divisor, $f$ is a log resolution in weak sense, but not a log resolution in the sense of Theorem 1.6.1(2). In order to obtain a log resolution, we need to do further blowups on $X'$ along the exceptional set of $f$ and that will induce an exceptional divisor with log discrepancy coefficient 1. However, this is not a rigorous proof of the fact that $(X, B)$ is not DLT.

The blowup $g \colon Y \to X$ along the origin $(0, 0, 0, 0)$ of $X$ is a log resolution. The exceptional set $E$ is a prime divisor isomorphic to $\mathbf{P}^1 \times \mathbf{P}^1$ and $C = \sum_{i=1}^{4} g_*^{-1} D_i + E$ is a normal crossing divisor. Since $g^*(K_X + B) = K_Y + C$, $(X, B)$ is LC.

54      *1 Algebraic Varieties with Boundaries*

(5) Take a smooth projective algebraic curve $C$ of genus 1 and two line bundles $L_1, L_2$ of negative degrees. Take $Y$ to be the total space of the vector bundle $L = L_1 \oplus L_2$ and denote by $C_1, C_2, E$ the subvarieties of $Y$ corresponding to subbundles $L_1 \oplus 0, 0 \oplus L_2, 0 \oplus 0$, respectively. Note that $E \cong C$. Denote $X = \mathrm{Spec}(\bigoplus_{m=0}^{\infty} H^0(C, L^{\otimes -m}))$, there is a natural birational morphism $f : Y \to X$ which contracts $E$ to a point $P = f(E)$. Write $B_i = f(C_i)$. Then $f^*(K_X + B_1 + B_2) = K_Y + C_1 + C_2$ and the pair $(X, B_1 + B_2)$ is not DLT but LC. Indeed, $X$ is not a rational singularity. The pairs $(B_i, 0)$ are also LC.

We introduce one more definition:

**Definition 1.11.6** A pair $(X, B)$ is $\overline{KLT}$ if it satisfies the following conditions:

(1) $(X, B)$ is LC.
(2) There is another effective **R**-divisor $B'$ such that $B' \leq B$ and $(X, B')$ is KLT.

In this situation, for any positive real number $\epsilon$ smaller than 1, $(X, (1-\epsilon) B + \epsilon B')$ is KLT. That is, $\overline{KLT}$ is the limit of KLT. For this reason, different from general LC pairs, it shares similar properties as a KLT pair.

Toric varieties provide good examples (see [31, 79] for details):

**Proposition 1.11.7** *Let $T$ be an algebraic torus and let $T \subset X$ be a toric variety, that is, a $T$-equivariant open immersion into a normal algebraic variety with a $T$-action. Consider the complement set $B = X \setminus T$ as a reduced divisor. Then the following assertions hold:*

*(1) The pair $(X, B)$ is LC. Moreover, it is $\overline{KLT}$.*
*(2) $X$ is **Q**-factorial if and only if the corresponding fan consists of simplicial cones.*

*Proof* (1) Take a $T$-equivariant resolution of singularities $f : Y \to X$ such that $f^{-1}(T) \cong T$ and $C = Y \setminus f^{-1}(T)$ is a normal crossing divisor.

Denote $\dim T = n$ and take coordinates $x_1, \ldots, x_n$ by pulling back from the standard embedding $T \subset \mathbf{A}^n$. The regular differential form $\theta = dx_1/x_1 \wedge \cdots \wedge dx_n/x_n$ on $T$ can be extended to a logarithmic differential form on $X$ and gives a global section of $K_X + B$ without zeros. Therefore, $K_X + B \sim 0$.

Similarly $\theta$ extends to a global section of $K_Y + C$ without zeros. Therefore, the equality $f^*(K_X + B) = K_Y + C$ holds, and hence $(X, B)$ is LC.

As $T$ is affine, there exists an effective Cartier divisor $B'$ with support $B$. For a sufficiently small real number $\epsilon > 0$, $(X, B - \epsilon B')$ is KLT, and hence $(X, B)$ is $\overline{KLT}$.

## 1.11 LC, DLT, and PLT Singularities for Pairs

(2) We may assume that $X$ is affine and its fan consists of a single cone. Irreducible components $B_i$ of $B$ correspond to points $P_i$ on 1-dimensional rays of this cone $\sigma$. The condition for $B_i$ becoming a $\mathbf{Q}$-Cartier divisor is that there exists a regular function on $X$ such that the corresponding divisor is a nonzero multiple of $B_i$. This is equivalent to saying that there exists a linear functional on $\sigma$ which takes value 1 at $P_i$ and 0 at all points on other rays, which is equivalent to $\sigma$ being simplicial. $\qquad\square$

The following is a corollary of Lemma 1.9.6.

**Corollary 1.11.8** *Let $(X, B)$ be an $n$-dimensional KLT pair and let $P$ be a point. Take sufficiently general effective Cartier divisors $D_1, \ldots, D_n, E$ passing through $P$ and a positive number $1 > \epsilon > 0$. Then there exists a sufficiently small number $\delta > 0$ such that the pair $(X, B + \sum(1 - \delta)D_i + \epsilon E)$ is KLT in a punctured neighborhood of $P$, but not LC at $P$.*

*Proof* As $D_1, \ldots, D_n, E$ are general outside $P$, take a log resolution $\bar{f} : Y \to (X, B)$ and write $\bar{f}^*(K_X + B) = K_Y + \bar{C}$, we may assume that $\bar{C} + \bar{f}^*(\sum D_i + E)$ is normal crossing outside $\bar{f}^{-1}(P)$. The coefficients of $D_1, \ldots, D_n, E$ in the pair $(X, B + \sum(1 - \delta)D_i + \epsilon E)$ are strictly smaller than 1 for $\delta > 0$, hence the pair is KLT in a punctured neighborhood of $P$.

On the other hand, take the log resolution $f$ and prime divisor $C_1$ as in Lemma 1.9.6, then the coefficient of $C_1$ in $f^*E$ is at least 1 and the coefficient of $C_1$ in $f^*(K_X + B + \sum D_i + \epsilon E)$ is strictly larger than 1. Hence $(X, B + \sum(1 - \delta)D_i + \epsilon E)$ is not LC at $P$ for sufficiently small $\delta > 0$. $\qquad\square$

### 1.11.2 The Subadjunction Formula

We will look at the behavior of singularities when restricting a given pair to lower dimensions.

First, we show Shokurov's connectedness lemma ([91, Theorem 17.4], [128]), which is a consequence of the vanishing theorem:

**Lemma 1.11.9** (*Connectedness lemma*) *Let $(X, B)$ be a pair of a normal variety and an $\mathbf{R}$-divisor such that $K_X + B$ is $\mathbf{R}$-Cartier, and let $f : (Y, C) \to (X, B)$ be a log resolution in weak sense. Write $C = C^+ - C^-$, where $C^+, C^-$ are effective $\mathbf{R}$-divisors with no common irreducible component. Then the natural homomorphism $\mathcal{O}_X \to f_*\mathcal{O}_{\llcorner C^+ \lrcorner}$ is surjective and the induced morphism $\mathrm{Supp}(\llcorner C^+ \lrcorner) \to f(\mathrm{Supp}(\llcorner C^+ \lrcorner))$ has connected geometric fibers.*

*Proof* Note that

$$-\llcorner C \lrcorner - (K_Y + C - \llcorner C \lrcorner) \equiv -f^*(K_X + B)$$

is $f$-nef and $f$-big. As the coefficients of $C - \llcorner C \lrcorner$ are contained in the open interval $(0, 1)$, by the vanishing theorem (Theorem 1.9.7),

$$R^1 f_*(\mathcal{O}_Y(-\llcorner C \lrcorner)) = 0.$$

Since $\llcorner C \lrcorner = \llcorner C^+ \lrcorner - \ulcorner C^- \urcorner$, the natural homomorphism

$$f_*(\mathcal{O}_Y(\ulcorner C^- \urcorner)) \to f_*(\mathcal{O}_{\llcorner C^+ \lrcorner}(\ulcorner C^- \urcorner))$$

is surjective. Since the support of the effective divisor $C^-$ is contained in the exceptional set, the natural homomorphism $f_*\mathcal{O}_Y \to f_*(\mathcal{O}_Y(\ulcorner C^- \urcorner))$ is bijective. In the commutative diagram

$$
\begin{array}{ccc}
\mathcal{O}_X \cong f_*\mathcal{O}_Y & \longrightarrow & f_*\mathcal{O}_{\llcorner C^+ \lrcorner} \\
\downarrow & & \downarrow \\
f_*(\mathcal{O}_Y(\ulcorner C^- \urcorner)) & \longrightarrow & f_*(\mathcal{O}_{\llcorner C^+ \lrcorner}(\ulcorner C^- \urcorner)),
\end{array}
$$

the left vertical arrow is bijective, the bottom horizontal arrow is surjective, and the right vertical arrow is injective, hence the top horizontal arrow is surjective. We conclude the proof. $\qquad\square$

**Corollary 1.11.10** *A DLT pair* $(X, B)$ *is PLT if and only if the irreducible components of* $\llcorner B \lrcorner$ *are disjoint from each other.*

*Proof* The sufficiency is easy. In order to show the converse direction, suppose that two irreducible components $B_1, B_2$ of $\llcorner B \lrcorner$ intersect. Take a log resolution $f : (Y, C) \to (X, B)$ as in Lemma 1.11.9, then the strict transforms $f_*^{-1} B_1, f_*^{-1} B_2$ are contained in the same connected component of the support of $\llcorner C^+ \lrcorner$. Then there exists an irreducible component of $\llcorner C^+ \lrcorner - f_*^{-1} B_1$ intersecting $f_*^{-1} B_1$. Blowing up along the intersection, the coefficient of the exceptional divisor is 1, which means that $(X, B)$ is not PLT. $\qquad\square$

**Corollary 1.11.11** *For a DLT pair* $(X, B)$, *every irreducible component of* $\llcorner B \lrcorner$ *is normal.*

*Proof* We may assume that $X$ is affine. Take $B_1$ to be an irreducible component of $\llcorner B \lrcorner$. Since there are sufficiently many regular functions on $X$, we can take a general effective $\mathbf{R}$-divisor $B'$ $\mathbf{R}$-linearly equivalent to $B - B_1$ such that $\llcorner B' \lrcorner = 0$. Indeed, take a sufficiently large integer $N$ and take general global sections $s_1, \ldots, s_N$ of the divisorial sheaf $\mathcal{O}_X(\llcorner B \lrcorner - B_1)$, then $\mathrm{div}(s_i) \sim \llcorner B \lrcorner - B_1$ and we may just take

$$B' = B - B_1 + \sum \mathrm{div}(s_i)/N - \llcorner B \lrcorner + B_1 = B - \llcorner B \lrcorner + \sum \mathrm{div}(s_i)/N.$$

## 1.11 LC, DLT, and PLT Singularities for Pairs 57

Since $s_i$ are taken to be general, a log resolution $f: Y \to X$ of $(X, B)$ is also a log resolution of $(X, B_1 + B')$. So $(X, B_1 + B')$ is still DLT. Write $K_Y + C = f^*(K_X + B_1 + B')$, then $\llcorner C^+ \lrcorner = f_*^{-1} B_1$. Therefore, Lemma 1.11.9 implies that $D$ is normal. □

**Remark 1.11.12** According to this corollary, the irreducible components of $\llcorner B \lrcorner$ have no "self-intersection." For example, if $X$ is a smooth complex algebraic variety and $B$ is a reduced divisor normal crossing in analytic sense but not simple normal crossing, then $(X, B)$ is not DLT. This is derived from the definition of normal crossing divisors in the definition of log resolutions.

Induction arguments on dimensions using the adjunction formula is compatible with the property of DLT. The reason is the following result:

**Theorem 1.11.13** (*Subadjunction formula*) *Let $(X, B)$ be a DLT pair and let $Z$ be an irreducible component of $\llcorner B \lrcorner$. Then we can naturally define an effective $\mathbb{R}$-divisor $B_Z$ on $Z$ satisfying*

$$(K_X + B)|_Z = K_Z + B_Z,$$

*and the pair $(Z, B_Z)$ is again DLT. Moreover, if $(X, B)$ is PLT in a neighborhood of $Z$, then $(Z, B_Z)$ is KLT.*

*Proof* Take a log resolution $f: (Y, C) \to (X, B)$ such that the coefficients of exceptional prime divisors in $C$ are less than 1. Write $W = f_*^{-1} Z$, $C_W = (C - W)|_W$ and $B_Z = (f|_W)_* C_W$. Here the coefficients of $C_W$ are at most 1, so are those of $B_Z$.

Here we claim that the coefficients of $B_Z$ are contained in the half-open interval $(0, 1]$. To see that $B_Z \geq 0$, after cutting $X$ by general hyperplanes, we may assume that $\dim X = 2$. In this case, $f: (Y, C) \to (X, B)$ factors through the minimal resolution of $X$ (see Proposition 1.13.8). Hence there exists a pair $(Y_1, C_1)$ and birational morphisms $f_1: Y \to Y_1$, $f_2: Y_1 \to X$ such that $f = f_2 \circ f_1$ and $K_{Y_1} + C_1 = f_2^*(K_X + B)$, and moreover $C_1 \geq 0$. Then $B_Z \geq 0$.

As $(K_Y + C)|_W = K_W + C_W$, we get $(K_X + B)|_Z = K_Z + B_Z$. Hence $K_Z + B_Z$ is $\mathbb{R}$-Cartier. Note that $f|_W$ is a log resolution of $(Z, B_Z)$ and $(f|_W)^*(K_Z + B_Z) = K_W + C_W$.

Recall that every irreducible component of $C$ with coefficient 1 is a strict transform of an irreducible component of $\llcorner B \lrcorner$. Take $D$ to be an irreducible component of $C_W$ with coefficient 1, then $D$ is contained in the intersection of $f_*^{-1} \llcorner B \lrcorner - W$ and $W$. Since $\mathrm{Exc}(f) \cup f_*^{-1} \llcorner B \lrcorner$ is a normal crossing divisor, $D$ is not contained in $\mathrm{Exc}(f)$. Therefore, $D$ is not contained in the exceptional set of $f|_W$ and hence $(Z, B_Z)$ is DLT.

The latter part is obvious. □

58    *1 Algebraic Varieties with Boundaries*

**Remark 1.11.14** It is possible that $B_Z \neq 0$ even if $B = Z$, that is, $K_Z$ might be smaller than expected, and this is why we use the word "sub." For example, consider the quadric surface $X$ defined by the equation $xy = z^2$ in the affine space $\mathbf{C}^3$ with coordinates $x, y, z$ and the divisor $Z$ on $X$ defined by the equation $x = z = 0$. Then the pair $(X, Z)$ is DLT and the subadjunction formula in this case is $(K_X + Z)|_Z = K_Z + \frac{1}{2}P$ (see Example 1.3.2).

For a pair $(X, B)$, a subvariety $Z$ of $X$ is called an *LC center* if there exists a log resolution $f : (Y, C) \to (X, B)$ such that there is an irreducible component $C_i$ of $\llcorner C^+ \lrcorner$ with $Z = f(C_i)$.

**Lemma 1.11.15** *Fix a log resolution* $f : (Y, C) \to (X, B)$ *of an LC pair* $(X, B)$. *Then the LC centers of the pair* $(X, B)$ *are exactly the images of irreducible components of intersections of several irreducible components of* $\llcorner C^+ \lrcorner$.

*Proof* Take the blowup $Y$ along an irreducible component of the intersection of several irreducible components of $\llcorner C^+ \lrcorner$, we get a new log resolution and the exceptional divisor has coefficient 1 in the new boundary. Hence the image is an LC center. On the other hand, by an easy computation, blowing up along other centers gives an exceptional divisor with coefficient strictly smaller than 1. By Theorem 1.6.4, any log resolution is dominated by a log resolution obtained in this way, which concludes the proof.                                       $\square$

In particular, when $(X, B)$ is DLT, there exists a log resolution $f : (Y, C) \to (X, B)$ with $\llcorner C^+ \lrcorner = f_*^{-1}\llcorner B \lrcorner$, hence an LC center is nothing but an irreducible component of the intersection of several irreducible components of $\llcorner B \lrcorner$. In other words, the reduced part of the boundary of the DLT pairs obtained by applying the subadjunction formula several times to $(X, B)$ are LC centers.

We extend the vanishing theorem to DLT pairs. Note that the condition "relatively ample" cannot be replaced by "relatively nef and relatively big" as DLT is not an open condition.

**Theorem 1.11.16** *Let $X$ be a normal algebraic variety, let $f : X \to S$ be a projective morphism, let $B$ be an $\mathbf{R}$-divisor on $X$, and let $D$ be a $\mathbf{Q}$-Cartier integral divisor on $X$. Assume that the pair $(X, B)$ is DLT and $D - (K_X + B)$ is relatively ample. Then for any positive integer $p$, $R^p f_*(\mathcal{O}_X(D)) = 0$.*

*Proof* Take a log resolution $g : (Y, C) \to (X, B)$ in strong sense and denote $h = f \circ g$. By the definition of DLT, we may assume that the coefficients of exceptional divisors in $C$ are strictly less than 1, note that here we use the fact

## 1.11 LC, DLT, and PLT Singularities for Pairs 59

that DLT is equivalent to WLT (see Remark 1.11.4). Take a sufficiently small effective $\mathbf{R}$-divisor $A$ supported on the exceptional set of $g$ such that $-A$ is $g$-ample, $\lfloor C + A \rfloor = \lfloor C \rfloor$, and $g^*D - (K_Y + C + A)$ is $h$-ample. Take a sufficiently small number $\epsilon > 0$ such that $g^*D - (K_Y + (1 - \epsilon)C + A)$ is again $h$-ample.

Write $D' - C' = g^*D - ((1 - \epsilon)C + A)$, where $D'$ is a divisor with integral coefficients and $C'$ is an $\mathbf{R}$-divisor with coefficients in the interval $(0, 1)$, in other words, take $D' = \lceil g^*D - ((1 - \epsilon)C + A) \rceil$. Since the support of $C'$ is a normal crossing divisor, by Theorem 1.9.3, for $p > 0$, $R^p g_*(\mathcal{O}_Y(D')) = R^p h_*(\mathcal{O}_Y(D')) = 0$. Therefore, for $p > 0$, $R^p f_*(g_*(\mathcal{O}_Y(D'))) = 0$. Since $g_*D' = D$ by definition and $D' \geq \lfloor g^*D \rfloor$ as the coefficients of $(1 - \epsilon)C + A$ are smaller than 1, we have $g_*(\mathcal{O}_Y(D')) = \mathcal{O}_X(D)$ and the theorem is proved. $\square$

Here we remark that we can give an alternative proof by applying Lemma 2.1.8 to replace $(X, B)$ by a KLT pair and then applying Theorem 1.9.3 directly.

### 1.11.3 Terminal and Canonical Singularities

In the end of this section, we introduce terminal and canonical singularities. These singularities are not considered in the main part of this book. However, they are important in applications and have a longer history than KLT, DLT, LC, et cetera in dimensions 3 and higher. Originally 3-dimensional algebraic geometry was successful because these singularities can be classified. However, classification of singularities is impossible in higher dimensions, and it is replaced by using log pairs and induction on dimensions.

**Definition 1.11.17** A normal algebraic variety $X$ is said to have *canonical singularities* if the following conditions are satisfied:

(1) $K_X$ is $\mathbf{Q}$-Cartier.
(2) For a resolution of singularities $f: Y \to X$, if write $f^*K_X = K_Y + C$, then $-C$ is effective.

Furthermore, $X$ is said to have *terminal singularities* if the following is satisfied:

(3) The support of $-C$ coincides with the divisorial part of $\text{Exc}(f)$.

In terms of discrepancy coefficients, the feature of terminal singularities (canonical singularities) is that all discrepancy coefficients are positive

60                    *1 Algebraic Varieties with Boundaries*

(nonnegative). It is easy to see that conditions (2) and (3) do not depend on the choice of resolutions of singularities.

The concept of terminal and canonical singularities can be also extended to pairs.

**Definition 1.11.18** A pair $(X, B)$ consisting of a normal algebraic variety $X$ and an effective $\mathbf{R}$-divisor $B$ on $X$ is said to have *canonical singularities* if the following conditions are satisfied:

(1) $K_X + B$ is $\mathbf{R}$-Cartier.
(2) For any resolution of singularities $f : Y \to X$, if write $f^*(K_Y + B) = K_Y + C$, then $-C + f_*^{-1}B$ is effective.

Furthermore, $(X, B)$ is said to have *terminal singularities* if the following is satisfied:

(3) The support of $-C + f_*^{-1}B$ coincides with the divisorial part of $\mathrm{Exc}(f)$.

In conditions (2) and (3), it is not sufficient to check for only one log resolution.

As will be explained in Section 2.5, discrepancy coefficients are nondecreasing under the MMP, hence the MMP preserves types of singularities. That is, when applying a birational map in the MMP to an algebraic variety with certain singularities, we get an algebraic variety with the same type of singularities. In other words, the MMP can be considered within the category of varieties having certain singularities. In particular, when considering the MMP starting from a smooth algebraic variety, everything is within the category of terminal singularities. Note that 2-dimensional terminal singularities without boundaries are just smooth, that is the reason why it is not necessary to consider singularities in the classical 2-dimensional MMP.

## 1.12 Minimality and Log Minimality

The minimality in the minimal model theory is defined by the minimality of canonical divisors. A log minimal model is the log version of a minimal model, where the log canonical divisor is minimized. The MMP is a process to find a "minimal model" which is a birational model with good properties for a given algebraic variety.

First, we define "minimality" by the property of singularities and numerical property of canonical divisors:

## 1.12 Minimality and Log Minimality 61

**Definition 1.12.1** (1) A projective morphism $f : X \to S$ from a normal algebraic variety to another algebraic variety is said to be relatively *minimal* over $S$ if it satisfies the following conditions (a), (b). It is said to be relatively *minimal in weak sense* over $S$ if it satisfies the following conditions (a'), (b).

(a) $X$ has **Q**-factorial terminal singularities.
(a') $X$ has canonical singularities.
(b) $K_X$ is relatively nef over $S$.

(2) A projective morphism $f : (X, B) \to S$ from a pair consisting of a normal algebraic variety and an **R**-divisor to another algebraic variety is said to be relatively *log minimal* over $S$ if it satisfies the following conditions (a), (b). It is said to be relatively *log minimal in weak sense* over $S$ if it satisfies the following conditions (a'), (b).

(a) $X$ is **Q**-factorial and the pair $(X, B)$ is DLT.
(a') The pair $(X, B)$ is LC.
(b) $K_X + B$ is relatively nef over $S$.

The minimality in weak sense defined above leads to the minimality of the canonical divisor $K_X$ and the log canonical divisor $K_X + B$:

**Proposition 1.12.2** (1) *Let* $f : X \to S$ *be a relatively minimal morphism in weak sense. Consider a projective morphism* $g : Y \to S$ *from another normal algebraic variety and birational projective morphisms* $f' : Z \to X$ *and* $g' : Z \to Y$ *from a third normal algebraic variety with* $f \circ f' = g \circ g'$. *If* $K_Y$ *is* **Q**-*Cartier, then the inequality* $(f')^* K_X \le (g')^* K_Y$ *holds. That is,* $K_X$ *is minimal in birational equivalence classes.*

(2) *Let* $f : (X, B) \to S$ *be a relatively log minimal morphism in weak sense. Consider a projective morphism* $g : (Y, C) \to S$ *from another pair of a normal algebraic variety and an* **R**-*divisor, and birational projective morphisms* $f' : Z \to X$ *and* $g' : Z \to Y$ *from a third normal algebraic variety with* $f \circ f' = g \circ g'$. *Furthermore, assume the following conditions:*

(a) *For each irreducible component* $B_i$ *of* $B$, *its strict transform* $C_i = g'_*(f')_*^{-1} B_i$ *is an irreducible component of* $C$. *If we denote the coefficients of* $B_i, C_i$ *by* $b_i, c_i$, *then the inequalities* $b_i \le c_i$ *hold.*
(b) *For each irreducible component* $C_j$ *of* $C$ *satisfying* $f'_*(g')_*^{-1} C_j = 0$, *its coefficient* $c_j$ *is 1.*

*If* $K_Y + C$ *is* **R**-*Cartier, then the inequality* $(f')^*(K_X + B) \le (g')^*(K_Y + C)$ *holds. That is,* $K_X + B$ *is minimal in birational equivalence classes.*

*Proof* (1) By the desingularization theorem we may assume that $Z$ is smooth. Write $(f')^* K_X = K_Z + E$, $(g')^* K_Y = K_Z + F$.

Since $X$ has canonical singularities, $-E$ is effective. That is, $K_X$ is smaller than $K_Z$. So the condition on singularities guarantees the minimality locally.

In order to see the global property, we apply the negativity lemma (Lemma 1.6.3). Write $F - E = G^+ - G^-$, where $G^+, G^-$ are effective **Q**-divisors with no common irreducible component. Our goal is to show $G^- = 0$. Suppose that $G^- \neq 0$. As $-E$ is effective, the support of $G^-$ is contained in the support of $F$, which is contracted by $g'$.

By Lemma 1.6.3, there exists a curve $C$ contracted by $g'$ such that $(G^- \cdot C) < 0$ and $(G^+ \cdot C) \geq 0$. Note that $((K_Z + F) \cdot C) = 0$. On the other hand, since $K_X$ is nef,

$$0 \leq ((K_Z + E) \cdot C) = ((E - F) \cdot C) = -(G^+ \cdot C) + (G^- \cdot C) < 0,$$

which is a contradiction. Therefore, $G^- = 0$ and $F - E$ is effective.

(2) We may assume that $f', g'$ are log resolutions. Write $(f')^*(K_X + B) = K_Z + E$ and $(g')^*(K_Y + C) = K_Z + F$.

Since $(X, B)$ is LC, the coefficients of $E$ are at most 1. Therefore, if we denote by $\bar{E}$ the sum of the strict transform $(f')_*^{-1} B$ and all exceptional divisors of $f'$ with given coefficients 1, then $(f')^*(K_X + B)$ is smaller than $K_Z + \bar{E}$. So the LC condition guarantees the minimality locally.

Let us look at the global property. Write $F - E = G^+ - G^-$, where $G^+, G^-$ are effective **R**-divisors with no common irreducible component. Our goal is to show $G^- = 0$. Suppose that $G^- \neq 0$.

Once it is shown that the support of $G^-$ is contracted by $g'$, the conclusion follows exactly as the proof of (1). In order to show that the support of $G^-$ is contracted by $g'$, for any prime divisor $R$ on $Z$, we are going to show that $R$ is not an irreducible component of $G^-$ if $g'_* R = Q$ is a prime divisor on $Y$.

If $f'_* R = P$ is a prime divisor on $X$, by assumption (a), the coefficient of $P$ in $B$ is not greater than that of $Q$ in $C$. This holds even if $P$ is not an irreducible component of $B$ in which case we just formally set the coefficient to be 0. Therefore, the coefficient of $R$ in $F - E$ is nonnegative and it is not an irreducible component of $G^-$.

If $f'_* R = 0$, by assumption (b), the coefficient of $Q$ in $C$ is 1 while that of $R$ in $E$ is at most 1. Therefore, the coefficient of $R$ in $F - E$ is nonnegative and it is not an irreducible component of $G^-$. $\qquad\square$

## 1.12 Minimality and Log Minimality 63

**Remark 1.12.3** (1) In the minimal model theory in classical algebraic surface theory, a minimal model is defined to be the minimal one under the following relation using birational morphisms: For two smooth projective algebraic surfaces $X, Y$, we define $X \leq Y$ if there exists a birational morphism $Y \to X$.

However, in dimensions 3 and higher, there are examples showing that such a definition does not work ([25, 26]). Therefore, in the minimal model theory discussed in this book, we consider projective algebraic varieties with singularities, and define the minimal model by the size of canonical divisors; the relation $X \leq Y$ between two birationally equivalent algebraic varieties is defined by the inequality $K_X \leq K_Y$. Here the inequality of divisors is by comparing the pullbacks on an appropriate birational model: We write $K_X \leq K_Y$ if $f^* K_X \leq g^* K_Y$ for birational projective morphisms $f \colon Z \to X$ and $g \colon Z \to Y$.

The relation $(X, B) \leq (Y, C)$ for log pairs is defined by $f^*(K_X + B) \leq g^*(K_Y + C)$ for birational projective morphisms $f \colon Z \to X$ and $g \colon Z \to Y$ together with two conditions of (2) of the above proposition.

Such change of viewpoint has already been observed in the log version of algebraic surfaces ([51]). The importance of considering the log version showed up at that time. Furthermore, extending to the log version is indispensable for the inductive proof of the existence of minimal models in this book.

(2) From the above proposition, one can see that the minimality in weak sense is equivalent to the minimality of canonical divisors. Furthermore, according to Corollary 3.6.10 which is derived from the main theorems of this book, minimal models are maximal among minimal models in weak sense under the relation defined by birational morphisms.

Looking at this locally, we can say that: Canonical singularities are characterized by the property that the canonical divisors are locally minimal. Furthermore, **Q**-factorial terminal singularities are maximal, among those with canonical divisors locally minimal, under the relation defined by birational morphisms.

For pairs, the log minimality in weak sense is equivalent to the minimality of log canonical divisors. But as a DLT blowup can be further blown up, it is impossible to construct a "maximal minimal model." However, if the minimal model is KLT, then we can construct a maximal minimal model by Corollary 3.6.10. This is a pair with **Q**-factorial terminal singularities.

Looking at this locally, we can say that: LC pairs are characterized by the property that the log canonical divisors are locally minimal. Furthermore,

64                     *1 Algebraic Varieties with Boundaries*

by looking at only KLT pairs, $\mathbf{Q}$-factorial terminal pairs are maximal, among pairs with log canonical divisors locally minimal, under the relation defined by birational morphisms.

Therefore, the theory requiring $\mathbf{Q}$-factorial terminal singularities can be regarded as "maximalist" and the theory requiring canonical singularities or LC singularities can be regarded as "minimalist." Models that are expected to be obtained using the minimal model program will be "maximalist."

(3) Let $\alpha\colon X \dashrightarrow Y$ be a birational map between normal algebraic varieties projective over $S$. $X, Y$ are said to be *crepant* or *$K$-equivalent* to each other if there are birational projective morphisms $f\colon Z \to X$, $g\colon Z \to Y$ from a third normal algebraic variety with $g = \alpha \circ f$ such that $f^*K_X = g^*K_Y$. Here the comparison of canonical divisors is by using rational differential forms identified by the birational map. By the above proposition, birationally equivalent minimal models are crepant to each other.

Furthermore, given effective $\mathbf{R}$-divisors $B, C$ on $X, Y$, assume that $K_X + B$ and $K_Y + C$ are $\mathbf{R}$-Cartier. The pairs $(X, B)$ and $(Y, C)$ are said to be *log crepant* or *$K$-equivalent* to each other if $f^*(K_X+B) = g^*(K_Y+C)$, or just *crepant* for simplicity. When considering minimal models with boundaries, only being birational is not enough, we should also pay attention to how to define the boundaries. This is settled in Section 2.5.5.

## 1.13 The 1-Dimensional and 2-Dimensional Cases

In this section, we describe known results including the finite generation of canonical rings in dimensions up to 2. Many of them are special phenomena which only happen in dimensions up to 2. In particular, we describe the classification of DLT pairs in dimension 2. We obtain a subadjunction formula from this, and apart from this formula, other results will not be used in subsequent sections. For a DLT pair in arbitrary dimension, its structure in codimension 2 can be considered by cutting down the dimension by general hyperplanes and reducing to the classification of DLT pairs in dimension 2.

### 1.13.1 The 1-Dimensional Case

First, we discuss the 1-dimensional case briefly. Take an algebraic curve $X$, that is, a smooth projective 1-dimensional algebraic variety. Denote its genus by $g$. If $g = 0$, then $X \cong \mathbf{P}^1$ and $R(X, K_X) \cong k$. If $g = 1$, then $K_X \sim 0$ and $R(X, K_X) \cong k[t]$. These cases are simple.

## 1.13 The 1-Dimensional and 2-Dimensional Cases 65

In the following we consider the case $g \geq 2$. This condition is equivalent to $X$ being of general type. It is also equivalent to that the degree of the canonical divisor is positive $\deg(K_X) > 0$ since the degree of the canonical divisor $K_X$ is $2g - 2$. The plurigenera are given by $\dim H^0(X, mK_X) = (2m - 1)(g - 1)$ for $m \geq 2$. As $K_X$ is ample, the canonical ring $R(X, K_X)$ is finitely generated and

$$X = \text{Proj } R(X, K_X).$$

$X$ is called a *hyperelliptic curve* if there exists a finite morphism $\pi : X \to \mathbf{P}^1$ of degree 2. The canonical linear system $|K_X|$ is always free, but it is very ample if and only if $X$ is not a hyperelliptic curve. When $X$ is a hyperelliptic curve,

$$|K_X| = \pi^*|\mathcal{O}_{\mathbf{P}^1}(g - 1)|,$$

where $\pi$ is the morphism corresponding to $|K_X|$. In this case, $|3K_X|$ is very ample ([44, IV.5]).

To be more specific, if $X$ is not a hyperelliptic curve, then the canonical ring is generated by the degree 1 part $H^0(X, K_X)$ (a theorem of Max Noether [5, p. 117]). On the other hand, if $X$ is a hyperelliptic curve, then degree up to 3 parts are required to generate the canonical ring.

### 1.13.2 Minimal Models in Dimension 2

In the following, we consider the 2-dimensional case. For details please refer to [10]. Let $X$ be an *algebraic surface*, that is, a 2-dimensional algebraic variety.

Numerical geometry is particularly effective on algebraic surfaces. This is because the intersection number becomes a symmetric bilinear form since prime divisors are the same as curves. The following powerful theorem is often used in algebraic surface theory. It can be used even for problems in higher dimensional algebraic geometry, by cutting by hyperplane sections and reducing to algebraic surfaces (see Lemma 1.6.3).

**Theorem 1.13.1** (*Hodge index theorem* [44, Theorem V.1.9]) *Let $A, B$ be Cartier divisors on a proper 2-dimensional algebraic variety $X$. If $(A^2) > 0$, $(A \cdot B) = 0$, and $B \not\equiv 0$, then $(B^2) < 0$.*

**Corollary 1.13.2** *Let $f : Y \to X$ be a resolution of singularities of an algebraic surface and let $D$ be a nonzero divisor on $Y$ supported in the exceptional set $\text{Exc}(f)$. Then $(D^2) < 0$. Therefore, if the exceptional divisors of $f$ are $E_1, \ldots, E_r$, then the matrix of intersection numbers $[(E_i \cdot E_j)]$ is negative definite.*

*Proof* We may assume that $X$ is projective. Take an ample divisor $H$ on $X$, then $(f^*H \cdot f^*H) > 0$ and $(f^*H \cdot D) = 0$. If $D \geq 0$, as $Y$ is projective, $D \neq 0$ implies $D \not\equiv 0$. Therefore, $(D^2) < 0$. In general, we can write $D = D^+ - D^-$ in terms of the positive part and the negative part, then $(D^2) \leq (D^+)^2 + (D^-)^2 < 0$. $\qquad \square$

In general, given a resolution of singularities $f : Y \to X$, the *dual graph* $\Gamma$ can be constructed from the exceptional set as the following:

(1) Take vertices $v_1, \ldots, v_r$ of $\Gamma$ corresponding to prime divisors $E_1, \ldots, E_r$ in $\mathrm{Exc}(f)$.
(2) Join $v_i, v_j$ with an edge if two distinct prime divisors $E_i, E_j$ intersect and associate the edge with weight $(E_i \cdot E_j)$.
(3) Associate each vertex $v_i$ with the self-intersection number $(E_i^2)$ as its weight.

First of all, we recall the minimality of algebraic surfaces. The definition of minimal models in algbraic surface theory is different from that in higher dimensional algebraic geometry. Hence here we use "minimal in the classical sense." Given two smooth algebraic surfaces $X, Y$, the relation $X \geq Y$ is defined by that there is a birational projective morphism $f : X \to Y$. An algebraic surface minimal under this relation is defined to be *minimal in the classical sense*.

A curve $C$ on $X$ is called a $(-1)$-*curve* if $C \cong \mathbf{P}^1$ and the self-intersection number $(C^2) = -1$. If we blow up a smooth algebraic surface $Y$ at a point $P$, then the exceptional set is a $(-1)$-curve. Conversely, a $(-1)$-curve can be contracted to a smooth point:

**Theorem 1.13.3** (*Castelnuovo's contraction theorem* [44, Theorem V.5.7]) *For a smooth algebraic surface $X$ and a $(-1)$-curve $C$ on $X$, there exists a birational projective morphism $f : X \to Y$ to another smooth algebraic surface such that $f(C)$ is a point and $f$ induces an isomorphism $X \setminus C \cong Y \setminus f(C)$.*

Minimality is characterized by the absence of $(-1)$-curve:

**Theorem 1.13.4** ([44, Proposition V.5.3]) *A smooth algebraic surface $X$ is minimal in the classical sense if and only if there is no $(-1)$-curve on $X$.*

**Corollary 1.13.5** *For a smooth projective algebraic surface $X$, its minimal model in the classical sense always exists.*

*Proof* In the case that $f : X \to Y$ is a contraction of a $(-1)$-curve, the Picard number decreases exactly by one: $\rho(X) = \rho(Y) + 1$. As the Picard number

## 1.13 The 1-Dimensional and 2-Dimensional Cases

is always positive, a minimal model in the classical sense can be obtained by taking contractions finitely many times. □

Minimal projective algebraic surfaces in the classical sense are classified into the following three types:

(1) A surface with $K_X$ nef.
(2) A $\mathbf{P}^1$-bundle over a curve.
(3) $\mathbf{P}^2$.

In this book, (1) is called a *minimal model*, and (2) or (3) is called a *Mori fiber space*. In case (1), the minimal model is unique, so it is the minimum one. On the other hand, in cases (2) and (3), the minimal model (in the classical sense) is not unique, so such a model is sometimes said to be *relatively minimal*, but to avoid confusion we will not use this terminology.

Combining the existence of resolution of singularities and Castelnuovo's contraction theorem, we get the *minimal resolution of singularities* of a normal algebraic surface. It is a minimal model in the relative setting, which is obtained by considering $\rho(Y/X)$ instead of $\rho(X)$:

**Corollary 1.13.6** ([44, Theorem V.5.8]) *Let $X$ be a normal algebraic surface. Then among all birational projective morphisms $g\colon Y \to X$ from smooth algebraic surfaces, there exists a unique minimal one in the classical sense.*

We also have the following *minimal log resolution of singularities* which is the log version of the minimal resolution of singularities:

**Proposition 1.13.7** *Let $(X, B)$ be a pair consisting of a normal algebraic surface and a reduced divisor. Then among all birational projective morphisms $g\colon Y \to X$ from smooth algebraic surfaces such that the sum of $f_*^{-1}B$ and the exceptional divisor $E$ is a normal crossing divisor, there exists a unique minimal one in the classical sense.*

For a projective algebraic curve $C$ on a smooth algebraic surface $X$, the following *genus formula* holds ([44, Example V.3.9.2]):

$$(K_X \cdot C) + (C^2) = 2\bar{g} - 2 \geq -2.$$

Here $\bar{g}$ is called the *virtual genus* of $C$, which is a nonnegative integer. Take $g$ to be the genus of the smooth projective curve $C^v$ obtained from taking the normalization of $C$, then $\bar{g} \geq g$. The difference $\bar{g} - g$ comes from the singularities of $C$. In particular, the equality holds if and only if $C$ is smooth.

Minimal resolution of singularities is characterized by relative nefness of the canonical divisor. This coincides with the definition of minimality in this book:

68            *1 Algebraic Varieties with Boundaries*

**Proposition 1.13.8** *(1) A birational projective morphism* $f: Y \to X$ *from a smooth algebraic surface to a normal algebraic surface is the minimal resolution of singularities if and only if* $K_Y$ *is relatively nef.*

*(2) Let* $f: Y \to X$ *be the minimal resolution of singularities of a normal algebraic surface. If we write* $f^* K_X = K_Y + C$, *then C is effective.*

*Proof* (1) If there is a $(-1)$-curve $C$ such that $f(C)$ is a point, then $(K_Y \cdot C) = -1$ and $K_Y$ is not relatively nef.

Conversely, if $K_Y$ is not relatively nef, then there is a curve $C$ such that $(K_Y \cdot C) < 0$ and $f(C)$ is a point. By the Hodge index theorem (Corollary 1.13.2), $(C^2) < 0$. On the other hand, by the genus formula, $(K_Y \cdot C) + (C^2) \geq -2$. Hence we have $((K_Y + C) \cdot C) = -2$, and hence $C \cong \mathbf{P}^1$ and $(C^2) = -1$. So $C$ is a $(-1)$-curve.

(2) Write $C = C^+ - C^-$, where $C^+$ and $C^-$ are effective divisors with no common irreducible component. If $C^- \neq 0$, then $(K_Y \cdot C^-) = -(C^+ \cdot C^-) + (C^- \cdot C^-) < 0$, which contradicts the fact that $K_Y$ is relatively nef. $\qquad\square$

For the Euler characteristic $\chi(\mathcal{O}_X) = \sum (-1)^i \dim H^i(X, \mathcal{O}_X)$ of a smooth projective algebraic surface $X$, we have *Noether's formula*

$$\chi(\mathcal{O}_X) = \frac{1}{12}((K_X^2) + c_2(X)).$$

Here $c_2(X)$ is the *second Chern class* of the tangent bundle of $X$ and $-K_X = c_1(X)$ is the *first Chern class*.

### 1.13.3 The Classification of Algebraic Surfaces

Let us consider the finite generation problem for canonical rings of smooth projective algebraic surfaces. The important thing here is that canonical rings are invariant under contractions of $(-1)$-curves: $f^*: R(X', K_{X'}) \cong R(X, K_X)$. Therefore, in the following we consider $X$ to be minimal.

In the classification of minimal models in the classical sense, for a Mori fiber space in case (2) or (3), its canonical ring is just $k$, and the finite generation is trivial. In the following we just consider case (1). The following content is a deep result called the *Kodaira–Enriques classification theory* for algebraic surfaces. In addition, Kodaira also classified (not necessarily algebraic) compact complex surfaces, but we will not discuss them here ([9]).

The Kodaira dimension $\kappa(X)$ takes value among $0, 1, 2$. When $\kappa(X) = 0$, there exists a positive integer $r$ such that $r K_X \sim 0$. If we take $r$ to be the smallest one with such property, then $r$ is among $1, 2, 3, 4, 6$. In particular, $R(X, K_X) \cong k[t^r]$.

When $\kappa(X) = 1$, there exists a surjective morphism $f: X \to Y$ to a smooth projective algebraic curve such that the generic fiber is an elliptic curve. The following *Kodaira's canonical bundle formula* holds:

$$K_X \sim_{\mathbf{Q}} f^*(K_Y + B).$$

Moreover, $\deg(K_Y + B) > 0$. Here $B$ is a $\mathbf{Q}$-divisor on $Y$ determined by types of singular fibers of $f$ and $\sim_{\mathbf{Q}}$ means $\mathbf{Q}$-linearly equivalent. Singular fibers are completely classified and the corresponding coefficients of $B$ are determined. Here the coefficients of $B$ are not necessarily contained in the open interval $(0, 1)$. This is because it also includes a part induced from the $J$-function $J: Y \to \mathbf{P}^1$ coming from the fibers of $f$. Anyway, there exists a positive integer $r$ such that $r K_X \sim f^*(r(K_Y + B))$ and $R(X, r K_X) \cong R(Y, r(K_Y + B))$. The latter is finitely generated as $r(K_Y + B)$ is an ample divisor, which implies that $R(X, K_X)$ is finitely generated.

Consider the case $\kappa(X) = 2$. A minimal model $X$ is of general type if and only if the self-intersection number of the canonical divisor is positive $(K_X^2) > 0$. For $m \geq 2$, by a vanishing theorem of Kodaira type, we have the following plurigenus formula:

$$\dim H^0(X, m K_X) = \frac{1}{2}m(m - 1)(K_X^2) + \chi(\mathcal{O}_X).$$

We discuss the canonical models. A curve $C$ on $X$ is called a $(-2)$-*curve* if $C \cong \mathbf{P}^1$ and $(C^2) = -2$. On a minimal surface of general type, a $(-2)$-curve is characterized by the condition $(K_X \cdot C) = 0$. This is because, on the one hand, $(C^2) < 0$ by the Hodge index theorem (Corollary 1.13.2) and on the other hand, $(K_X \cdot C) + (C^2) \geq -2$ by the genus formula. According to *Artin's contraction theorem* ([6] or Theorem 1.13.10), we can contract all $(-2)$-curves by a birational morphism; there exists a birational morphism $g: X \to Y$ to a normal algebraic surface such that the exceptional set of $g$ coincides with the union of all $(-2)$-curves. $Y$ is called the *canonical model*.

The canonical divisor $K_Y$ of $Y$ is a Cartier divisor and $K_X = g^* K_Y$. Therefore, there is an isomorphism $g^*: R(Y, K_Y) \cong R(X, K_X)$. Since $K_Y$ is ample, the canonical ring $R(X, K_X)$ is finitely generated and $Y = \operatorname{Proj} R(X, K_X)$. This is the proof of the finite generation of canonical rings in dimension 2 by Mumford ([107]). More precisely, on the canonical model, $|5K_Y|$ is very ample ([17]).

### 1.13.4 Rational Singularities

For a minimal model $X$ of general type, its canonical model $Y$ has *canonical singularities* because the birational morphism $g: X \to Y$ is crepant

($K_X = f^*K_Y$). Canonical singularities in dimension 2 are known to be the same as *rational double points*, that is, rational singularities of multiplicity 2. Such singularities were investigated in many different situations historically. They are also called *Du Val singularities, Klein singularities, simple singularities*, or *ADE singularities*. Here we summarize the classification of 2-dimensional canonical singularities:

**Theorem 1.13.9** *Let $P \in X$ be a canonical singularity in dimension 2.*

(1) *Take $f: Y \to X$ to be the minimal resolution of singularity, then the exceptional set $\mathrm{Exc}(f)$ is a normal crossing divisor whose irreducible components are all $(-2)$-curves and the dual graph defined by their intersections is among the* Dynkin diagrams *of type $A_n, D_n, E_6, E_7, E_8$ (see Figure 1.2).*

*Conversely, on a smooth algebraic surface, a normal crossing divisor whose irreducible components are all $(-2)$-curves with dual graph of type $A_n, D_n, E_6, E_7, E_8$ can be contracted to a canonical singularity by a birational projective morphism.*

(2) *When the base field is $\mathbf{C}$, there exists an analytic neighborhood of $P$ isomorphic to the neighborhood of the origin of the hypersurface in $\mathbf{C}^3$ defined by one of the following equations:*

$$A_n : x^2 + y^2 + z^{n+1} = 0, \quad n \geq 1;$$
$$D_n : x^2 + y^2 z + z^{n-1} = 0, \quad n \geq 4;$$
$$E_6 : x^2 + y^3 + z^4 = 0;$$

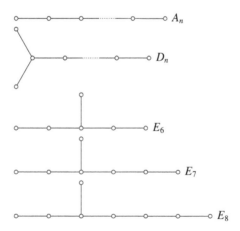

Figure 1.2 Dynkin diagrams.

## 1.13 The 1-Dimensional and 2-Dimensional Cases

$$E_7 : \ x^2 + y^3 + yz^3 = 0;$$
$$E_8 : \ x^2 + y^3 + z^5 = 0.$$

*Here $(x, y, z)$ are coordinates of $\mathbf{C}^3$.*

*(3) When the base field is $\mathbf{C}$, it is analytically isomorphic to the singularity of the image of the origin of the quotient space $\mathbf{C}^2/G$ by a finite subgroup $G$ of $\mathrm{SL}(2, \mathbf{C})$.*

More generally, rational singularities on algebraic surfaces are defined by Artin ([7]). Please refer to the original paper for the proof. The theorem is characteristic free:

**Theorem 1.13.10** *Let $X$ be a smooth algebraic surface and let $E_i$ ($i = 1, \ldots, r$) be projective curves on $X$ such that the union $E = \bigcup E_i$ is connected. Assume that the matrix of intersections $[(E_i \cdot E_j)]$ is negative definite. Then the following assertions hold:*

*(1) There exists a smallest effective integral divisor $F = \sum e_i E_i \neq 0$ satisfying the property that $(F \cdot E_i) \leq 0$ for all $i$. It is called the* fundamental cycle.
*(2) The inequality $(K_X \cdot F) + (F^2) \geq -2$ holds.*
*(3) If the equality $(K_X \cdot F) + (F^2) = -2$ holds, then there exists a birational projective morphism $f \colon X \to Y$ to a normal algebraic surface and the exceptional set $\mathrm{Exc}(f)$ coincides with $E$. In this case, the singularity of $Y$ is called a* rational singularity.
*(4) Rational singularities are $\mathbf{Q}$-factorial. Moreover, $R^1 f_* \mathcal{O}_X = 0$. Conversely, a normal singularity on an algebraic surface $Y$ with a resolution of singularity $f \colon X \to Y$ satisfying $R^1 f_* \mathcal{O}_X = 0$ is a rational singularity.*

The condition $R^1 f_* \mathcal{O}_X = 0$ is independent of the choice of resolutions of singularities since for $g \colon X' \to X$ a blowup of a smooth algebraic surface at a point, $R^1 g_* \mathcal{O}_{X'} = 0$ and $g_* \mathcal{O}_{X'} \cong \mathcal{O}_X$ hold.

**Example 1.13.11** (1) On a smooth algebraic surface, a curve satisfying $C \cong \mathbf{P}^1$ and $(C^2) < 0$ can be contracted to a rational singularity.
(2) Dual graphs obtained by taking resolutions of singularities of 2-dimensional DLT pairs (see Figure 1.3) can be contracted to rational singularities.

**Proposition 1.13.12** *Let $X$ be a normal algebraic surface with at most rational singularities and let $f \colon Y \to X$ be a resolution of singularities. Then prime divisors in the exceptional set of $f$ are all isomorphic to $\mathbf{P}^1$ and the dual graph*

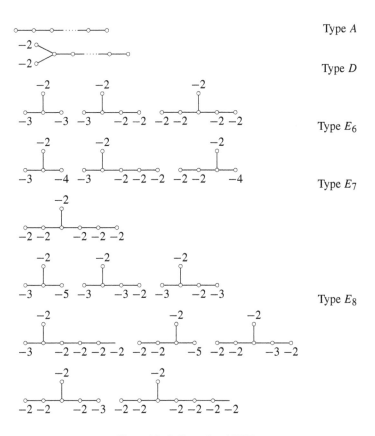

Figure 1.3 2-dimensional DLT.

is a tree. Here a tree is a graph with all edges having weight one and with no cycles.

*Proof* Since $R^1 f_* \mathcal{O}_Y = 0$, $\lim_E H^1(E, \mathcal{O}_E) = 0$ by [44, Theorem III.11.1]. Here the limit is the inverse limit for all subschemes $E$ supported on the exceptional set of $f$. Since the exceptional set of $f$ is 1-dimensional, for any effective divisor $E$ supported in $\mathrm{Exc}(f)$, we have $H^1(E, \mathcal{O}_E) = 0$. This concludes the proof. □

**Remark 1.13.13** According to a theorem of Grauert ([33]), for a smooth complex analytic surface $X$ and projective curves $E_i$ ($i = 1, \ldots, r$) on $X$ such that the union $E = \bigcup E_i$ is connected and the matrix of intersections $[(E_i \cdot E_j)]$ is negative definite, there always exists a proper birational morphism $f : X \to Y$ to a normal complex analytic surface such that the exceptional set

## 1.13 The 1-Dimensional and 2-Dimensional Cases

of $f$ coincides with $E$. However, $Y$ does not necessarily admit an algebraic structure and $f$ is not necessarily algebraic.

### 1.13.5 The Classification of DLT Surface Singularities I

Numerical geometry becomes easy for normal algebraic surfaces. Even for **R**-divisors which are not **R**-Cartier, intersection numbers and pullback by a morphism can be well defined.

Let $X$ be a normal algebraic surface and let $D$ be an **R**-divisor on $X$. Take a resolution of singularities $f : Y \to X$ and denote by $E_i$ $(i = 1, \dots, r)$ the exceptional divisors. *Mumford's numerical pullback* $f^* D = f_*^{-1} D + \sum e_i E_i$ is defined as the following ([106]): The coefficients $e_i$ are the solution of the equations $(f^* D \cdot E_i) = 0$ for all $i$, which are uniquely determined since $[(E_i \cdot E_j)]$ is negative definite. If moreover $D$ is effective, we can see that $f^* D$ is again effective.

For two **R**-divisors $D$ and $D'$, their intersection number can be defined by $(D \cdot D') = (f^* D \cdot f^* D')$.

From now on, we work on the classification of 2-dimensional DLT pairs. Here, in all discussions, we assume that the base field is of characteristic 0. There is also a classification in positive characteristics ([51]).

As the definition of pullback extends to all **R**-divisors, for a pair $(X, B)$ we can define the concept such as KLT and DLT without assuming that $K_X + B$ is **R**-Cartier. Therefore, in the following, this assumption is removed. However, as will be shown later in this section, it turns out that $K_X + B$ automatically becomes **R**-Cartier.

First, we generalize the vanishing theorem slightly. For algebraic surfaces, the normal crossing condition which is important in Theorem 1.9.7 can be removed:

**Proposition 1.13.14** *Let $X$ be a smooth projective algebraic surface defined over an algebraically closed field of characteristic 0, let $f : X \to S$ be a projective morphism to another algebraic variety, and let $D$ be a relatively nef and relatively big **R**-divisor on $X$. Then $R^1 f_*(\mathcal{O}_X(K_X + \ulcorner D \urcorner)) = 0$.*

*Proof* Take a log resolution $g : Y \to X$ of $(X, D)$. By Theorem 1.9.7, $R^1(f \circ g)_*(\mathcal{O}_Y(K_Y + \ulcorner g^* D \urcorner)) = R^1 g_*(\mathcal{O}_Y(K_Y + \ulcorner g^* D \urcorner)) = 0$. Then, arguing by spectral sequence, we get $R^1 f_*(g_*(\mathcal{O}_Y(K_Y + \ulcorner g^* D \urcorner))) = 0$. In the exact sequence

$$0 \to g_*(\mathcal{O}_Y(K_Y + \ulcorner g^* D \urcorner)) \to \mathcal{O}_X(K_X + \ulcorner D \urcorner) \to Q \to 0,$$

74  *1 Algebraic Varieties with Boundaries*

the cokernal $Q$ of the natural homomorphism has 0-dimensional support, hence it does not have higher cohomologies. Therefore, the proof is completed.

$\square$

DLT pairs have rational singularities:

**Proposition 1.13.15** *Let $(X, B)$ be a 2-dimensional DLT pair defined over an algebraically closed field of characteristic 0. Then $X$ has rational singularities. If $(X, B)$ is only LC, then $X$ has rational singularities at points in the support of $B$.*

*Proof* Since $(X, B)$ is DLT, $(X, 0)$ is again DLT. Here note that the condition $K_X + B$ being **R**-Cartier is removed in the definition of DLT. As $(X, 0)$ has no boundary, it is KLT. Take the minimal resolution of singularities $f : Y \to X$ and write $f^* K_X = K_Y + C$. As it is the minimal resolution, $C$ is effective. Since $(X, 0)$ is KLT, $\ulcorner -C \urcorner = 0$. Applying Proposition 1.13.14 to $D = -f^* K_X$, we get $R^1 f_* \mathcal{O}_Y = R^1 f_* (\mathcal{O}_Y(\ulcorner -C \urcorner)) = 0$.

For the latter assertion, when the pair $(X, B)$ is LC, $(X, 0)$ is KLT at points in the support of $B$.

$\square$

Rationality of singularities implies **Q**-factoriality:

**Proposition 1.13.16** *Algebraic surfaces defined over the complex number field with only rational singularities are **Q**-factorial.*

*Proof* Take a resolution of singularities $f : Y \to X$. Consider $Y$ as a complex analytic variety, consider its sheaves in the classical topology instead of the Zariski topology. Then there exists an *exponential exact sequence*

$$0 \to \mathbf{Z}_Y \to \mathcal{O}_Y \to \mathcal{O}_Y^* \to 0.$$

Here the map $\mathcal{O}_Y \to \mathcal{O}_Y^*$ is defined by the exponential function $z \mapsto e^{2\pi i z}$. Note that such kind of exact sequence does not exist in the Zariski topology.

By assumption, $R^1 f_* \mathcal{O}_Y = 0$, hence the map $R^1 f_* \mathcal{O}_Y^* \to R^2 f_* \mathbf{Z}_Y$ is injective.

For any divisor $D$ on $X$, its numerical pullback $f^* D$ is a **Q**-divisor, so we can take a positive integer $m$ such that $m f^* D$ is integral. Note that $\mathcal{O}_Y(m f^* D)$ determines an element in $R^1 f_* \mathcal{O}_Y^*$ whose image in $R^2 f_* \mathbf{Q}_Y$ is 0 since $(m f^* D \cdot E) = 0$ for every $f$-exceptional curve $E$. Therefore, there exists a positive integer $m'$ such that the image of $\mathcal{O}_Y(mm' f^* D)$ in $R^2 f_* \mathbf{Z}_Y$ is 0. This induces an isomorphism

$$\mathcal{O}_Y(mm' f^* D) \cong \mathcal{O}_Y.$$

## 1.13 The 1-Dimensional and 2-Dimensional Cases 75

The global section of the left-hand side corresponding to 1 of the right-hand side determines a rational function $h$ on $Y$ such that $\text{div}(h)_Y = -mm' f^* D$. Hence $\text{div}(h)_X = -mm' D$ which means that $mm' D$ is Cartier. $\square$

As 2-dimensional DLT pairs are rational singularities, they are **Q**-factorial, and hence numerical pullback is actually the same as pullback. For an LC pair, the same holds true on the support of the boundary.

Next, we show that the KLT or LC property is preserved under covering:

**Lemma 1.13.17** *Let $f: Y \to X$ be a finite surjective morphism étale in codimension 1 between normal algebraic varieties defined over an algebraically closed field of characteristic 0.*

*Let $B$ be an effective **R**-divisor on $X$ such that $K_X + B$ is **R**-Cartier and write $f^*(K_X + B) = K_Y + C$. Then the pair $(X, B)$ is LC if and only if the pair $(Y, C)$ is LC. The same holds true for KLT pairs.*

*Proof* As $f$ is étale in codimension 1, $C$ is effective. Take a log resolution $g: X' \to X$ of $(X, B)$ and take $Y'$ to be the normalization of $X'$ in the function field $k(Y)$. Denote the induced morphisms by $h: Y' \to Y$ and $f': Y' \to X'$. Write $g^*(K_X + B) = K_{X'} + B'$ and $h^*(K_Y + C) = K_{Y'} + C'$.

First, we show that $(X, B)$ is LC assuming that $(Y, C)$ is LC. Take an arbitrary prime divisor $D$ contracted by $g$ and denote its coefficient in $B'$ by $d$. Take a prime divisor $E$ on $Y'$ such that $f'(E) = D$ and denote the ramification index of $E$ with respect to $f'$ by $r$. Then the coefficient of $E$ in $(f')^* D$ and $K_{Y'} - (f')^* K_{X'}$ are $r$ and $r - 1$, respectively. Therefore, take $e$ to be the coefficient of $E$ in $C'$, we get the relation

$$dr = r - 1 + e.$$

Since $e \le 1$ by assumption, we get $d \le 1$. Moreover, if $e < 1$, then $d < 1$.

Conversely, we show that $(Y, C)$ is LC assuming that $(X, B)$ is LC. By using the result we just proved in the first part, we may replace $Y$ by taking the Galois closure and assume that the field extension $k(Y)/k(X)$ is Galois from the beginning. As the Galois group $G$ acts on $Y$, we now take $h: Y' \to Y$ to be a $G$-equivariant log resolution. For example, a canonical resolution (Remark 1.6.2(4)) is automatically $G$-equivariant. The quotient space $X' = Y'/G$ has quotient singularities. Denote by $g: X' \to X$ and $f': Y' \to X'$ the induced morphisms. Take a prime divisor $E$ contracted by $h$ and define $D, e, d$ in the same way as the first part. Although $X'$ is not smooth, we still have $dr = r - 1 + e$. Since $d \le 1$ by assumption, we get $e \le 1$. Moreover, if $d < 1$, then $e < 1$. $\square$

**Remark 1.13.18** Here we give some remarks about the topology of algebraic varieties defined over the complex number field. In general, the topology of algebraic varieties is the Zariski topology, but when the base field is the complex number field, the classical Euclidean topology is also useful. For example, the exponential exact sequence that appeared in Proposition 1.13.16 makes sense only in the latter topology.

As an open subset in the Zariski topology is large, it admits nontrivial structure itself, on the other hand, classical topology has polydisks as a base and its local structure is trivial. Since there are many open subsets, even the constant sheaf has nontrivial cohomology groups.

For algebraic varieties defined over the complex number field, many definitions and results hold both for the Zariski topology and the classical topology. Furthermore, in many cases they can be generalized to nonalgebraic complex analytic varieties. For example, the definitions of DLT pairs and LC pairs can be generalized using resolutions of complex analytic singularities. The same is true for DLT pairs having rational singularities. The fact that LC and KLT are preserved by étale in codimension 1 coverings can be also generalized since it is a consequence of the ramification formula.

The construction of *index* 1 *covers* can be also generalized. For example, for an effective divisor $D$ on a complex analytic variety $X$ such that there is an isomorphism $\mathcal{O}_X(rD) \cong \mathcal{O}_X$, take a regular function $h$ such that $\operatorname{div}(h) = rD$, take the normalization of the subvariety defined by the equation $z^r = h$ in the trivial line bundle $X \times \mathbf{C}$ over $X$, we get the index 1 cover. Here $z$ is the coordinate in the fiber direction. When $D$ is not effective, we can consider a similar construction in $X \times \mathbf{P}^1$.

However, as stated in Remark 1.1.2, we should take care of the concept of normal crossing divisor. We should also take care of $\mathbf{Q}$-factoriality. A complex analytic variety $X$ is *analytically* $\mathbf{Q}$-*factorial* if for any analytic neighborhood $U$ of any point $P \in X$ and any codimension 1 subvariety $D$ defined on $U$, there exists a neighborhood $U'$ of $P$ in $U$, a positive integer $r$, and a regular function $h$ on $U'$ such that $\operatorname{div}_{U'}(h) = r(D \cap U')$. As the algebraic $\mathbf{Q}$-factoriality is a condition for globally defined prime divisors, analytical $\mathbf{Q}$-factoriality is a stronger condition.

## 1.13.6 The Classification of DLT Surface Singularities II

We describe the classification of DLT pairs for algebraic surfaces. The results are established in a sufficiently small analytic neighborhood near the singularity.

## 1.13 The 1-Dimensional and 2-Dimensional Cases

First, consider the structure near points in the support of the boundary:

**Theorem 1.13.19** ([61]) *Let $X$ be an algebraic surface defined over the complex number field and let $B$ be a reduced divisor on $X$. Assume that $(X, B)$ is DLT. Then for any point $P \in X$ in the support of $B$, there exists an analytic neighborhood $U$ such that one of the following assertions holds:*

*(1) $U$ is smooth and $B|_U$ is a normal crossing divisor in complex analytic sense.*

*(2) $U$ has a cyclic quotient singularity of type $\frac{1}{r}(1, s)$ and $B|_U$ is irreducible. Here $r, s$ are coprime positive integers. In more detail, there exists a neighborhood $U_0$ of the origin of the affine space $\mathbf{C}^2$ with coordinates $x, y$, a group action by $G = \mathbf{Z}/(r)$ as $x \mapsto \zeta x$, $y \mapsto \zeta^s y$ such that the pair $(U, B|_U)$ is analytically isomorphic to $(U_0/G, B_0/G)$. Here $\zeta$ is a primitive $r$th root of 1 and $B_0 = \mathrm{div}(x)$. In this case, $(U, B|_U)$ is PLT.*

*Conversely, pairs satisfying (1) or (2) are DLT.*

*Proof* Take a sufficiently small analytic neighborhood $U$ of $P$ and take an analytic irreducible component $B_1$ of $B \cap U$. We may assume that $B_1$ remains irreducible when replacing $U$ by smaller neighborhoods. Here note that it is possible that an (algebraic) irreducible component of $B$ containing $B_1$ and passing $P$ is strictly bigger than $B_1$ when restricting to $U$.

Since $X$ has rational singularities, it is analytically $\mathbf{Q}$-factorial. Hence the divisor $B_1$ on $U$ is $\mathbf{Q}$-Cartier. Take $r_1$ to be the smallest positive integer such that $r_1 B_1$ is Cartier. Then we may assume that $\mathcal{O}_U(r_1 B_1) \cong \mathcal{O}_U$. Take $\pi_1 \colon Y_1 \to U$ to be the index 1 cover. As $\pi_1$ is étale in codimension 1, by Lemma 1.13.17, $(Y_1, \pi_1^* B)$ is LC.

If one of the analytic irreducible components of $\pi_1^* B$ is not Cartier, note that $Y_1$ has again rational singularities, we can construct an index 1 cover $\pi_2 \colon Y_2 \to Y_1$ again. Therefore, we can construct a finite cover $\pi \colon Y \to U$ étale in codimension 1 such that any analytically irreducible component of $C = \pi^* B$ is Cartier. By construction, $Q = \pi^{-1}(P)$ is one point.

We will show that $Y$ is smooth. Suppose not, take the minimal resolution of singularities $g \colon Z \to Y$. Take $C_j$ to be an analytically irreducible component of $C$, as $C_j$ is Cartier, $g^* C_j$ is an integral divisor. Note that the support of $g^* C_j$ contains the exceptional set of $g$.

Take $s$ to be the number of such $C_j$. If $s \geq 2$, then any exceptional divisor of $g$ has coefficients at least 1 in $g^* C_1$ and $g^* C_2$. Since $K_Z \leq g^* K_Y$, this contradicts the fact that $(Y, C)$ is LC.

Now $s = 1$. Take $E_1, \ldots, E_r$ to be the exceptional divisors of $g$. Since $Y$ has rational singularities, the dual graph of the exceptional divisors of $g$ is a tree.

Since $(Y, C)$ is LC, we get $g^*C_1 = g_*^{-1}C_1 + \sum E_i$ and $K_Z = g^*K_Y$. Since $C_1$ is analytically irreducible, set-theoretically $g_*^{-1}C_1$ intersects the support of $\sum E_i$ at one point. If the graph of $g_*^{-1}C_1 + \sum E_i$ is not a tree, then we need more blowups to get a log resolution of $(Y, C)$, but this procedure will produce an exceptional divisor with log discrepancy coefficient at least 2, which is a contradiction.

On the other hand, if the graph of $g_*^{-1}C_1 + \sum E_i$ is a tree, then there exists an irreducible component $E_1$ intersecting $g_*^{-1}C_1 + \sum_{i \neq 1} E_i$ at just one point. But by $(K_Z \cdot E_1) = 0$ we get $(E_1^2) = -2$, which contradicts to $(g^*C_1 \cdot E_1) = 0$.

In summary, we showed that $Y$ is smooth. By a similar argument, we can show that $C$ is normal crossing. Note that $Y \setminus Q$ is connected and simply connected, so it coincides with the universal covering of $U \setminus P$. In particular, $\pi : Y \to U$ is a Galois covering. Take $G$ to be the Galois group.

Embed $Y$ into the affine space $\mathbf{C}^2$ with coordinates $x, y$ such that $Q$ is the origin. Since $(Y, C)$ is LC and $Q$ is contained in the support of $C$, we may assume that the equation of $C$ is $xy = 0$ or $x = 0$. By construction, $C$ is invariant under the action of $G$.

If the equation of $C$ is $x = 0$, then $B \cap U$ is analytically irreducible, and hence $G$ is the Galois group of an index 1 cover which is isomorphic to $\mathbf{Z}/(r_1)$. We get into case (2) by diagonalizing the generator of $G$. Here if $r, s$ are not coprime, then there is a nontrivial subgroup of $G$ with fixed locus outside $Q$, which contradicts the fact that $\pi : Y \to U$ is étale in codimension 1.

Consider the case that the equation of $C$ is $xy = 0$. First, consider the case that every irreducible component of $C$ is invariant under the action of $G$. By choosing coordinates properly, the log canonical form $dx/x \wedge dy/y$ is invariant under the action of $G$, and determines a log canonical form $\theta \in H^0(U, K_U + B)$ on the quotient space $Y/G \cong U$. Since $\theta$ has no zeros, $K_U + B$ is Cartier on $U$. Suppose that $U$ is not smooth, take $h : V \to U$ to be the minimal resolution of singularities and write $h^*(K_U + B) = K_V + B_V$, then the coefficients of $B_V$ are integers. Since $h^*K_U \geq K_V$, the coefficients of $B_V$ are at least 1. This contradicts the fact that $(X, B)$ is DLT. Hence $U$ is smooth and we get into case (1).

Next, suppose that there exists an element in $G$ exchanging irreducible components of $C$. Then $B \cap U$ is again analytically irreducible. Hence the DLT pair $(U, B)$ is PLT. Take $G'$ to be the subgroup of $G$ consisting of all elements preserving irreducible components of $C$, then $G_1 = G/G' \cong \mathbf{Z}/(2)$ and the log canonical divisor $K_{Y'} + C'$ on $Y' = Y/G'$ is Cartier. Here $C'$ is the image of $C$, which is a reduced divisor with two irreducible components. If $Y'$ is not smooth, take $g' : Z' \to Y'$ to be the minimal resolution of singularities and write $(g')^*(K_{Y'} + C') = K_{Z'} + C'_{Z'}$, then the coefficients of $C'_Z$ all equal to 1. The action of $G_1$ on $Y'$ extends to $Z'$ and induces a birational morphism

## 1.13 The 1-Dimensional and 2-Dimensional Cases

$h: V = Z'/G_1 \to U = Y'/G_1$. This is not necessarily the minimal resolution of singularities, but if write $h^*(K_U + B) = K_V + B_V$, then by the ramification formula, the coefficients of $B_V$ all equal to 1, which contradicts the fact that $(U, B)$ is PLT. Therefore, $Y'$ is smooth. Then $G' = \{1\}$ and the action of $G_1$ exchanging irreducible components of $C$ is étale in codimension 1, which is absurd. $\qquad\square$

As an application in arbitrary dimension, we can show the subadjunction formula for DLT pairs (see Theorem 1.11.13):

**Corollary 1.13.20** *Let $(X, B)$ be a DLT pair and let $Z$ be an irreducible component of $\llcorner B \lrcorner$. Define the **R**-divisor $B_Z$ on $Z$ by $(K_X + B)|_Z = K_Z + B_Z$. Take an irreducible component $P$ of $B_Z$ with coefficient $p$. Denote by $b_i$ the coefficients of irreducible components of $B$ containing $P$. Then there exist positive integers $m_i, r$ such that*

$$p = \frac{r - 1 + \sum b_i m_i}{r}.$$

*Proof* As we can check the coefficient of $P$ on its generic point, we may assume that $\dim X = 2$ and $P$ is a point. The coefficient remains the same when $X$ is considered as a complex analytic variety, hence we just need to consider two cases in Theorem 1.13.19 applied to $(X, \llcorner B \lrcorner)$. Case (1) is trivial, we only consider case (2).

Let $Y = \mathbf{C}^2$, $W = \mathrm{div}(x)$, $G = \mathbf{Z}/(r)$, $X = Y/G$, and $Z = W/G$. Denote the projection by $\pi : Y \to X$. Take the origin $Q \in Y$ and denote $P = \pi(Q)$. In the DLT pair $(X, B)$, $B = Z + \sum b_i B_i$. Take $C_i = \pi^* B_i$ and $m_i = (C_i \cdot W)_Q$ which are local intersection numbers at $Q$. When $B_i$ passes through $P$, $m_i$ is a positive integer.

Since the covering $\pi : Y \to X$ is étale outside the origin, $\pi^*(K_X + Z) = K_Y + W$. On the other hand, $\pi|_W : W \to Z$ is ramified over $Q$ with index $r$, hence $\pi^* P = rQ$, $K_W = (\pi|_W)^* K_Z + (r - 1)Q$. On the smooth variety $Y$ we have the usual adjunction formula $(K_Y + W)|_W = K_W$. Then the assertion follows. $\qquad\square$

Next we consider points outside the boundary:

**Theorem 1.13.21** ([61]) *Let $X$ be an algebraic surface defined over the complex number field. Assume that the pair $(X, 0)$ is DLT. Then any point $P \in X$ is a quotient singularity. That is, there exists an analytic neighborhood $U$ of $P$ which is analytically isomorphic to the quotient of a neighborhood of the origin $(0, 0)$ of $\mathbf{C}^2$ by the linear action of a finite subgroup $G$ of the general linear group $\mathrm{GL}(2, \mathbf{C})$.*

*Conversely, if $X$ has quotient singularities, then $(X, 0)$ is DLT.*

# 80    *1 Algebraic Varieties with Boundaries*

*Proof* Since $B = 0$, $(U, 0)$ is KLT. First, take the index 1 cover $\pi_1 : Y_1 \to U$ of $K_X$. Since $(Y_1, 0)$ is also KLT and $K_{Y_1}$ is Cartier, $Y_1$ has canonical singularities. Therefore, $Y_1 = U_0/G_1$, where $U_0$ is a neighborhood of the origin of $\mathbf{C}^2$ and $G_1$ is a finite subgroup of SL$(2, \mathbf{C})$. Now $U_0 \setminus \{0\}$ is the universal cover of $U \setminus \{P\}$ and we get the conclusion.

The converse statement follows from the ramification formula and holds for any dimension (Proposition 1.10.6). $\qquad\square$

Birational geometry of algebraic surfaces works for arbitrary characteristics. The classification theorem of minimal models works under certain modification ([18, 19, 110]). The theory of rational singularities remains true, also the contraction theorem remains true ([6, 7]). The dual graph of the resolution of singularities of a DLT pair is completely classified, which is the same as in characteristic 0 ([51]; Figure 1.3). However, in characteristic 0 the singularity can be determined by the dual graph of the resolution of singularity, which turns out to be a quotient singularity, but on the other hand, in positive characteristics it is only known to be a rational singularity and the structure of the singularity is not determined only by the dual graph of the resolution of singularity, the classification seems to be more complicated. In addition, [51] is the origin where the author was involved in the minimal model theory.

## 1.13.7 The Zariski Decomposition

Finally, we state the Zariski decomposition theorem for divisors on algebraic surfaces:

**Theorem 1.13.22** *Let $D$ be an integral divisor on a smooth projective surface $X$. Assume that there exists a positive integer $m$ such that $|mD| \neq \emptyset$. Then there exists an effective $\mathbf{Q}$-divisor $N$ satisfying the following conditions:*

*(1) $P = D - N$ is nef.*
*(2) $(P \cdot E_i) = 0$ for every $i$, where $E_1, \ldots, E_m$ are irreducible components of $N$.*
*(3) The matrix $[(E_i \cdot E_j)]$ is negative definite.*

*Moreover, $N$ is uniquely determined by the above conditions.*

Such a decomposition $D = P + N$ is called the *Zariski decomposition* of $D$ ([144]).

**Proposition 1.13.23** *Let $X$ be a smooth projective surface and let $f : X \to Y$ be a morphism to a minimal model in the classical sense. Assume that $K_Y$ is nef. Set $N = K_X - f^* K_Y$, then $K_X = f^* K_Y + N$ is the Zariski decomposition.*

## 1.13 The 1-Dimensional and 2-Dimensional Cases 81

That is, we can say that the Zariski decomposition indeed gives the minimal model without taking a birational model. This is the reason why Zariski decomposition has drawn a lot of attention.

**Example 1.13.24** We give an example of a log minimal model in dimension 2. The correspondence between Zariski decompositions and log minimal models holds in general ([51]).

Consider an irreducible curve $B$ of degree 4 with three ordinary cusp singularities on the projective plane $X = \mathbf{P}^2$. Here an ordinary cusp singularity is a singularity analytically equivalent to the singularity given by the equation $x^2 - y^3 = 0$ at the origin. By the genus formula, $B$ is a rational curve, that is, its normalization is isomorphic to $\mathbf{P}^1$. Let $f : Y \to X$ be the minimal log resolution of the pair $(X, B)$ and let $C_0 = f_*^{-1} B$ be the strict transform. Let $P_i$ ($i = 1, 2, 3$) be the three singular points on $B$. Over each point there are three exceptional divisors $E_{ij}$ ($i, j = 1, 2, 3$) on $Y$. It is easy to calculate the intersection numbers $(C_0^2) = -2$ and $(E_{ij}^2) = -j$. $C = C_0 + \sum_{i,j} E_{ij}$ is a normal crossing divisor with all irreducible components isomorphic to $\mathbf{P}^1$. The dual graph is shown in Figure 1.4.

The Zariski decomposition $K_Y + C = P + N$ is given by

$$P = K_Y + C_0 + \sum_i \left( E_{i1} + \frac{1}{2} E_{i2} + \frac{2}{3} E_{i3} \right), \quad N = \sum_i \left( \frac{1}{2} E_{i2} + \frac{1}{3} E_{i3} \right).$$

Here $P$ is nef and big with $(P^2) = 1/2$.

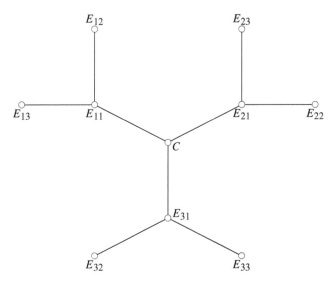

Figure 1.4 Dual graph of the resolution of singularities.

Denote by $g: Y \to Z$ the contraction of six curves $E_{i2}, E_{i3}$ ($i = 1, 2, 3$) in the support of $N$ and $D = g_* C$. Then $K_Z + D$ is ample and $P = g^*(K_Z + D)$. The pair $(Z, D)$ is a minimal model of the DLT pair $(Y, C)$ which is also the canonical model.

In Chapter 2, we will generalize the definition of Zariski decomposition in a weak sense for pseudo-effective **R**-divisors in any dimension, which is called the "divisorial Zariski decomposition."

## 1.14 The 3-Dimensional Case

Let us consider the 3-dimensional case. In this situation, results in higher dimensional algebraic geometry discussed in subsequent chapters are necessary. Indeed, higher dimensional algebraic geometry starts from dimension 3. However, there are also special phenomena and results that only appear in dimension 3. We will describe them briefly as a comparison to results in dimensions up to 2. The results in this section will not be used in subsequent sections.

The MMP, including the existence of flips, the termination of flips, and the abundance conjecture which will be discussed in Chapter 2, is completely understood in dimension 3 even for the log version.

As a consequence of the minimal model theory, the following theorem holds:

**Theorem 1.14.1** *Let $X$ be a smooth projective 3-dimensional algebraic variety over a field of characteristic 0. Then there exists a projective algebraic variety $X'$ with at most* **Q**-*factorial terminal singularities and a birational map $f: X \dashrightarrow X'$ surjective in codimension 1 such that one of the following assertions holds:*

(1) $X'$ *is a* minimal model. *That is, the canonical divisor $K_{X'}$ is nef.*
(2) $X'$ *admits a* Mori fiber space *structure. That is, there exists a surjective morphism $g: X' \to Y$ to a normal algebraic variety $Y$ with $\dim Y < \dim X$ and connected geometric fibers such that $-K_X$ is $g$-ample and $\rho(X/Y) = 1$.*

**Remark 1.14.2** (1) $f$ is not necessarily a morphism and $X'$ is not necessarily smooth, this is a feature in dimensions 3 and higher.
(2) $X'$ has terminal singularities means that the pair $(X', 0)$ with divisor 0 has terminal singularities. The concept of terminal singularities was originally defined by Reid in dimension 3 ([121]). This was the starting point of higher dimensional minimal model theory. However, log terminal

*1.14 The 3-Dimensional Case* 83

singularities for algebraic surfaces already appeared before this ([51]). In dimension 2, terminal singularities are impossible to be aware of since they are automatically smooth.

(3) Any terminal singularity can appear in some minimal model. Terminal singularities in dimension 3 are isolated singularities and are completely classified (Theorem 1.14.5). For example, for two coprime positive integers $r, b$ with $b < r$, a cyclic quotient singularity of type $\frac{1}{r}(1, -1, b)$ is a terminal singularity (see Example 1.10.5 for the notation). The *Cartier index* of a singularity $P \in X$ is the minimal positive integer $m$ such that $m K_X$ is Cartier in a neighborhood of $P$. For example, the Cartier index of a cyclic quotient singularity of type $\frac{1}{r}(1, -1, b)$ is $r$. In particular, there are minimal models with arbitrarily large Cartier indices.

(4) The existence of flips in dimension 3 was proved by Mori via an almost complete classification of small contractions ([102]). As will be discussed in Chapter 3, the existence of flips in arbitrary dimension is proved in a completely different way by induction on dimensions, where the generalization to the log version is essential.

(5) The termination of flips in dimension 3 was proved by Shokurov ([127]). The termination of log flips in dimension 3 was proved in [65]. The termination of flips remains open in arbitrary dimension.

The abundance theorem holds in dimension 3 ([63, 95–97]):

**Theorem 1.14.3** *Let $X$ be a 3-dimensional minimal model. That is, $X$ is a projective algebraic variety with terminal singularities and $K_X$ is nef. Then there exists a positive integer $m$ such that the pluricanonical system $|m K_X|$ is free. Associated with this, there exists a surjective morphism $f : X \to Y$ to a normal projective algebraic variety with connected geometric fibers such that $K_X \sim_{\mathbf{Q}} f^* H$ for an ample $\mathbf{Q}$-divisor $H$ on $Y$. By definition, $\dim Y = \kappa(X)$. In particular, the canonical ring is finitely generated.*

**Remark 1.14.4** (1) The log version of the abundance conjecture in dimension 3 was also proved ([78]).

(2) As will be shown in Chapter 3, the finite generation of canonical rings is much weaker that the abundance theorem.

Terminal singularities in dimension 3 are completely classified as complex analytic singularities ([101, 121, 124]):

**Theorem 1.14.5** *Let $X$ be a 3-dimensional algebraic variety defined over the complex number field with terminal singularities and take $P \in X$ to be a singular point. Then $(X, P)$ is an isolated singularity. Take $r$ to be the Cartier*

84              *1 Algebraic Varieties with Boundaries*

*index, then there exists an analytic neighborhood of $P$ isomorphic to the neighborhood of the image of the origin of one of the following singularities:*

(1) *A cyclic quotient singularity of type $\frac{1}{r}(a, -a, 1)$. Here $r, a$ are coprime positive integers (see Example 1.10.5 for the notation).*

(2) General type: *The quotient space of the hypersurface in $\mathbf{C}^4$ defined by the equation $xy + f(z^r, w) = 0$ at the origin by the cyclic group $\mathbf{Z}/(r)$. In other words, the prime divisor in a 4-dimensional quotient singularity defined by*

$$\left\{ (x, y, z, w) \in \frac{1}{r}(a, -a, 1, 0) \,\middle|\, xy + f(z^r, w) = 0 \right\}.$$

*Here $r, a$ are coprime positive integers and $f$ has no constant term or $w$ term.*

    *The following (3), (4) are also prime divisors in 4-dimensional quotient singularities.*

(3) Special type:

$$\left\{ (x, y, z, w) \in \frac{1}{2}(1, 0, 1, 1) \,\middle|\, x^2 + y^2 + f(z, w) = 0 \right\}, \quad f \in \mathfrak{m}^4, r = 2;$$

$$\left\{ (x, y, z, w) \in \frac{1}{2}(1, 0, 1, 1) \,\middle|\, x^2 + f(y, z, w) = 0 \right\},$$
$$f \in \mathfrak{m}^3 \setminus \mathfrak{m}^4, f_3 \ne y^3, r = 2;$$

$$\left\{ (x, y, z, w) \in \frac{1}{3}(0, 1, 2, 2) \,\middle|\, x^2 + f(y, z, w) = 0 \right\},$$
$$f \in \mathfrak{m}^3, f_3 = y^3 + z^3 + w^3, y^3 + zw^2, \text{ or } y^3 + z^3, r = 3;$$

$$\left\{ (x, y, z, w) \in \frac{1}{2}(1, 0, 1, 1) \,\middle|\, x^2 + y^3 + yf(z, w) + g(z, w) = 0 \right\},$$
$$f \in \mathfrak{m}^4, g \in \mathfrak{m}^4 \setminus \mathfrak{m}^5, r = 2.$$

*Here $\mathfrak{m}$ is the maximal ideal of the origin.*

(4) Exceptional type:

$$\left\{ (x, y, z, w) \in \frac{1}{4}(1, 3, 1, 2) \,\middle|\, x^2 + y^2 + f(z^2, w) = 0 \right\}, \quad r = 4.$$

*Here $f$ has no constant term or $w$ term.*

    The exceptional type is different since $f$ is not invariant under the group action.

**Example 1.14.6** A terminal singularity appearing as the target of a divisorial contraction from a smooth 3-dimensional algebraic variety is either smooth or among one of the following cases:

## 1.14 The 3-Dimensional Case

(1) A cyclic quotient singularity of type $\frac{1}{2}(1,1,1)$.
(2) The hypersurface defined by the equation $xy + zw = 0$ in $\mathbf{C}^4$.
(3) The hypersurface defined by the equation $xy + z^2 + w^3 = 0$ in $\mathbf{C}^4$.

In cases (2) and (3), $K_X$ is Cartier.

More complicated terminal singularities appear when taking divisorial contractions from singular 3-dimensional algebraic varieties. Conversely, for the equation of each singularity above, we can construct a divisorial contraction $f : Y \to X$ explicitly by a *weighted blowing up* of $X$ (see [128, Appendix]).

Let $X$ be a 3-dimensional minimal projective algebraic variety. When $\kappa(X) = 3$, we want to have a formula for plurigenera. Being of general type for $X$ is equivalent to that the self-intersection of the canonical divisor on a minimal model is positive $(K_X^3) > 0$ (Theorem 1.5.12). However, as $K_X$ is not necessarily Cartier, $(K_X^3)$ is in general only a rational number.

By the finite generation of canonical rings, we can define the *canonical model* $Y = \operatorname{Proj} R(X, K_X)$. There exists a birational morphism $g : X \to Y$ such that $K_X = g^* K_Y$ which is the same as in dimension 2. Here this equality is in the following sense: For an integer $m$, $m K_X$ is Cartier if and only if $m K_Y$ is Cartier, moreover, in this case the equality $m K_X = g^*(m K_Y)$ holds. In particular, $|m K_X|$ is free if and only if $|m K_Y|$ is free.

In order to state *Reid's plurigenus formula* in [124], we introduce the concept of baskets of singularities. Take $\{P_1, \ldots, P_t\}$ to be the set of singular points of $X$. Each singular point $(X, P_i)$ is associated with a set of couples of integers $\{\frac{1}{r_{ij}}(1, -1, b_{ij})\}$ which is called the *basket*. Here $r_{ij}, b_{ij}$ are coprime positive integers with $b_{ij} < r_{ij}$. For example, when $(X, P_i)$ is a cyclic quotient singularity of type $\frac{1}{r}(1, -1, b)$, its basket just consists of one couple $\{\frac{1}{r}(1, -1, b)\}$, which coincides with the type of the quotient singularity. In general, a 3-dimensional terminal singularity can be locally deformed into several cyclic quotient singularities, in which case its basket is the collection of types of those cyclic quotient singularities. The Cartier index $r_i$ of $(X, P_i)$ coincides with the least common multiple of $r_{ij}$ in its basket. By considering baskets, terminal singularities can be replaced by a set of virtual cyclic quotient singularities.

Reid's plurigenus formula for $m \geq 2$ is the following:

$$\dim H^0(X, m K_X) = \frac{1}{12} m(m - 1)(2m - 1)(K_X^3) + (1 - 2m)\chi(\mathcal{O}_X)$$
$$+ \sum_{i,j} \left( \frac{r_{ij}^2 - 1}{12 r_{ij}}(m - \bar{m}) + \sum_{k=0}^{\bar{m}-1} \frac{\overline{b_{ij}k} \cdot (r_{ij} - \overline{b_{ij}k})}{2 r_{ij}} \right).$$

Here $\overline{m}$ denotes the residue of $m$ modulo $r_{ij}$ ([124]). This formula is a sum of a polynomial in $m$ and a periodic correction term with respect to $m$ (see [144]). The correction term runs over the baskets of all singularities. As plurigenera are birational invariants, the left-hand side is the same as the starting smooth model, but the right-hand side can be only computed on a minimal model with singularities. In other words, when computing plurigenera on a smooth model, the singularities of its minimal model appear, which is a surprising phenomenon.

Also we have the following formula ([59]):

$$\chi(\mathcal{O}_X) = -\frac{1}{24}(K_X \cdot c_2(X)) + \sum_{i,j} \frac{r_{ij}^2 - 1}{24 r_{ij}}.$$

Here, since $X$ has only isolated singularities, the intersection number $(K_X \cdot c_2(X))$ can be defined properly.

**Remark 1.14.7** In this book, we will show the finite generation of canonical rings. However, it is impossible to find a bound of the degrees of generators depending only on the dimension. This can already be observed in dimension 3.

Let $P$ be a singular point on a minimal model $X$. If $m$ is not divisible by the Cartier index $r$ of $P$, then $P$ is a basepoint of $|mK_X|$. Hence for arbitrary large $m$, we can construct examples such that $|mK_X|$ is not free.

For example, if $\dim X = 3$ and $P$ is a cyclic quotient singularity of type $\frac{1}{r}(a, -a, 1)$, then the canonical ring cannot be generated by elements of degree less than $r$. This is a completely different phenomenon from that in dimensions up to 2, because singularities appear in minimal models in dimensions 3 and higher.

# 2

# The Minimal Model Program

The purpose of this chapter is to formulate the minimal model program (MMP). The basepoint-free theorem and the cone theorem are two main pillars of the MMP, which are known results at the time of [76]. We will also discuss subsequent developments as an effective version of the basepoint-free theorem, the MMP with scaling, the lengths of extremal rays, the divisorial Zariski decomposition, and the Shokurov polytopes. The extension theorem obtained by using multiplier ideal sheaves is an important result that leads to the newest developments of the MMP discussed in Chapter 3.

Numerical geometry plays an important role in the minimal model theory. But unlike Kleiman's criterion, the basepoint-free theorem and the cone theorem do not hold for arbitrary schemes. A feature of the minimal model theory is that the canonical divisor plays a special role.

## 2.1 The Basepoint-Free Theorem

The basepoint-free theorem is one of the two pillars supporting the minimal model theory. It is an important consequence of the vanishing theorem of cohomologies.

For algebraic surfaces, minimal models are obtained by applying Castelnuovo's contraction theorem repeatedly. Contracting topological spaces is always possible. Also, as in Grauert's theorem, contraction maps in complex geometry are known to exist by only assuming numerical conditions. However, as in Artin's theorem, contraction maps in algebraic geometry are more subtle. In order to construct a contraction map in algebraic geometry, one needs a basepoint-free linear system. By using the basepoint-free theorem, one can construct a free linear system in a general setting.

88                           2 *The Minimal Model Program*

### 2.1.1 Proof of the Basepoint-Free Theorem

**Theorem 2.1.1** (*Basepoint-free theorem*) *Let $(X, B)$ be a KLT (Kawamata log terminal) pair, let $f : X \to S$ be a projective morphism, and let $D, E$ be Cartier divisors on $X$. Assume the following conditions:*

*(1) $D$ is relatively nef.*
*(2) There exists a positive integer $m_1$ such that $m_1 D + E - (K_X + B)$ is relatively nef and relatively big.*
*(3) $E$ is effective and there exists a positive integer $m_2$ such that for any positive integer $m \geq m_2$, the natural homomorphism $f_*(\mathcal{O}_X(mD)) \to f_*(\mathcal{O}_X(mD + E))$ is an isomorphism.*

*Then there exists a positive integer $m_3$ such that for any integer $m \geq m_3$, $mD$ is relatively free. That is, the natural homomorphism $f^* f_*(\mathcal{O}_X(mD)) \to \mathcal{O}_X(mD)$ is surjective.*

**Remark 2.1.2** For a given divisor, assuming that its numerical equivalence class is in the closure of the ample cone, that is, assuming that it is nef, to show that it is semi-ample is beyond the limit of Kleiman's criterion. The basepoint-free theorem can be generalized in many different directions, but it is not true if one completely removes the condition on singularities and the condition about canonical divisors. This reflects the complicated geometry of algebraic varieties.

*Proof* **Step 0** As the assertion is relative over $S$, we may assume that $S$ is affine. Then the assertion of the theorem says that the natural homomorphism $H^0(X, mD) \otimes \mathcal{O}_X \to \mathcal{O}_X(mD)$ is surjective, in other words, the linear system corresponding to $H^0(X, mD)$ has no basepoints. Note that when $S$ is not a point, $H^0(X, mD)$ may be infinite-dimensional.

**Step 1** We may assume that $m_1 D + E - (K_X + B)$ is relatively ample and $B$ is a **Q**-divisor.

Indeed, by assumption (2), we can write $m_1 D + E - (K_X + B) = A + B'$ for a relatively ample **R**-divisor $A$ and an effective **R**-divisor $B'$ by Kodaira's lemma. Then for a real number $\epsilon$ with $0 < \epsilon \leq 1$, $m_1 D + E - (K_X + B + \epsilon B')$ is relatively ample, and if $\epsilon$ is sufficiently small, $(X, B + \epsilon B')$ is again KLT. We can just replace $B$ by $B + \epsilon B'$. As ampleness is an open condition, we can adjust the coefficients of $B$ to become rational numbers.

**Step 2** Under the same assumption, we consider a weaker version of the basepoint-free theorem, which is independently called the *nonvanishing*

## 2.1 The Basepoint-Free Theorem

*theorem.* According to the historical order, we will give the proof of the nonvanishing theorem later in this section and show the basepoint-free theorem assuming the nonvanishing theorem in this step.

**Theorem 2.1.3** (*Nonvanishing theorem*) *Assume the same conditions as in Theorem 2.1.1. Then there exists a positive integer $m_3'$ such that for any integer $m \geq m_3'$, $f_*(\mathcal{O}_X(mD)) \neq 0$.*

Fix an integer $m \geq m_3'$, suppose that the linear system $|mD|$ corresponding to $H^0(X, mD)$ has a basepoint. Take a general divisor $M \in |mD|$. Take a log resolution $g: Y \to X$ of $(X, B+E+M)$ in strong sense, denote $h = f \circ g$, and write $g^*(K_X + B) = K_Y + C$. We may take a decomposition $g^*M = M_1 + M_2$, where $|M_1|$ is free and $M_2$ is the fixed part of $|g^*M|$.

We can construct an effective $\mathbf{Q}$-divisor $C'$ such that $\text{Exc}(g) \cup \text{Supp}(C + g^*E + M_2) = \text{Supp}(C')$ and $-C'$ is $g$-ample. The construction of $C'$ is as follows: First, by the definition of log resolution in strong sense, we can take an effective $\mathbf{Q}$-divisor $C''$ such that $\text{Exc}(g) = \text{Supp}(C'')$ and $-C''$ is $g$-ample, then we can perturb it to extend the support by the openness of ampleness.

We can take a sufficiently small positive rational number $\epsilon$ such that $g^*(m_1 D + E - (K_X + B)) - \epsilon C'$ is $h$-ample and $\ulcorner -C - \epsilon C' \urcorner = \ulcorner -C \urcorner \geq 0$.

One key point of the proof is to consider the following *threshold*:

$$c = \sup\{t \in \mathbf{R} \mid \llcorner tM_2 - g^*E + C + \epsilon C' \lrcorner \leq 0\}.$$

This is a kind of *LC (log canonical) threshold*. By definition, the maximal coefficient of $cM_2 - g^*E + C + \epsilon C'$ is exactly 1. By perturbing the coefficients of $C'$ while preserving the ampleness of $-C'$, we may assume that there is exactly one prime divisor attaining the maximal coefficient 1.

This idea of breaking the balance of coefficients by perturbing the coefficients of $\mathbf{Q}$-divisors is called *tiebreaking*. This is the advantage of considering $\mathbf{Q}$-divisors and $\mathbf{R}$-divisors instead of only integral divisors.

Denote by $Z$ the prime divisor with coefficient 1 in $cM_2 - g^*E + C + \epsilon C'$. As the coefficients of $C + \epsilon C'$ are less than 1, $Z$ is contained in the support of $M_2$. Hence $g(Z)$ is contained in the base locus $\text{Bs} |mD|$. Write

$$cM_2 - g^*E + C + \epsilon C' = F + Z.$$

By construction, $F$ does not contain the prime divisor $Z$ and $\ulcorner -F \urcorner \geq 0$.

Let $m'$ be an integer, and as $mg^*D \equiv_S M_1 + M_2$, we get the following equation which plays the trick:

$$m'g^*D - F - Z - K_Y \equiv_S (m' - cm)g^*D + cM_1 + g^*E - (K_Y + C + \epsilon C').$$

If $m' \geq m_1 + cm$, as $M_1$ is free, then the right-hand side is $h$-ample. Applying Theorem 1.9.3, we get

$$H^1(Y, m'g^*D + \ulcorner -F \urcorner - Z) = 0.$$

Therefore, the natural homomorphism

$$H^0(Y, m'g^*D + \ulcorner -F \urcorner) \to H^0(Z, m'g^*D|_Z + \ulcorner -F \urcorner|_Z)$$

is surjective. Here the restriction can be defined, since $F$ does not contain the prime divisor $Z$, and $D$ can be replaced by a (not necessarily effective) linearly equivalent divisor which does not contain the prime divisor $Z$.

On the other hand, as the negative coefficient part of $F + g^*E$ comes from the negative coefficient part of $C$, its support is contained in the exceptional set of $g$. Therefore, the support of $\max\{\ulcorner -F \urcorner - g^*E, 0\}$ is contained in the exceptional set of $g$, and hence there are natural injective homomorphisms

$$H^0(X, m'D) \to H^0(X, m'D + \ulcorner -g_*F \urcorner) \to H^0(X, m'D + E).$$

If $m' \geq m_2$, then the terms on both ends coincide by assumption (3), hence these three terms coincide. Therefore, $H^0(Y, m'g^*D) \to H^0(Y, m'g^*D + \ulcorner -F \urcorner)$ is bijective.

Define the boundary $B_Z = (F + \ulcorner -F \urcorner)|_Z$ on $Z$, then the pair $(Z, B_Z)$ is KLT. Let us check that the projective morphism $h|_Z \colon Z \to S$ and the Cartier divisors $g^*D|_Z, \ulcorner -F \urcorner|_Z$ on $Z$ satisfy conditions of the theorem. Obviously, (1) holds, and (2) also holds since $m'g^*D|_Z - F|_Z - K_Z$ is relatively ample. Consider the following commutative diagram:

$$
\begin{array}{ccc}
H^0(Y, m'g^*D) & \longrightarrow & H^0(Y, m'g^*D + \ulcorner -F \urcorner) \\
\downarrow & & \downarrow \\
H^0(Z, m'g^*D|_Z) & \longrightarrow & H^0(Z, m'g^*D|_Z + \ulcorner -F \urcorner|_Z).
\end{array}
$$

If $m' \geq m_2$, then the top horizontal arrow is bijective. Moreover, if $m' \geq m_1 + cm$, then the right vertical arrow is surjective. Hence the bottom horizontal arrow is surjective and (3) holds.

By applying the nonvanishing theorem to $Z$, there exists a positive integer $m_3''$ such that if $m' \geq m_3''$, then $H^0(Z, m'g^*D|_Z) \neq 0$. Here we may assume that $m_3'' \geq \max\{m_2, m_1 + cm\}$. By the above commutative diagram, this implies that $g(Z)$ is not contained in the base locus of $|m'D|$. If $m'$ is a multiple of $m$ which is fixed in the beginning of the theorem, then we have a strict inclusion of base loci $\mathrm{Bs}\, |m'D| \subsetneq \mathrm{Bs}\, |mD|$.

Fix a prime number $p$ and take $m, m'$ to be powers of $p$. As there is no strictly decreasing sequence of closed subsets in $X$, by repeating the

## 2.1 The Basepoint-Free Theorem

above argument, for any sufficiently large power $p^t$, $\mathrm{Bs}\,|p^t D| = \emptyset$. This argument is called *Noetherian induction*. For another prime number $q$, by the same argument, there exists a sufficiently large positive integer $s$ such that $\mathrm{Bs}\,|q^s D| = \emptyset$. As $p^t$ and $q^s$ are coprime, there exists a positive integer $m_3$ such that for any integer $m \geq m_3$ there exist positive integers $a, b$ such that $m = ap^t + bq^s$. In this case, $\mathrm{Bs}\,|mD| = \emptyset$. Therefore, assuming the nonvanishing theorem, we have proved the basepoint-free theorem.

**Step 3** We will show the nonvanishing theorem by induction on $\dim X$. The method is similar to the proof of the basepoint-free theorem, but we create basepoints artificially.

It suffices to show the nonvanishing on the generic fiber of $f$, hence we may assume that $S = \mathrm{Spec}\, k$.

The statement of the nonvanishing theorem is that for all sufficiently large $m$, $H^0(X, mD + E) \neq 0$. By Theorem 1.10.8, for any integers $p > 0$ and $m \geq m_1$, $H^p(X, mD+E) = 0$, and hence the equality $\dim H^0(X, mD+E) = \chi(X, mD + E)$ holds. The latter is a polynomial in $m$, so it suffices to show that it is not identically 0.

In general, proving the existence of global sections is a difficult problem. In our situation we reduce the problem to a problem for the Euler–Poincaré characteristic which is a topological quantity, and prove it as follows.

First, consider the case $D \equiv 0$. In this case, as $E - (K_X + B)$ is nef and big, by Theorem 1.10.8, for any integers $p > 0$, $H^p(X, E) = 0$. Since $E$ is effective, $\chi(X, E) = \dim H^0(X, E) \neq 0$, and the proof is finished.

**Step 4** Finally we show the nonvanishing theorem in the case $D \not\equiv 0$.

As in Step 1, we may assume that $m_1 D + E - (K_X + B) = A$ is an ample $\mathbb{Q}$-divisor. Take a positive integer $a$ such that $aA$ is a Cartier divisor.

Since $D \not\equiv 0$, there exists a curve $\Gamma$ such that $(D \cdot \Gamma) > 0$. Denote by $I_\Gamma$ the ideal sheaf of $\Gamma$, replacing $a$ by a sufficiently large multiple, we may assume that $\mathcal{O}_X(aA) \otimes I_\Gamma$ is generated by global sections. Denote $d = \dim X$, by taking the intersection of zeros of $d - 1$ general sections of this sheaf, we get $A^{d-1} \equiv c\Gamma + \Gamma'$. Here $c > 0$ and $\Gamma'$ is an effective linear sum of curves distinct from $\Gamma$. Hence $(D \cdot A^{d-1}) > 0$. We can take a sufficiently large integer $m$ such that the self-intersection number $(mD + aA)^d$ of the ample divisor $mD + aA$ is larger than $a^d (d + 1)^d$.

By the Serre vanishing theorem, there exists an integer $k_1$ such that for all integers $k \geq k_1$ and $p > 0$, $H^p(X, k(mD + aA)) = 0$. Therefore, $\dim H^0(X, k(mD+aA)) = \chi(X, k(mD+aA))$ is a polynomial in $k$ of degree $d$, and the coefficient of the leading term is larger than $a^d (d + 1)^d / d!$.

# 92                    2 The Minimal Model Program

Fix a smooth point $P$ in $X$ not contained in the support of $E + B$. Take $\mathfrak{m}_P$ the maximal ideal of $\mathcal{O}_{X,P}$, then $\text{length}(\mathcal{O}_{X,P}/\mathfrak{m}_P^{a(d+1)k})$ is a polynomial in $k$ of degree $d$, and the coefficient of the leading term is $a^d(d+1)^d/d!$. Hence for any sufficiently large $k$,

$$H^0(X, \mathcal{O}_X(k(mD+aA))) \otimes \mathfrak{m}_P^{a(d+1)k}) \neq 0.$$

Therefore, there exists an element $M \in |k(mD+aA)|$ such that $\text{mult}_P M \geq a(d+1)k$. This is called the *concentration method*.

From now on, the proof is the same as that of the basepoint-free theorem. Take a log resolution $g: Y \to X$ of $(X, B+E+M)$ in strong sense and write $g^*(K_X+B) = K_Y+C$. Note that here we can first take the blowup at $P$, then construct $g$ by further blowups after this blowup. We can take an effective $\mathbf{Q}$-divisor $C'$ such that $\text{Exc}(g) \cup \text{Supp}(g^*(B+E+M)) = \text{Supp}(C')$ and $-C'$ is $g$-ample. We can take a sufficiently small positive rational number $\epsilon$ such that $g^*(m_1 D+E-(K_X+B))-(d+1)\epsilon C'$ is ample and $\ulcorner -C-\epsilon C' \urcorner = \ulcorner -C \urcorner \geq 0$. Consider the threshold

$$c = \sup\{t \in \mathbf{R} \mid \llcorner tg^*M - g^*E + C + \epsilon C' \lrcorner \leq 0\}.$$

By perturbing the coefficients of $C'$, we may assume that there is exactly one prime divisor $Z$ attaining the maximal coefficient 1.

Take $C_0$ to be the strict transform of the exceptional divisor of the first blowup at $P$. Here $C_0$ and $Z$ may or may not coincide. The coefficient of $C_0$ in $tg^*M - g^*E + C + \epsilon C'$ is larger than $a(d+1)kt - (d-1)$, hence $c < d/a(d+1)k$ by definition. Write

$$cg^*M - g^*E + C + \epsilon C' = F + Z.$$

By construction, $F$ does not contain the prime divisor $Z$ and $\ulcorner -F \urcorner \geq 0$.

Take an integer $m'$ such that $m' \geq m_1 + ckm$, as $1 - ack > 1/(d+1) > 0$,

$$m'g^*D - F - Z - K_Y$$
$$\equiv (m' - m_1 - ckm)g^*D$$
$$+ (1-ack)(m_1 g^*D + g^*E - (K_Y + C + \epsilon C'/(1-ack)))$$

is ample. By Theorem 1.10.8,

$$H^1(Y, m'g^*D + \ulcorner -F \urcorner - Z) = 0,$$

hence the natural homomorphism

$$H^0(Y, m'g^*D + \ulcorner -F \urcorner) \to H^0(Z, m'g^*D|_Z + \ulcorner -F \urcorner|_Z)$$

## 2.1 The Basepoint-Free Theorem

is surjective. Also, for

$$H^0(Y, m'g^*D) \to H^0(Y, m'g^*D + \ulcorner -F \urcorner) \to H^0(Y, m'g^*D + g^*E),$$

the three terms coincide if $m' \geq m_2$.

Denote $B_Z = (F + \ulcorner -F \urcorner)|_Z$, then the pair $(Z, B_Z)$ is KLT. The projective morphism $h|_Z \colon Z \to S$ and the Cartier divisors $g^*D|_Z, \ulcorner -F \urcorner|_Z$ on $Z$ satisfy the conditions of the nonvanishing theorem. Here recall that $S$ is assumed to be a point. By applying the nonvanishing theorem to $Z$, there exists a positive integer $m_3''$ such that if $m' \geq m_3''$, then $H^0(Z, m'g^*D|_Z) \neq 0$. Here we may assume that $m_3'' \geq \max\{m_2, m_1 + ckm\}$. Hence by surjectivity, $H^0(X, m'D) \neq 0$. The proof of the nonvanishing theorem is finished. $\qquad\square$

**Remark 2.1.4** The original basepoint-free theorem only treats the case $E = 0$, but the proof is exactly the same ([76]). As in the proof, even if we assume that $E = 0$ in the beginning, $F \neq 0$ appears naturally after taking a log resolution. Therefore, it is natural to consider $E \neq 0$ in the beginning. Also, in the statement of the nonvanishing theorem, $E$ appears in the beginning ([127]). In order to apply the basepoint-free theorem to the abundance theorem, according to [28], we show the general form with $E$.

In the former half of the above proof, we apply the vanishing theorem to a linear system which appears naturally in the process, while in the latter half, we apply the vanishing theorem to a linear system which is artificially constructed. The latter method is called *Shokurov's concentration method* of singularities.

The argument of proving the basepoint-free theorem by using the vanishing theorem was originally developed in [56]. In dimension 3, the nonvanishing theorem follows easily from the *Riemann–Roch theorem*. Later this argument was applied by Shokurov ([127]) to the proof of the nonvanishing theorem, and hence the basepoint-free theorem is proved in any dimension. Furthermore, it was shown in [54] that this argument can be applied to the proof of the cone theorem using the rationality theorem discussed in Section 2.3. It was also used in the establishment of the abundance conjecture ([57]). So this argument has been found to have a wide range of applications, and is known as the *X-method*.

### 2.1.2 Paraphrasings and Generalizations

The following corollary is an equivalent statement of the basepoint-free theorem:

**Corollary 2.1.5** *Suppose that $f \colon (X, B) \to S$, $D$, $E$ satisfy the assumptions of Theorem 2.1.1. Then there exists a projective morphism $g \colon Z \to S$ from a* *normal algebraic variety, a surjective projective morphism $h \colon X \to Z$ with connected geometric fibers such that $f = g \circ h$, and a $g$-ample Cartier divisor $H$ such that $h^*H \sim D$.*

*Proof* By the basepoint-free theorem, there exists a positive integer $m_3$ such that for $m \geq m_3$, $mD$ is $f$-free. Denote by $\phi_m = h'_m \circ h_m \colon X \to Z_m \to Z'_m$ the *Stein factorization* of the associated morphism over $S$, and denote by $g_m \colon Z_m \to S$ the induced morphism. By construction, there exists a $g_m$-ample Cartier divisor $H_m$ on $Z_m$ such that $mD \sim h_m^* H_m$.

Note that, for a curve $C$ on $X$ such that $f(C)$ is a point on $S$, $h_m(C)$ is a point on $Z_m$ if and only if $(D \cdot C) = 0$. By Zariski's main theorem, there exists an isomorphism $k_m \colon Z_m \to Z_{m+1}$ such that $k_m \circ h_m = h_{m+1}$. Take $H = k_m^* H_{m+1} - H_m$, then $h_m^* H \sim D$. $\qquad\square$

**Corollary 2.1.6** *Let $(X, B)$ be a KLT pair and let $f \colon X \to S$ be a projective morphism. Assume that $K_X + B$ is $f$-nef and $B$ is an $f$-big $\mathbf{Q}$-Cartier divisor. Then there exists a projective morphism $g \colon Z \to S$ from a normal algebraic variety, a surjective projective morphism $h \colon X \to Z$ with connected geometric fibers such that $f = g \circ h$, and a $g$-ample $\mathbf{Q}$-Cartier divisor $H$ such that $h^*H \sim_{\mathbf{Q}} K_X + B$.*

*Proof* The assertion is local, so we may assume that $S$ is affine. As the coefficients of $B$ are in $\mathbf{Q}$, there exists a positive integer $m_1$ such that $D = m_1(K_X + B)$ is an $f$-nef Cartier divisor. As $B$ is $f$-big, there exists an $f$-ample $\mathbf{Q}$-Cartier divisor $A$ and an effective $\mathbf{Q}$-Cartier divisor $E$ such that we can write $B = A + E$. For a sufficiently small positive rational number $\epsilon$, $(X, (1 - \epsilon)B + \epsilon E)$ is KLT. Also, $D - (K_X + (1 - \epsilon)B + \epsilon E) = (m_1 - 1)(K_X + B) + \epsilon A$ is $f$-ample. We get the conclusion by the above corollary. $\qquad\square$

**Remark 2.1.7** The condition that the coefficients of $B$ are rational numbers can be removed by using the cone theorem (Corollary 2.4.13).

The following lemma is useful when generalizing statements for KLT pairs to DLT (divisorially log terminal) pairs.

**Lemma 2.1.8** ([86]) *Let $(X, B)$ be a DLT pair, let $f \colon X \to S$ be a projective morphism, let $H$ be a relatively ample divisor on $X$, and let $\epsilon$ be a positive real number. Assume that $S$ is quasi-projective. Then there exists an ample divisor $A$ on $S$ and an effective $\mathbf{R}$-divisor $B'$ on $X$ such that $B + \epsilon(H + f^*A) \sim_{\mathbf{R}} B'$ and the pair $(X, B')$ is KLT.*

## 2.1 The Basepoint-Free Theorem 95

*Proof* We can choose an ample divisor $A$ on $S$ such that $H + f^*A$ is ample on $X$. Take a log resolution $g \colon (Y, C) \to (X, B)$ in strong sense, denote $h = f \circ g$. By the definition of DLT, we may assume that the coefficients of the exceptional divisors in $C$ are strictly less than 1; note that here we use the fact that DLT is equivalent to WLT (see Remark 1.11.4). Take a sufficiently small effective **Q**-divisor $E$ supported on the exceptional set of $g$ such that $-E$ is $g$-ample, $\llcorner C + E \lrcorner = \llcorner C \lrcorner$, and $g^*(H + f^*A) - E$ is ample on $Y$.

Write $B = \sum b_i B_i$ where $B_i$ are distinct prime divisors and write $g_*^{-1} B = \sum b_i B_i'$ the strict transform on $Y$. We can choose a positive integer $m$ such that for every $i$, the divisorial sheaf $\mathcal{O}_Y(B_i' + m(g^*(H + f^*A) - E))$ is generated by global sections. By taking a general global section, we can find a general effective divisor $D_i' \sim B_i' + m(g^*(H + f^*A) - E)$. Take a sufficiently small positive real number $\delta$ and take

$$C' = C - \delta \sum b_i B_i' + \delta \sum b_i D_i' + m\delta \sum b_i E$$
$$\sim_{\mathbf{R}} C + m\delta \sum b_i (g^*(H + f^*A)).$$

Note that the support of $C'$ is a normal crossing divisor as $D_i'$ are general, and the coefficients of $C'$ are less than 1 as $\delta$ is sufficiently small. Then we can take $B' = g_* C' = (1 - \delta)B + \delta \sum b_i g_* D_i' \sim_{\mathbf{R}} B + m\delta \sum b_i (H + f^*A)$. Note that $K_X + B'$ is **R**-Cartier and $f^*(K_X + B') = K_Y + C'$, which implies that $(X, B')$ is KLT. $\qquad\square$

Now we can show the basepoint-free theorem for DLT pairs:

**Corollary 2.1.9** (*Basepoint-free theorem*) *Let* $(X, B)$ *be a DLT pair, let* $f \colon X \to S$ *be a projective morphism, and let* $D, E$ *be Cartier divisors on* $X$. *Assume the following conditions:*

*(1) $D$ is relatively nef.*

*(2) There exists a positive integer $m_1$ such that $m_1 D + E - (K_X + B)$ is relatively ample.*

*(3) $E$ is effective and there exists a positive integer $m_2$ such that for any integer $m \geq m_2$, the natural homomorphism $f_*(\mathcal{O}_X(mD)) \to f_*(\mathcal{O}_X(mD + E))$ is an isomorphism.*

*Then there exists a positive integer $m_3$ such that for any integer $m \geq m_3$, $mD$ is relatively free. That is, the natural homomorphism $f^* f_*(\mathcal{O}_X(mD)) \to \mathcal{O}_X(mD)$ is surjective.*

*Proof* As the assertion is local, we may assume that $S$ is affine. Take $B' \sim_{\mathbf{R}} B + \epsilon(H + f^*A)$ as in Lemma 2.1.8 such that $(X, B')$ is KLT. If $\epsilon$ is sufficiently small, then $m_1 D + E - (K_X + B')$ is still relatively ample. Then the corollary follows. $\qquad\square$

## 2.2 An Effective Version of the Basepoint-Free Theorem

The basepoint-free theorem states that a multiple of a certain Cartier divisor is free. Its effective version shows how large this multiple can be taken in practice. The proof is not just a refinement of the proof of the basepoint-free theorem, but it actually needs to use the conclusion of the noneffective version on the existence of such a morphism.

**Theorem 2.2.1** (*Effective basepoint-free theorem* [84]) *Let $(X, B)$ be a KLT pair consisting of an n-dimensional algebraic variety and an $\mathbf{R}$-divisor, let $E$ be an effective Cartier divisor on $X$, let $D$ be a Cartier divisor on $X$, and let $f : X \to S$ be a projective morphism. Assume the following conditions hold:*

*(1) $D$ is $f$-nef and $D + E - (K_X + B)$ is $f$-nef and $f$-big.*
*(2) The natural homomorphism*

$$f_* \mathcal{O}_X(mD) \to f_* \mathcal{O}_X(mD + E)$$

*is bijective for any positive integer $m$.*

*Then for any $m \geq 2n + 3$, $|m^{n+1}D|$ is $f$-free, that is, the natural homomorphism*

$$f^* f_* \mathcal{O}_X(m^{n+1}D) \to \mathcal{O}_X(m^{n+1}D)$$

*is surjective. Therefore, there exists a positive integer $m_0$ depending only on the dimension n such that $|mD|$ is $f$-free for any $m \geq m_0$.*

*Proof* We may assume that $S$ is affine. Note that in this case "ample over $S$" or "free over $S$" is simply the same as "ample" or "free." By slightly perturbing the coefficients of $B$, we may assume that $D + E - (K_X + B)$ is ample.

By Corollary 2.1.5 of the basepoint-free theorem, there exists a normal algebraic variety $Y$, a surjective projective morphism $g : X \to Y$ over $S$ with connected geometric fibers, and a relatively ample Cartier divisor $H$ on $Y$ such that $D = g^* H$. Denote by $h : Y \to S$ the morphism such that $f = h \circ g$ and $d = \dim Y$. Take $X_s, Y_s$ to be the fibers over a general point $s$ in $h(Y)$ and denote $d_s = \dim Y_s \leq d \leq n$.

## 2.2 An Effective Version of the Basepoint-Free Theorem

First, we show the effective version of the nonvanishing theorem (see Theorem 2.1.3). By the vanishing theorem, for $m > 0$,

$$h^0(X_s, mD) = h^0(X_s, mD + E) = \chi(X_s, mD + E).$$

By the nonvanishing theorem, the latter is a nonzero polynomial of degree $d_s$ and has at most $d_s$ distinct roots.

We claim that for $m \geq 2d_s + 3$, $H^0(X_s, mD) \neq 0$. Take $d_s + 1$ pairs $\{i, m-i\}$ $(1 \leq i \leq d_s + 1)$. If $H^0(X_s, mD) = 0$, then either $H^0(X_s, iD) = 0$ or $H^0(X_s, (m-i)D) = 0$, which implies that $\chi(X_s, mD + E)$ has at least $d_s + 1$ roots, a contradiction. Therefore, $H^0(X_s, mD) \neq 0$ for $m \geq 2d_s + 3$.

The theorem can be reduced to the following lemma:

**Lemma 2.2.2** *Fix any $m \geq 2d + 3$ and take an irreducible component $\bar{Z}$ of the base locus*

$$\mathrm{Bs}\,|mH| = \mathrm{Supp}(\mathrm{Coker}(h^* h_* \mathcal{O}_Y(mH) \to \mathcal{O}_Y(mH))).$$

*Then for any $k \geq 2d + 3$, $\bar{Z} \not\subset \mathrm{Bs}\,|kmH|$.*

Let us continue the proof by assuming Lemma 2.2.2. Note that $g^{-1}(\mathrm{Bs}\,|mH|) = \mathrm{Bs}\,|mD|$. Fix any $m \geq 2d + 3$, Lemma 2.2.2 shows that the dimension of the base locus $\mathrm{Bs}\,|m^j D|$ is decreasing for $1 \leq j \leq d + 1$. When $j = d + 1$, the base locus becomes an empty set, which concludes the former assertion. The interesting point of this proof is that, no matter how many irreducible components there are in the base locus, the above lemma can be applied to every irreducible component of the base locus at the same time to cut down the dimension of the base locus by multiplying by $m$.

Take two distinct prime numbers $p, q \geq 2d + 3$. Then $|p^{d+1}D|$ and $|q^{d+1}D|$ are free. There exists a positive integer $m_0$ such that any integer $m \geq m_0$ can be expressed as $m = ap^{d+1} + bq^{d+1}$ for some $a, b \in \mathbf{Z}_{>0}$ and hence $|mD|$ is free. This proves the latter assertion. $\qquad\square$

*Proof of Lemma 2.2.2* Note that $h(\bar{Z})$ is a subset of $h(Y)$, and they may not coincide. Denote $Z = g^{-1}(\bar{Z})$. The proof is by applying the argument in the basepoint-free theorem to the pullback of the neighborhood of the generic point of $\bar{Z}$ by $g$.

First, we construct singularities in a neighborhood of $Z$. Denote $\bar{d} = \dim \bar{Z}$. Take $d - \bar{d} + 1$ divisors corresponding to general global sections of $\mathcal{O}_Y(mH)$, say, $\bar{M}_i$ $(1 \leq i \leq d - \bar{d})$ and $\bar{N}$. Note that the supports of $\bar{M}_i$ and $\bar{N}$ all contain $\mathrm{Bs}\,|mH|$, but they are free outside $\mathrm{Bs}\,|mH|$. Take sufficiently small numbers $\epsilon, \delta > 0$ and take $M = (1 - \delta)\sum_i g^* \bar{M}_i + \epsilon g^* \bar{N}$. Take a neighborhood $\bar{U}$ of the generic point of $\bar{Z}$ and denote $U = g^{-1}(\bar{U})$. By shrinking $\bar{U}$, we may

98                          2 *The Minimal Model Program*

assume that it does not intersect irreducible components of Bs $|mH|$ other than $\bar{Z}$. Here note that $\bar{Z}$ is not necessarily contained in $\bar{U}$.

Since $\bar{Z}$ has codimension $d - \bar{d}$, by Corollary 1.11.8, we may assume that the pair $(X, B + M)$ is KLT on $U \setminus \mathrm{Bs}\,|mD|$ but not LC in the neighborhood of any point in $U \cap Z$ by taking $\epsilon, \delta$ appropriately and shrinking $U$.

Take a log resolution $\mu \colon X' \to (X, B + M)$ in strong sense and write $\mu^*(K_X + B) = K_{X'} + B'$. The coefficients of $B'$ are all less than 1. Take an effective divisor $F$ supported on the exceptional set of $\mu$ such that $-F$ is $\mu$-ample. Take a sufficiently small positive number $\epsilon'$ such that the coefficients of $B' + \epsilon'F$ are all less than 1 and $\mu^*(D + E - (K_X + B)) - \epsilon'F$ is ample.

Consider the LC threshold $c$ as the following: On $\mu^{-1}(U)$, the coefficients $B' + c\mu^*M + \epsilon'F$ are all no greater than 1, and at least one coefficient is exactly 1. Here $c < 1$ as $(X, B + M)$ is not LC on $U \cap Z$, and note that outside $\mu^{-1}(U)$ the coefficients may be greater than 1. Write

$$\mu^*(K_X + B + cM) + \epsilon'F = K_{X'} + F_0 + B''.$$

Here $F_0$ is the sum of all irreducible components with coefficient 1 intersecting $\mu^{-1}(U)$. By perturbing the coefficients of $F$ and shrinking $\bar{U}$, we may assume that $F_0$ is irreducible, $g \circ \mu(F_0) = \bar{Z}$, and $\llcorner B'' \lrcorner \le 0$ on $\mu^{-1}(U)$.

For a positive integer $m'$, consider the following exact sequence

$$0 \to \mathcal{O}_{X'}(\mu^*(m'D + E) - \llcorner B'' \lrcorner - F_0) \to \mathcal{O}_{X'}(\mu^*(m'D + E) - \llcorner B'' \lrcorner)$$
$$\to \mathcal{O}_{F_0}((\mu^*(m'D + E) - \llcorner B'' \lrcorner)|_{F_0}) \to 0.$$

If $m' \ge c((d - \bar{d})(1 - \delta) + \epsilon)m + 1$, then

$$\mu^*(m'D + E) - B'' - F_0 - K_{X'}$$
$$\equiv (m' - c((d - \bar{d})(1 - \delta) + \epsilon)m - 1)\mu^*D + \mu^*(D + E - (K_X + B)) - \epsilon'F$$

is ample. By the vanishing theorem, higher cohomologies of the first and the third terms in the exact sequence vanish, and the following natural homomorphism is surjective:

$$H^0(X', \mu^*(m'D + E) - \llcorner B'' \lrcorner) \to H^0(F_0, (\mu^*(m'D + E) - \llcorner B'' \lrcorner)|_{F_0}).$$

On the other hand, as the support of $(B'')^-$ is contained in the exceptional set of $\mu$, we have an injective homomorphism

$$H^0(X', \mu^*(m'D + E) - \llcorner B'' \lrcorner) \to H^0(X, m'D + E)$$

and a bijective homomorphism

$$H^0(Y, m'H) \to H^0(X, m'D + E).$$

Therefore, for $m' \geq (d - \bar{d} + 1)m$, the image of the natural homomorphism

$$H^0(Y, m'H) \to H^0(\bar{Z}, m'H|_{\bar{Z}}) \to H^0(F_0, \mu^*(m'D + E)|_{F_0})$$

contains $H^0(F_0, (\mu^*(m'D + E) - \llcorner B'' \lrcorner)|_{F_0})$.

Recall that $\llcorner B'' \lrcorner \leq 0$ on $\mu^{-1}(U)$. This means that there might be some irreducible component of $B''$ with coefficient greater than 1, but it does not intersect $\mu^{-1}(U)$.

Take a general point $t$ in $h(\bar{Z})$ and denote by $F_{0,t}$ the fiber of $F_0$ over $t$. Since $\llcorner B'' \lrcorner \leq 0$ on $\mu^{-1}(U)$, we have injective homomorphisms

$$H^0(F_{0,t}, (\mu^* g^*(m'H))|_{F_{0,t}}) \to H^0(F_{0,t}, (\mu^*(m'D + E))|_{F_{0,t}})$$
$$\to H^0(F_{0,t}, (\mu^*(m'D + E) - \llcorner B'' \lrcorner)|_{F_{0,t}}).$$

By the vanishing theorem, for $m' \geq (d - \bar{d} + 1)m$,

$$h^0(F_{0,t}, (\mu^*(m'D + E) - \llcorner B'' \lrcorner)|_{F_{0,t}}) = \chi(F_{0,t}, (\mu^*(m'D + E) - \llcorner B'' \lrcorner)|_{F_{0,t}}).$$

As $H$ is ample on $Y$, this is a nonzero polynomial of degree at most $\bar{d}$, and hence has at most $\bar{d}$ zeros.

Note that if the image of $H^0(Y, m'H) \to H^0(\bar{Z}, m'H|_{\bar{Z}})$ is not 0, then $\bar{Z}$ is not contained in the base locus of $|m'H|$. Hence

$$\bar{Z} \subset \mathrm{Bs}\, |(d - \bar{d} + j)mH|$$

could be true only for at most $\bar{d}$ values of $j \geq 1$. Using a similar argument as in the proof of the effective basepoint-free theorem, for $k \geq 2d + 3$,

$$\bar{Z} \not\subset \mathrm{Bs}\, |kmH|. \qquad \square$$

## 2.3 The Rationality Theorem

The rationality theorem which will be proved in this section is a key point of the proof of the cone theorem discussed in Section 2.4.

The first part of the rationality theorem shows that a certain threshold is a rational number. It concludes the existence of extremal rays in the cone theorem. The second part gives an estimate of the denominator of the threshold, which concludes the discreteness of extremal rays. As we will explain in Section 2.4, the discreteness of extremal rays can be proved alternatively by the estimate of the lengths of extremal rays. The latter argument uses the theorem on the existence of rational curves which depends on methods of algebraic geometry in positive characteristics.

100          *2 The Minimal Model Program*

The proof of the cone theorem uses the argument in the basepoint-free theorem. It was developed in [54], which completed the formulation of the MMP. Reid ([119]) observed that the rationality statement is the key for the generalization of Mori's cone theorem ([100]) to singular varieties. We will prove it by using the argument of the basepoint-free theorem.

**Theorem 2.3.1** (*Rationality theorem* [76, Theorem 3.1.1]) *Let $(X, B)$ be a KLT pair where $B$ is a $\mathbf{Q}$-divisor, let $f : X \to S$ be a projective morphism, and let $A$ be a relatively nef and relatively big Cartier divisor. Assume that $K_X + B$ is not relatively nef. Then the threshold*

$$r = \max\{t \in \mathbf{R} \mid A + t(K_X + B) \text{ is relatively nef}\}$$

*is a rational number. Moreover, denote by a the minimal positive integer such that $a(K_X + B)$ is Cartier and denote by b the maximal dimension of fibers of $f$, if we write $r/a = p/q$ as an irreducible fraction, then*

$$q \le a(b + 1).$$

*Proof* The proof is by exploring the proof of the basepoint-free theorem in more detail. To get a contradiction, we assume that either $r$ is not a rational number or $r$ is a rational number but $q > a(b + 1)$.

**Step 1** Clearly $r$ is a positive real number. We may assume that $S$ is affine. We will show that we may assume that $A$ is free and relatively ample.

By Theorem 2.1.1, we can take sufficiently large integers $m, n$ such that $a < mr$, $(mn, q) = 1$ (if $r$ is rational), and $A' = n(mA + a(K_X + B))$ is free and big. Then the threshold

$$r' = \max\{t \in \mathbf{R} \mid A' + t(K_X + B) \text{ is relatively nef}\}$$

satisfies the relation $mnr = an + r'$. So $r$ is rational if and only if $r'$ is rational. Moreover, if $r'$ is rational and we write $r'/a = p'/q'$ as an irreducible fraction, then $q = q'$. So after replacing $A$ by $A'$, we may assume that $A$ is free.

Moreover, from the construction of $A'$ and the fact that $a < mr$, we can see that if $(A \cdot C) = 0$ for a relative curve $C$, then $((K_X + B) \cdot C) = 0$. So as in the proof of Corollary 2.1.5, there exists a projective morphism $f_0 \colon X \to X_0$ over $S$ induced by $A$ such that $A = f_0^*(A_0)$ and $K_X + B = f_0^*(K_{X_0} + B_0)$, where $A_0$ is a relatively ample Cartier divisor on $X_0$, $B_0 = (f_0)_* B$ and $a(K_{X_0} + B_0)$ is Cartier. Here note that $f_0$ is birational as $A$ is relatively big. Then after replacing $X, B, A$ by $X_0, B_0, A_0$, we may assume that $A$ is relatively ample.

**Step 2** The following lemma plays a similar role as the nonvanishing theorem in the proof of the basepoint-free theorem. Here our assumption is

## 2.3 The Rationality Theorem

101

more general than that in [76] and the proof is irrelevant to the nonvanishing theorem.

**Lemma 2.3.2** ([76, Lemma 3.1.2]) *Let $(X, B)$ be a projective KLT pair, let $D_1, D_2, E$ be Cartier divisors on $X$, let $d$ be a positive integer, and let $r', s$ be positive real numbers. For integers $x, y$, denote $D(x, y) = x D_1 + y D_2$. Assume the following conditions:*

*(1) $E$ is effective.*
*(2) There exists a positive integer $y_1$ such that if $x > 0$, $y \geq y_1$, and $y - r'x < s$, then $D(x, y) + E - (K_X + B)$ is nef and big and the natural homomorphism $H^0(X, D(x, y)) \to H^0(X, D(x, y) + E)$ is bijective.*
*(3) The polynomial $\chi(X, D(x, y) + E)$ in two variables $x, y$ is of degree at most $d$ and not identically $0$.*
*(4) Either $r'$ is irrational, or $r'$ is rational and $qs > d + 1$, where we write $r' = p/q$ as an irreducible fraction.*

*Then there exists a positive integer $y_2$ such that $H^0(X, D(x, y) + E) \neq 0$ if $y - r'x < s$ and $y \geq y_2$.*

*Proof* If $r'$ is irrational, then there are infinitely many couples of positive integers $(x, y)$ such that $0 < y - r'x < s/(d + 1)$. If $r'$ is rational, then as $p, q$ are coprime, there are infinitely many couples of positive integers $(x, y)$ such that $y - r'x = 1/q < s/(d + 1)$. So in either case, there are infinitely many couples of positive integers $(x, y)$ such that $0 < y - r'x < s/(d + 1)$. We may assume that $y \geq y_1$ in each couple.

For any such a couple $(x_0, y_0)$, consider the polynomial $\chi(X, D(mx_0, my_0) + E)$ in $m$. For any integer $m$ such that $1 \leq m \leq d + 1$, $my_0 - mr'x_0 < s$ holds, and hence $D(mx_0, my_0) + E - (K_X + B)$ is nef and big. By the vanishing theorem, higher cohomologies vanish and

$$\chi(X, D(mx_0, my_0) + E) = \dim H^0(X, D(mx_0, my_0) + E).$$

On the other hand, if $H^0(X, D(mx_0, my_0) + E) = 0$ for all $1 \leq m \leq d + 1$, then $\chi(X, D(x, y) + E)$, which is a polynomial of degree at most $d$ in two variables, is identically $0$ on the line $y_0 x - x_0 y = 0$. By construction, there are infinitely many such lines and $\chi(X, D(x, y) + E)$ cannot be identically $0$ on all such lines since $y_0/x_0$ takes infinitely many values. Hence there exists a couple $(x, y)$ such that $x > 0$, $y \geq y_1$, $0 < y - r'x < s$, and $H^0(X, D(x, y) + E) \neq 0$.

If such a positive integer $y_2$ in the assertion does not exist, then there are infinitely many couples of positive integers $(x', y')$ such that $y' - r'x' < s$, $x' > dx$, $y' \geq y_1 + dy$, and $H^0(X, D(x', y') + E) = 0$. Since

$$H^0(X, D(x, y)) \cong H^0(X, D(x, y) + E) \neq 0,$$

we know that $H^0(X, D(x' - mx, y' - my) + E) = 0$ for $0 \leq m \leq d$. So again by the vanishing theorem, $\chi(X, D(\bar{x}, \bar{y}) + E)$ is identically 0 on infinitely many lines $y(\bar{x} - x') - x(\bar{y} - y') = 0$, a contradiction. This concludes the lemma. $\square$

**Step 3** If $r$ is rational, then by the assumption that $q > a(b + 1)$, we may take a sufficiently small positive real number $\delta$ such that $q(1 - \delta) > a(b + 1)$. If $r$ is irrational, just take any $0 < \delta < 1$. Take $E, d, r', s$ to be 0, $b$, $r/a$, $(1 - \delta)/a$, respectively, and take $D(x, y) = xA + ay(K_X + B)$.

Applying Lemma 2.3.2 to the generic fiber of $f$, we know that there exists a couple of positive integers $(x, y)$ such that

$$0 < ay - rx < 1 - \delta \quad \text{and} \quad H^0(X, D(x, y)) \neq 0.$$

Here we can have $ay - rx > 0$ by the first paragraph of the proof of Lemma 2.3.2. Note that since $S$ is affine, the nonvanishing of $H^0$ on the generic fiber implies the nonvanishing on $X$.

Fix such a couple $(x, y)$. Since $ay - rx > 0$, $D(x, y)$ is not relatively nef, and therefore the linear system $|D(x, y)|$ has basepoints. Taking a general element $M \in |D(x, y)|$, we are going to apply the argument in the basepoint-free theorem to kill the base locus and get a contradiction.

Take a log resolution $g: Y \to X$ of $(X, B + M)$ in strong sense and write $h = f \circ g$ and $g^*(K_X + B) = K_Y + C$. Take the decomposition $g^*M = M_1 + M_2$ where $|M_1|$ is free and $M_2$ is the fixed part of $|g^*M|$.

Take an effective divisor $C'$ such that $\mathrm{Exc}(g) \cup \mathrm{Supp}(C + M_2) = \mathrm{Supp}(C')$ and $-C'$ is $g$-ample. Take a sufficiently small positive real number $\epsilon$ such that $\delta g^*A - r\epsilon C'$ is $h$-ample and $\ulcorner -C - \epsilon C' \urcorner = \ulcorner -C \urcorner \geq 0$. Consider the following threshold:

$$c = \sup\{t \in \mathbf{R} \mid \llcorner tM_2 + C + \epsilon C' \lrcorner \leq 0\}.$$

We may assume that there exists exactly one prime divisor $Z$ attaining the maximal coefficient 1 in $cM_2 + C + \epsilon C'$. Note that $g(Z)$ is contained in the base locus $\mathrm{Bs} |D(x, y)|$. Write

$$cM_2 + C + \epsilon C' = F + Z.$$

Here the support of $F$ does not contain the prime divisor $Z$ and $\ulcorner -F \urcorner \geq 0$.

For a couple of integers $(x', y')$, consider

$$g^*D(x', y') - F - Z - K_Y$$
$$\equiv (x' - cx)g^*A + (ay' - acy)g^*(K_X + B) + cM_1 - (K_Y + C + \epsilon C').$$

## 2.3 The Rationality Theorem

This **R**-divisor is $h$-ample if $x' > cx$, $y' > cy + 1/a$, and $r(x' - cx) \geq a(y' - cy) - 1 + \delta$ (that is, $ay' - rx' \leq c(ay - rx) + 1 - \delta$). In particular, the last one is satisfied if $ay' - rx' < 1 - \delta$.

Therefore, if $x' > cx$, $y' > cy + 1/a$, and $ay' - rx' < 1 - \delta$, then by Theorem 1.9.1,

$$H^1(Y, g^*D(x', y') + \ulcorner -F \urcorner - Z) = 0$$

and the natural homomorphism

$$H^0(Y, g^*D(x', y') + \ulcorner -F \urcorner) \to H^0(Z, g^*D(x', y')|_Z + \ulcorner -F \urcorner|_Z)$$

is surjective. On the other hand, the natural map

$$H^0(Y, g^*D(x', y')) \to H^0(Y, g^*D(x', y') + \ulcorner -F \urcorner)$$

is bijective. By the commutative diagram

$$
\begin{array}{ccc}
H^0(Y, g^*D(x', y')) & \longrightarrow & H^0(Y, g^*D(x', y') + \ulcorner -F \urcorner) \\
\downarrow & & \downarrow \\
H^0(Z, g^*D(x', y')|_Z) & \longrightarrow & H^0(Z, (g^*D(x', y') + \ulcorner -F \urcorner)|_Z),
\end{array}
$$

the bottom horizontal arrow is surjective.

Denote $B_Z = (F + \ulcorner -F \urcorner)|_Z$, then the pair $(Z, B_Z)$ is KLT. We may apply Lemma 2.3.2 to the generic fiber of $h|_Z : Z \to S$. Here we take $D_1, D_2, E$ to be the restrictions of $g^*A$, $ag^*(K_X + B)$, $\ulcorner -F \urcorner$, and take $d, r', s$ to be $b, r/a$, $(1 - \delta)/a$, respectively. It can be checked that the conditions of Lemma 2.3.2 are satisfied, where (1), (2), and (4) are clear and (3) follows from the fact that the dimensions of the fibers of $g(Z) \to S$ are at most $b$. So by Lemma 2.3.2, $g(Z)$ is not contained in Bs $|D(x', y')|$ if $ay' - rx' < 1 - \delta$ and $y'$ is sufficiently large.

Now consider a couple $(x', y')$ satisfying $0 < ay' - rx' < 1 - \delta$ with $y'$ sufficiently large defined in the following way. If $r$ is irrational, take a sufficiently large integer $l$ such that

$$x' = \llcorner aly/r \lrcorner = lx + \llcorner l(ay - rx)/r \lrcorner, \qquad y' = ly,$$

and $ay' - rx' < 1 - \delta$; if $r$ is rational, take a sufficiently large integer $l$ and take

$$x' = x + lq, \qquad y' = y + lp.$$

Note that $A$ is free and in the latter case $l(qA + ap(K_X + B))$ is free by the basepoint-free theorem, hence

$$\text{Bs}\,|D(x', y')| \subset \text{Bs}\,|D(x, y)|.$$

104                      *2 The Minimal Model Program*

To summarize, starting from a couple $(x, y)$ such that $0 < ay - rx < 1 - \delta$ and $H^0(X, D(x, y)) \neq 0$, we constructed a couple $(x', y')$ such that $0 < ay' - rx' < 1 - \delta$ and $\mathrm{Bs}\,|D(x', y')| \subsetneq \mathrm{Bs}\,|D(x, y)|$. Applying Noetherian induction as in the proof of the basepoint-free theorem, there exists a couple of positive integers $(x'', y'')$ such that $0 < ay'' - rx'' < 1 - \delta$ and $D(x'', y'')$ is free. This implies that $x''A + ay''(K_X + B)$ is relatively nef, which contradicts the maximality of $r$. $\qquad\square$

## 2.4 The Cone Theorem

The cone theorem, together with the basepoint-free theorem, are two main pillars for the minimal model theory. Higher dimensional minimal model theory began from the introduction of the concept of extremal rays in [100]. Birational geometry becomes visible by looking at cones and polyhedra in finite-dimensional real vector spaces.

The cone theorem states that the cone of curves is locally a rational polyhedral cone in the part which intersects negatively with the canonical divisor. This statement splits into two parts: the existence and the discreteness of extremal rays. The discreteness of extremal rays can be proved by the rationality theorem proved in Section 2.3, or the boundedness of the lengths of extremal rays which will be proved in Section 2.8. We will introduce both arguments, the former stays in characteristic 0, while the latter uses positive characteristic methods.

### 2.4.1 The Contraction Theorem

In general, a subset $\mathscr{C}$ in a finite-dimensional real vector space $V$ is called a *cone* if it is invariant under the map $v \mapsto tv$ for any $t \in \mathbf{R}^*$. It is said to be *convex* if for any $v, v' \in \mathscr{C}$ and any $t \in [0, 1]$, $tv + (1 - t)v' \in \mathscr{C}$. Consider a closed convex cone $\mathscr{C}$. A closed subset $F$ of $\mathscr{C}$ is called a *face* if there exists $u \in V^*$ such that $\mathscr{C} \subset V_{u \geq 0}$ and $F = \mathscr{C}_{u=0}$. Here $V_{u \geq 0} = \{v \in V \mid (u, v) \geq 0\}$ and $\mathscr{C}_{u=0} = \{v \in \mathscr{C} \mid (u, v) = 0\}$. In particular, a half line which is a face is called an *extremal ray*. Here $u$ is called the *supporting function* of $F$.

Given two projective morphisms $f \colon X \to S$ and $g \colon Y \to S$ from normal algebraic varieties, a projective morphism $h \colon X \to Y$ over $S$ is called a *contraction morphism* if the natural homomorphism $\mathcal{O}_Y \to h_* \mathcal{O}_X$ is bijective. In other words, $h$ is surjective and with connected geometric fibers. Here $h$ is a morphism over $S$ means that $g \circ h = f$. A contraction morphism is also called an *algebraic fiber space*. Usually the former is used for birational morphisms,

## 2.4 The Cone Theorem

and the latter is mainly used in the case dim $Y < \dim X$. However, these can often be handled in the same way.

Consider a face $F$ of the closed cone of curves $\overline{\mathrm{NE}}(X/S)$ and a contraction morphism $h: X \to Y$. Then $h$ is called the *contraction morphism associated to $F$* if the following conditions are satisfied:

- For a curve $C$ on $X$ such that $f(C)$ is a point, $h(C)$ is a point if and only if $[C] \in F$, in this case we say that $C$ is *contracted* by $h$.
- The smallest closed convex cone containing the equivalence classes of curves contracted by $h$ coincides with $F$.

In particular, $h$ is not an isomorphism if $F \neq 0$.

By *Zariski's main theorem*, the contraction morphism $h$ is determined by the face $F$ and is independent of the choice of the supporting function.

The following contraction theorem is a consequence of the basepoint-free theorem (Theorem 2.1.1):

**Theorem 2.4.1** (*Contraction theorem*) *Let $(X, B)$ be a KLT pair, let $f: X \to S$ be a projective morphism and let $F$ be a face of $\overline{\mathrm{NE}}(X/S)$. Assume that the supporting function $u$ of $F$ is defined over the rational number field and the function on $N_1(X/S)$ corresponding to $K_X + B$ takes negative values on $F \setminus \{0\}$. Then the following assertions hold:*

*(1) The contraction morphism $h: X \to Y$ associated to $F$ exists.*
*(2) The smallest linear subspace containing $F$ coincides with the image of the injective map $N_1(X/Y) \to N_1(X/S)$ and $F$ coincides with the image of $\overline{\mathrm{NE}}(X/Y)$.*
*(3) $-(K_X + B)$ is h-ample.*
*(4) For a Cartier divisor $D$ on $X$, if its corresponding function on $N_1(X/S)$ is identically $0$ on $N_1(X/Y)$, then there exists a Cartier divisor $E$ on $Y$ such that $D \sim h^* E$.*
*(5) $\rho(X/S) = \rho(Y/S) + \rho(X/Y)$.*

*Proof* (1) After taking a multiple of $u$, we may assume that it corresponds to a Cartier divisor $L$. Then $L$ is relatively nef.

As $K_X + B$ is negative on $F \setminus \{0\}$, $\overline{\mathrm{NE}}(X/S)_{K_X+B\geq 0} \cap F = \{0\}$. On the compact subset $(\overline{\mathrm{NE}}(X/S)_{K_X+B\geq 0}\setminus\{0\})/\mathbf{R}_{>0}$ in the $(\rho(X/S)-1)$-dimensional sphere $(N_1(X/S)\setminus\{0\})/\mathbf{R}_{>0}$, the quotient of functions $(K_X+B)/L$ takes finite values, and hence it is bounded. Therefore, there exists a sufficiently small positive real number $\epsilon$ such that $L - \epsilon(K_X + B)$ is positive on $\overline{\mathrm{NE}}(X/S) \setminus \{0\}$. By Kleiman's criterion, $L - \epsilon(K_X + B)$ is relatively ample.

Applying the basepoint-free theorem, after replacing $L$ by a multiple of a positive integer, we may assume that the natural homomorphism

$f^* f_* \mathcal{O}_X(L) \to \mathcal{O}_X(L)$ is surjective. Correspondingly, we get a projective morphism $\bar{h} \colon X \to \mathbf{P}_S(f_* \mathcal{O}_X(L))$ over $S$. Here the latter is the projective scheme over $S$ corresponding to the coherent sheaf $f_* \mathcal{O}_X(L)$. By definition, $\bar{h}^* \mathcal{O}_{\mathbf{P}_S(f_* \mathcal{O}_X(L))}(1) \cong \mathcal{O}_X(L)$.

Take the *Stein factorization* of $\bar{h}$, we get a surjective morphism $h \colon X \to Y$ to a normal algebraic variety and a finite morphism $Y \to \mathbf{P}_S(f_* \mathcal{O}_X(L))$. Take $g \colon Y \to S$ to be the induced morphism.

We claim that $h$ is the contraction morphism associated to $F$. First, for a curve $C$ on $X$, if $h(C)$ is a point, then $\mathcal{O}_X(L) \otimes \mathcal{O}_C \cong \mathcal{O}_C$ and hence $(L \cdot C) = 0$, which implies that $[C] \in F$.

Second, take $F'$ to be the closed convex cone spanned by equivalence classes of curves contracted by $h$, then $F' = \overline{\mathrm{NE}}(X/Y) \subset F$. Assume, to the contrary, that $F' \neq F$, then there exists a Cartier divisor $L'$ on $X$ which is positive on $F' \setminus \{0\}$ but negative at some point of $F$. Note that $L = h^* L''$ for some $g$-ample Cartier divisor $L''$. Since $L'$ is $h$-ample, for any sufficiently large $m$, $L' + mh^* L''$ is $f$-ample. In this case, $L' + mh^* L''$ is positive on $F \setminus \{0\}$, but this is a contradiction since $h^* L''$ is identically 0 on $F \setminus \{0\}$.

(2) and (3) follow directly from (1).

(4) Since $D$ is $h$-nef and $D - (K_X + B)$ is $h$-ample, the basepoint-free theorem (Theorem 2.1.1) can be applied to $D$ and $h$, which implies that there exists a positive integer $m_1$ such that $mD$ is $h$-free for $m \geq m_1$. The corresponding morphism $X \to Y'$ over $Y$ coincides with $h$ since $mD \equiv 0$ over $Y$. That is, there exists a Cartier divisor $E_m$ on $Y$ such that $mD \sim h^* E_m$. We can conclude (4) by taking $E = E_{m_1+1} - E_{m_1}$.

(5) From (4), we get the following exact sequence

$$0 \to N^1(Y/S) \to N^1(X/S) \to N^1(X/Y) \to 0,$$

which concludes (5). □

**Remark 2.4.2** The phenomenon in (4) suggests that the fibers of a contraction morphism are special varieties similar to $\mathbf{P}^1$. That is because, for example, on elliptic curves there exist many Cartier divisors which are numerically 0 but not 0. Later in Corollary 2.8.4, we will prove that the fibers of a contraction morphism are covered by rational curves. However, rational curves with bad singularities could have similar Cartier divisors as in the case of elliptic curves, so we may expect a much stronger statement.

### 2.4.2 The Cone Theorem

The shape of the closed cone of curves $\overline{\mathrm{NE}}(X/S)$ varies, but according to the following cone theorem, if restricted to the area taking negative values on the

## 2.4 The Cone Theorem

canonical divisor, then locally it is generated by finitely many extremal rays. By the contraction theorem, those extremal rays are associated with contraction morphisms.

**Theorem 2.4.3** (*Cone theorem*) *Let $(X, B)$ be a KLT pair and let $f : X \to S$ be a projective morphism. Fix a relatively ample divisor $A$ and a positive real number $\epsilon$. Then there are finitely many extremal rays $R_i$ of the closed cone of curves $\overline{\mathrm{NE}}(X/S) \subset N_1(X/S)$ such that*

$$\overline{\mathrm{NE}}(X/S) = \overline{\mathrm{NE}}(X/S)_{K_X+B+\epsilon A \geq 0} + \sum R_i.$$

*This equation means that the smallest convex cone containing all terms on the right-hand side is the left-hand side. Moreover, after removing unnecessary terms in the sum, for each $i$, $K_X + B$ is negative on $R_i \setminus \{0\}$, and there exists a contraction morphism $h_i : X \to Y_i$ associated to the extremal ray $R_i$.*

*Proof* We do induction on $\rho(X/S) = \dim N_1(X/S)$. In the proof, the relative setting plays an important role.

**Step 1** We may assume that $\epsilon$ is a rational number. We will show that we may also assume that $B$ is a **Q**-divisor.

We may write $K_X + B = \sum r_i D_i$, where $D_1, \ldots, D_t$ are Cartier divisors. We may approximate real numbers $r_i$ by rational numbers $r'_i$ such that $\sum_i (r_i - r'_i) D_i + \epsilon A/3$ is relatively ample.

As $B' = \sum r'_i D_i - K_X$ is not necessarily effective, we can write $B' = (B')^+ - (B')^-$. Here $(B')^+, (B')^-$ are effective **Q**-divisors with no common irreducible component. If taking $r'_i - r_i$ sufficiently small, then the coefficients of $(B')^-$ are sufficiently small, and there exists an effective **Q**-divisor $B''$ with sufficiently small coefficients such that $\epsilon A/3 - (B')^- \sim_{\mathbf{Q}} B''$.

Also by taking $r'_i - r_i$ sufficiently small, we may assume that $(X, (B')^+ + B'')$ is again KLT. Once we proved the assertion for $(X, (B')^+ + B'')$, the assertion for $(X, B)$ follows from the fact that

$$\overline{\mathrm{NE}}(X/S)_{K_X+(B')^++B''+\epsilon A/3 \geq 0} \subset \overline{\mathrm{NE}}(X/S)_{K_X+B+\epsilon A \geq 0}.$$

Therefore, we may assume that $B$ is a **Q**-divisor.

**Step 2** If $\rho(X/S) = 1$, then there is nothing to prove. So we assume that $\rho(X/S) > 1$ in the following. Also we may assume that $K_X + B$ is not relatively nef.

For any relatively ample **Q**-divisor $H$, by the rationality theorem (Theorem 2.3.1), the threshold

$$r_H = \max\{t \in \mathbf{R} \mid H + t(K_X + B) \text{ is relatively nef}\} \in \mathbf{Q}$$

108     2 *The Minimal Model Program*

determines a **Q**-divisor $L_H = H + r_H(K_X + B)$. By construction, $L_H$ is relatively nef but not relatively ample. We know that

$$F_H = \overline{\mathrm{NE}}(X/S)_{L_H=0}$$

is a face of the closed cone of curves and satisfies the contraction theorem (Theorem 2.4.1). Denote by $h_H : X \to Y_H$ the corresponding contraction morphism.

**Step 3** Take $\mathscr{C}$ to be the smallest closed convex cone containing $\overline{\mathrm{NE}}(X/S)_{K_X+B\geq 0}$ and all faces $F_H$ with $L_H \not\equiv_S 0$. (Here we remark that such $L_H$ always exists as $\rho(X/S) > 1$.) We will show that the closed cone of curves $\overline{\mathrm{NE}}(X/S)$ coincides with $\mathscr{C}$. Note that in this step there might be infinitely many $F_H$.

To the contrary, assume that $\mathscr{C} \neq \overline{\mathrm{NE}}(X/S)$. Then under this assumption, there exists a **Q**-divisor $M$ such that $(M \cdot v) > 0$ for all $v \in \mathscr{C} \setminus \{0\}$ and $(M \cdot v_0) < 0$ for some $v_0 \in \overline{\mathrm{NE}}(X/S)$. Moreover, $M$ cannot be numerically equivalent to a multiple of $K_X + B$ over $S$. Indeed, suppose that $M \equiv_S t(K_X + B)$ for some $t \neq 0$, then $t < 0$ as $M$ and $-(K_X + B)$ are positive on some $F_H \setminus \{0\}$; on the other hand, $(M \cdot v_0) < 0$ implies that $v_0 \in \overline{\mathrm{NE}}(X/S)_{K_X+B\geq 0} \setminus \{0\} \subset \mathscr{C} \setminus \{0\}$, a contradiction.

The dual closed convex cone $(\overline{\mathrm{NE}}(X/S)_{K_X+B\geq 0})^*$ of $\overline{\mathrm{NE}}(X/S)_{K_X+B\geq 0}$ is just the closed convex cone spanned by $\overline{\mathrm{Amp}}(X/S)$ and $K_X + B$, because the dual of the latter is $\overline{\mathrm{NE}}(X/S)_{K_X+B\geq 0}$.

As $M$ is positive on $\overline{\mathrm{NE}}(X/S)_{K_X+B\geq 0} \setminus \{0\}$, it is an interior point of $(\overline{\mathrm{NE}}(X/S)_{K_X+B\geq 0})^*$. Therefore, we can write $M = H + t(K_X + B)$ for some relatively ample **Q**-divisor $H$ and some positive rational number $t$.

Since $M$ is not relatively nef, $r_H < t$. On the other hand, since $L_H = H + r_H(K_X + B) \not\equiv_S 0$ (as $M$ is not numerically equivalent to a multiple of $K_X + B$ over $S$), we have $F_H \subset \mathscr{C}$ and hence $M$ is positive on $F_H \setminus \{0\}$. This is a contradiction.

**Step 4** Take $\mathscr{C}_1$ to be the closed convex cone containing $\overline{\mathrm{NE}}(X/S)_{K_X+B\geq 0}$ and all extremal rays of the form $R_H = F_H$. We will show that $\overline{\mathrm{NE}}(X/S) = \mathscr{C}_1$. Note that in this step there might be infinitely many extremal rays $R_H$.

For a face $F_H$ with $\dim F_H \geq 2$, we may apply Step 3 to $F_H = \overline{\mathrm{NE}}(X/Y_H) \subset \overline{\mathrm{NE}}(X/S)$. Since $(F_H)_{K_X+B\geq 0} = \{0\}$, $F_H$ is generated by lower dimensional faces.

**Step 5** We will show the *discreteness* of extremal rays by applying the estimate of denominators $q \leq a(b + 1)$ in the rationality theorem, that is, to

## 2.4 The Cone Theorem

show that there are only finitely many extremal rays negative on $K_X + B + \epsilon A$. Here $a, b$ are from the notation in the rationality theorem.

For each extremal ray $R_i$, take the associated contraction morphism $h_i \colon X \to Y_i$. Since $-(K_X + B)$ is $h_i$-ample, there is a unique element $v_i \in R_i$ with $(a(K_X + B) \cdot v_i) = -1$.

Take relatively ample Cartier divisors $H_1, \ldots, H_{\rho(X/S)-1}$ such that together with $a(K_X + B)$ they form a basis of $N^1(X/S)$. Since $\dim N_1(X/Y_i) = 1$, we can define $r_{ij}$ such that $H_j + r_{ij}(K_X + B) \equiv 0$ over $Y_i$. Applying the rationality theorem (Theorem 2.3.1) to $h_i$, we can express $r_{ij}/a = p_{ij}/q_{ij}$ as an irreducible fraction and $q_{ij} \le a(b+1)$. Therefore, $(a(b+1))! \, (H_j \cdot v_i) \in \mathbf{Z}$.

Take a sufficiently large number $N$ such that $NA - H_j$ is $f$-ample for all $j$. If we only look at extremal rays $R_i$ such that $((K_X + B + \epsilon A) \cdot v_i) < 0$, then

$$(H_j \cdot v_i) < (NA \cdot v_i) < N/a\epsilon,$$

and hence there are only finitely many possible values for $(H_j \cdot v_i)$. This means that there are only finitely many extremal rays generated by such $v_i$.

**Step 5'** Let us give another proof of the discreteness of extremal rays by applying the estimate of the lengths of extremal rays instead of the rationality theorem.

Keep the notation in the last step. By Corollary 2.8.4, there exists an $h_i$-relative curve $C_i$ such that $(-(K_X + B) \cdot C_i) \le 2b$. If we only look at extremal rays $R_i$ such that $((K_X + B + \epsilon A) \cdot C_i) < 0$, then $(A \cdot C_i) < 2b/\epsilon$. As the degree of $C_i$ is bounded, there exists a scheme of finite type $W$ and a closed subscheme $V$ of $X \times W$ such that all $C_i$ appear as fibers of the projection $\phi \colon V \to W$. Therefore, there are only finitely many numerical equivalence classes of those $C_i$.

Also we can use the following argument. Since $(-a(K_X + B) \cdot C_i) \le 2ab$ and $(H_j \cdot C_i) \in \mathbf{Z}$, we have $(2ab)! \, (H_j \cdot v_i) \in \mathbf{Z}$. Then we can argue the same as the end of Step 5. $\square$

**Remark 2.4.4** (1) The contraction theorem was first proved in the case that $X$ is smooth, $B = 0$, and $\dim X \le 3$ ([100]). The proof is by completely classifying the contraction morphisms. The classification shows for the first time that even if we start from a smooth $X$, the image $Y$ of the contraction morphism may have singularities, which is different from the surface case. The general contraction theorem was proved in a completely different way as an application of the basepoint-free theorem ([54, 55]).

110                     *2 The Minimal Model Program*

(2) The cone theorem was first proved in the case that $X$ is smooth and $B = 0$ (Mori [100]). The proof efficiently uses Frobenius morphisms which is a special method of algebraic geometry in positive characteristics (Theorem 2.7.2 discussed later is an application of this method). However, the deformation theory used in this method is difficult to be generalized to algebraic varieties with singularities, so it is limited as in the minimal model theory we cannot avoid dealing with algebraic varieties with singularities. Therefore, a completely different proof was developed by extending that of the basepoint-free theorem ([54]).

(3) In Step 5' of the proof, it might be possible to get a stronger estimate of the lengths of extremal rays $(-(K_X + B) \cdot C) \leq b + 1$. This is still an open problem.

(4) When considering an extremal ray $R$ in this book, we always assume that the log canonical divisor $K_X + B$ takes negative values on the complement set of the origin $R \setminus \{0\}$. Such an extremal ray is called a $(K_X + B)$-negative extremal ray.

**Corollary 2.4.5** *Keep the assumption of Theorem 2.4.3. Assume that $S$ is quasi-projective. Assume that $B$ is **R**-Cartier and relatively big. Then there are only finitely many extremal rays in the closed cone of curves $\overline{\mathrm{NE}}(X/S) \subset N_1(X/S)$ which take negative values on $K_X + B$.*

*Proof* Write $B = A + E$ for some relatively ample **R**-divisor $A$ and some effective **R**-divisor $E$. We may take a sufficiently small positive real number $\epsilon$ such that $(X, (1 - \epsilon)B + \epsilon E)$ is KLT. By Theorem 2.4.3, there are only finitely many $(K_X + (1 - \epsilon)B + \epsilon E + \epsilon A)$-negative extremal rays. Therefore, the same holds true for $K_X + B$. $\qquad\square$

It is easy to extend the cone theorem to DLT pairs:

**Corollary 2.4.6** *Let $(X, B)$ be a DLT pair and let $f : X \to S$ be a projective morphism. Assume that $S$ is quasi-projective. Fix a relatively ample divisor $A$ and a positive real number $\epsilon$. Then there exist finitely many extremal rays $R_i$ of $\overline{\mathrm{NE}}(X/S) \subset N_1(X/S)$ such that*

$$\overline{\mathrm{NE}}(X/S) = \overline{\mathrm{NE}}(X/S)_{K_X+B+\epsilon A \geq 0} + \sum R_i.$$

*Moreover, after removing unnecessary terms in the sum, for each $i$, $K_X + B$ is negative on $R_i \setminus \{0\}$ and there exists a contraction morphism $h_i : X \to Y_i$ associated to the extremal ray $R_i$.*

*Proof* By Lemma 2.1.8, there is $B' \sim_{\mathbf{R},S} B + \frac{1}{2}\epsilon A$ such that $(X, B')$ is KLT. The corollary can be reduced to the cone theorem. $\qquad\square$

### 2.4.3 Contraction Morphisms in Dimensions 2 and 3

In the following, we describe the contraction morphism associated to an extremal ray in dimension up to 3. First, let us consider the surface case.

**Example 2.4.7** Consider the case that $X$ is smooth, $S = \operatorname{Spec} k$, $B = 0$, and $\dim X = 2$. Here the base field $k$ is algebraically closed of arbitrary characteristic. The contraction morphism $\phi \colon X \to Y$ associated to an extremal ray $R$ can be classified as the following ([100]).

(a) There exists a $(-1)$-curve $C \subset X$ (see the context before Theorem 1.13.3) such that $R = \mathbf{R}_+[C]$. $Y$ is smooth, $\phi(C) = P$ is a point, and $\phi$ is the blowup of $Y$ at $P$. Conversely, a $(-1)$-curve always generates an extremal ray.

(b) $\phi \colon X \to Y$ is a $\mathbf{P}^1$-bundle over a smooth curve $Y$ and $R = \mathbf{R}_+[C]$ for any fiber $C$. In this case, $X$ is called a *ruled surface*. Conversely, if $X$ admits a $\mathbf{P}^1$-bundle structure, then its fiber determines an extremal ray.

(c) $X \cong \mathbf{P}^2$, $Y = \operatorname{Spec} k$, and $R$ is generated by the equivalence class of a line on $\mathbf{P}^2$.

It is important that, in each case, the extremal ray is generated by a curve isomorphic to $\mathbf{P}^1$.

As we will see in Section 3.9, the theory of extremal rays can be extended to algebraically nonclosed base field $k$. Take the base change $\bar{X} = X \times \operatorname{Spec} \bar{k}$ to the algebraic closure, the classification can be generalized as the following.

(a') There exist disjoint $(-1)$-curves $C_1, \ldots, C_t$ on $\bar{X}$ such that a certain multiple of their sum $C = m \sum C_i$ is defined over $k$ and $R = \mathbf{R}_+[C]$. $Y$ is smooth, $\phi(C) = P$ is a point, and $\phi$ is the blowup of $Y$ at $P$. Here the residue field of $P$ is an extension field of $k$.

(b') $\phi \colon X \to Y$ is a morphism to a smooth curve $Y$ and $R = \mathbf{R}_+[C]$ for any fiber $C$. In this case, every fiber is isomorphic to a curve of degree 2 in $\mathbf{P}^2$ and $X$ is called a *conic surface*.

(c') $-K_X$ is ample and $\rho(X) = 1$. Here $\rho(X) = \dim N^1(X)$ is the *Picard number*. Generally, a smooth projective surface with ample anti-canonical divisor is called a *del Pezzo surface*. There is a classical classification of del Pezzo surfaces.

We describe an example with infinitely many extremal rays.

**Example 2.4.8** (Nagata's example) By the cone theorem, there are only finitely many $(K_X + B + \epsilon A)$-negative extremal rays, but when taking the

112                    *2 The Minimal Model Program*

limit as $\epsilon \to 0$, it is possible to have infinitely many extremal rays. Here the base field $k$ is algebraically closed of characteristic 0.

Given two curves $C_1, C_2$ of degree 3 on the projective plane $\mathbf{P}^2$ intersecting at nine distinct points $P_1, \ldots, P_9$. The rational function $h$ defined by $\mathrm{div}(h) = C_1 - C_2$ determines a rational map $\bar{h} \colon \mathbf{P}^2 \dashrightarrow \mathbf{P}^1$. The indeterminacy locus of $\bar{h}$ is $\{P_1, \ldots, P_9\}$. Blowing up along those points $f \colon X \to \mathbf{P}^2$ resolves the indeterminacy points and gives a morphism $g = \bar{h} \circ f$.

For a smooth curve $C$ of degree 3 passing through these nine points, its strict transform $F = f_*^{-1} C$ becomes a smooth fiber of $g$, and $K_X = -F$. In particular, $F$ is an elliptic curve. The exceptional set of $f$ consists of nine $(-1)$-curves $E_i$ ($i = 1, \ldots, 9$), which are sections of $g$.

The generic fiber $F_\eta$ of $g$ is an elliptic curve defined over the field $k(\mathbf{P}^1)$. Take $Q_i = E_i \cap F_\eta$. Consider the additive group structure on $F_\eta$ with $Q_1$ as the origin. If $C_1, C_2$ are chosen generally, $Q_2$ is not a torsion point with respect to the addition, that is, $m Q_2 \neq Q_1$ for all positive integer $m$. Take $G_m$ to be the closure of $m Q_2$ in $X$, which is a section of $g$. Then $G_m \cong \mathbf{P}^1$ and $(K_X \cdot G_m) = -1$. That is, $G_m$ is a $(-1)$-curve. In this case, there are infinitely many extremal rays.

Take $S = \{(P_1, \ldots, P_9) \in (\mathbf{P}^2)^9 \mid P_i \neq P_j \ (i \neq j)\}$. The projection $\mathbf{P}^2 \times S \to S$ naturally admits nine sections. Take $\tilde{f} \colon \mathcal{X} \to \mathbf{P}^2 \times S$ to be the blowup along those sections, then the above constructed $X$ is a fiber of the smooth morphism $\pi \colon \mathcal{X} \to S$. That is, $\pi$ is a *deformation family* of $X$.

As $(-1)$-curves are preserved by small deformations, for each $m$ there exists a non-empty open set $U_m$ and a closed subvariety $\tilde{G}_m$ of $\pi^{-1}(U_m)$ such that $\tilde{G}_m \cap X = G_m$ and on each fiber $X_s = \pi^{-1}(s)$ ($s \in U_m$), $\tilde{G}_m \cap X_s$ is a $(-1)$-curve. In the case that the base field $k$ is the complex number field, the intersection $U = \bigcap U_m$ is not empty, and for each $s \in U$, $X_s$ has infinitely many extremal rays.

Generally, if there exists a non-empty open set such that a property holds for each point in this set, then we say that this property holds for *general* points; if a property holds for each point in the intersection of countably infinitely many non-empty open sets, like the above $U$, then we say that this property holds for *very general* points. So a very general fiber of $\pi$ has infinitely many extremal rays.

The 3-dimensional case is as follows.

**Example 2.4.9** Consider the case that $X$ is smooth, $S = \mathrm{Spec}\, k$, $B = 0$, and $\dim X = 3$. The contraction morphism $\phi \colon X \to Y$ associated to an extremal ray $R$ can be classified as the following ([100]). Here the base field $k$ is algebraically closed of characteristic 0.

## 2.4 The Cone Theorem 113

(1) The exceptional set of $\phi$ is a prime divisor $E$ and $\phi$ is the blowup of $Y$ along $\phi(E)$. However, $Y$ is not necessarily smooth. $E$ and $\phi$ are classified as the following.

(a) $\phi(E) = P$ is a point, $E \cong \mathbf{P}^2$, and $\mathcal{O}_E(E) \cong \mathcal{O}_{\mathbf{P}^2}(-1)$. In this case, $Y$ is smooth.

(b) $\phi(E) = P$ is a point, $E \cong \mathbf{P}^2$, and $\mathcal{O}_E(E) \cong \mathcal{O}_{\mathbf{P}^2}(-2)$. If $k = \mathbf{C}$, then $(Y, P)$ is analytically isomorphic to the cyclic quotient singularity of type $\frac{1}{2}(1, 1, 1)$.

(c) $\phi(E) = P$ is a point, $E$ is isomorphic to the quadratic surface in $\mathbf{P}^3$ defined by the equation $xy + zw = 0$, and $\mathcal{O}_E(E) \cong \mathcal{O}_E(-1)$. $E$ is isomorphic to $\mathbf{P}^1 \times \mathbf{P}^1$. If $k = \mathbf{C}$, then the singularity $(Y, P)$ is analytically isomorphic to the hypersurface singularity defined by $xy + zw = 0$ in $\mathbf{C}^4$.

(d) $\phi(E) = P$ is a point, $E$ is isomorphic to the quadratic surface in $\mathbf{P}^3$ defined by the equation $xy + z^2 = 0$, and $\mathcal{O}_E(E) \cong \mathcal{O}_E(-1)$. If $k = \mathbf{C}$, then the singularity $(Y, P)$ is analytically isomorphic to the hypersurface singularity defined by $xy + z^2 + w^3 = 0$ in $\mathbf{C}^4$.

(e) $\phi(E) = C$ is a smooth projective curve, $\phi|_E : E \to C$ is a $\mathbf{P}^1$-bundle, and $(E \cdot F) = -1$ for each fiber $F$. In this case, $Y$ is smooth.

(2) $Y$ is a smooth projective surface and the geometric generic fiber of $\phi$ is isomorphic to $\mathbf{P}^1$. Every fiber of $\phi$ is isomorphic to a conic curve in $\mathbf{P}^2$, hence $X$ is called a *conic bundle*.

(3) $Y$ is a smooth projective curve, the geometric generic fiber of $\phi$ is a del Pezzo surface.

(4) $Y$ is a point and $X$ is a *Fano variety* of Picard number $\rho(X) = 1$. Generally, a projective algebraic variety $X$ is called a Fano variety if $-K_X$ is ample. Three-dimensional smooth Fano varieties are classified ([48, 49, 103, 104]).

### 2.4.4 The Cone Theorem for the Space of Divisors

By dualizing the contraction theorem and the cone theorem, we can describe them in terms of the space of divisors as the following. The paraphrase is powerful when considering changing of birational models. For example, when the nef cones of two birational models adjoin along a face of both cones, the phenomenon of *wall crossing* is important, and can be described appropriately in the space of divisors. Here a wall is a face of codimension 1, which is the dual concept of an extremal ray.

114                    2 *The Minimal Model Program*

**Theorem 2.4.10** *Let $(X, B)$ be a KLT pair and let $f : X \to S$ be a projective morphism. Fix a relatively ample divisor $A$ and a positive real number $\epsilon$. Assume that $K_X + B + \epsilon A$ is not $f$-nef. Take $R_i$ ($i = 1, \ldots, N$) to be all $(K_X + B + \epsilon A)$-negative extremal rays and take $h_i : X \to Y_i$ to be the contraction morphism associated to $R_i$. Take the rational linear functional $l_i$ on $N^1(X/S)$ defined by $l_i(u) = (u \cdot v_i)$, where $v_i \in R_i$ is a nonzero rational point. Then the following assertions hold:*

*(1) $l_i$ is nonnegative on $\overline{\mathrm{Amp}}(X/S)$.*
*(2) $G_i = \{u \in \overline{\mathrm{Amp}}(X/S) \mid l_i(u) = 0\}$ is a face of codimension 1 in $\overline{\mathrm{Amp}}(X/S)$ which coincides with $h_i^* \overline{\mathrm{Amp}}(Y_i/S)$.*
*(3) Take $F$ to be the face of $\overline{\mathrm{NE}}(X/S)$ generated by several extremal rays $R_{i_1}, \ldots, R_{i_r}$ and take $h : X \to Y$ to be the associated contraction morphism. Then*

$$G = \bigcap_{j=1}^{r} G_{i_j} = \{u \in \overline{\mathrm{Amp}}(X/S) \mid (u \cdot v) = 0 \text{ for all } v \in F\}$$

*is $h^* \overline{\mathrm{Amp}}(Y/S)$.*
*(4) For any $f$-ample $\mathbf{R}$-divisor $H$, take*

$$t_0 = \min\{t \mid K_X + B + \epsilon A + tH \text{ is } f\text{-nef}\},$$

*then there exists a face $G$ of the form $G = \bigcap_{j=1}^r G_{i_j}$ such that $[K_X + B + \epsilon A + t_0 H]$ is a relative interior point of $G$. In other words, it is contained in $h^* \mathrm{Amp}(Y/S)$.*

*Proof* (1) This follows from $R_i \subset \overline{\mathrm{NE}}(X/S)$.

(2) By the contraction theorem, $G_i = h_i^* \overline{\mathrm{Amp}}(Y_i/S)$ is of codimension 1.

(3) This is a consequence of the contraction theorem.

(4) By definition, $u = [K_X + B + \epsilon A + t_0 H]$ is the supporting function of a face $F$ of $\overline{\mathrm{NE}}(X/S)$ such that $K_X + B$ is negative on $F \setminus \{0\}$. By the cone theorem, such a face is generated by extremal rays, say $R_{i_1}, \ldots, R_{i_r}$, which implies that $u$ is contained in $G = \bigcap_{j=1}^r G_{i_j}$. As $u$ is the supporting function of $F$, $u$ is an interior point of $G$. $\qquad\square$

**Remark 2.4.11** In other words, the cone theorem can be explained as the following: Imagine the nef cone as a planet and $[K_X + B] \in N^1(X/S)$ as a satellite moving around it. Suppose that the nef cone is opaque. First, if $[K_X + B] \in \overline{\mathrm{Amp}}(X/S)$, then we can observe nothing and hence the statement is void. If $[K_X + B] \notin \overline{\mathrm{Amp}}(X/S)$, then we can observe the front side $V$ of the surface $\partial \overline{\mathrm{Amp}}(X/S)$ of the nef cone. The back side $\partial \overline{\mathrm{Amp}}(X/S) \setminus V$ cannot be observed, just like the back of the moon.

## 2.4 The Cone Theorem

When we look at the planet from a slightly closer observation point $[K_X + B + \epsilon A] \in N^1(X/S)$, the surface $V$ looks like a polyhedron consisting of finitely many faces $G_i$. If we move the observation point to the limit $[K_X + B]$ as $\epsilon \to 0$, in the case with infinitely many extremal rays, there turns out to be infinitely many faces converging to the horizon.

As a corollary, we get the basepoint-free theorem for **R**-divisors:

**Corollary 2.4.12** *Let $(X, B)$ be a KLT pair, let $f \colon X \to S$ be a projective morphism, and let $D$ be an **R**-Cartier divisor. Assume that $D$ is $f$-nef and $D - (K_X + B)$ is $f$-nef and $f$-big. Then there exists a projective morphism $g \colon Z \to S$ from a normal algebraic variety, a projective surjective morphism $h \colon X \to Z$ with connected geometric fibers such that $f = g \circ h$, and a $g$-ample **R**-Cartier divisor $H$ on $Z$ such that $h^* H \sim_{\mathbf{R}} D$.*

*Proof* We may assume that $S$ is affine. Indeed, since the factorizations $X \to Z \to S$ over affine open subsets $S$ are uniquely determined by $D$, they glue together. We may assume that $D$ is not $f$-ample. Since $D - (K_X + B)$ is $f$-big, we may write $D - (K_X + B) = A + E$ for some $f$-ample **R**-Cartier divisor $A$ and some effective **R**-Cartier divisor $E$. Since $D - (K_X + B)$ is also $f$-nef, for any sufficiently small positive real number $\epsilon$, $L = D - (K_X + B) - \epsilon E$ is $f$-ample.

By taking $\epsilon$ sufficiently small, we may assume that $(X, B + \epsilon E)$ is KLT. We can apply Theorem 2.4.10(4) to this pair. Since $D$ is not $f$-ample, $K_X + B + \epsilon E = D - L$ is not $f$-nef. Therefore, for a sufficiently small $\delta > 0$, $K_X + B + \epsilon E + \delta L$ is not $f$-nef. Consider

$$t_0 = \min\{t \mid K_X + B + \epsilon E + \delta L + tL \text{ is } f\text{-nef}\},$$

then $t_0 = 1 - \delta$ and $K_X + B + \epsilon E + \delta L + t_0 L = D$. Then the conclusion follows from Theorem 2.4.10(4). $\qquad\square$

As a corollary of the above corollary, we can show the existence of *canonical models* when the boundary is big:

**Corollary 2.4.13** *Let $(X, B)$ be a KLT pair and let $f \colon X \to S$ be a projective morphism. Assume that $K_X + B$ is $f$-nef and $B$ is an $f$-big **R**-Cartier divisor. Then there exists a projective morphism $g \colon Z \to S$ from a normal algebraic variety, a projective surjective morphism $h \colon X \to Z$ with connected geometric fibers such that $f = g \circ h$, and a $g$-ample **R**-Cartier divisor $H$ on $Z$ such that $h^* H \sim_{\mathbf{R}} K_X + B$.*

*Proof* We may assume that $S$ is affine. Take $D = K_X + B$. Then $D$ is $f$-nef. As $B$ is $f$-big, we may write $B = A + E$ for some $f$-ample **R**-Cartier divisor

116                 *2 The Minimal Model Program*

$A$ and some effective $\mathbf{R}$-Cartier divisor $E$. Take a sufficiently small $\epsilon > 0$ such that $(X, (1 - \epsilon)B + \epsilon E)$ is KLT. Then, $D - (K_X + (1 - \epsilon)B + \epsilon E) = \epsilon A$ is $f$-ample, and we can apply Corollary 2.4.12.        □

## 2.5 Types of Contraction Morphisms and the Minimal Model Program

The MMP is an operation to simplify a given pair consisting of a variety and a boundary divisor by applying birational maps repeatedly. The pair we consider is assumed to be KLT or DLT, and the variety is assumed to be $\mathbf{Q}$-factorial and projective over the base variety. This condition is preserved under the operation of the MMP.

Such an operation in the MMP is constructed by the contraction morphism associated to an extremal ray. There are three types of contraction morphisms: divisorial contractions, small contractions, and Mori fiber spaces.

The goal of the MMP is to obtain either a minimal model (a pair with relatively nef log canonical divisor) or another different class called a Mori fiber space.

### 2.5.1 Classification of Contraction Morphisms

First, consider the case that the contraction morphism associated to an extremal ray is a birational morphism contracting a divisor:

**Theorem 2.5.1** *Let $(X, B)$ be a DLT pair and let $f : X \to S$ be a projective morphism. Assume that $X$ is $\mathbf{Q}$-factorial. Let $R$ be an extremal ray of $\overline{NE}(X/S)$ such that $K_X + B$ takes negative values on $R \setminus \{0\}$ and take $h : X \to Y$ to be the contraction morphism associated to $R$. Assume that $h$ is a birational morphism and its exceptional set contains a prime divisor. Then the following assertions hold:*

*(1) $-(K_X + B)$ is $h$-ample.*
*(2) $\rho(X/Y) = 1$ and $\rho(X/S) = \rho(Y/S) + 1$.*
*(3) The exceptional set of $h$ is a prime divisor, say $E$.*
*(4) $Y$ is $\mathbf{Q}$-factorial.*
*(5) We can write $K_X + B = h^*(K_Y + B_Y) + eE$, $e > 0$. Here $B_Y = h_* B$.*
*(6) $(Y, B_Y)$ is DLT. Moreover, if $(X, B)$ is KLT, then $(Y, B_Y)$ is KLT.*

*Proof* (1), (2) follow directly from the contraction theorem.

## 2.5 Types of Contraction Morphisms and the Minimal Model Program 117

(3) Let $E$ be a prime divisor contained in the exceptional set of $h$. Since $X$ is $\mathbf{Q}$-factorial, $E$ is $\mathbf{Q}$-Cartier. Since $E$ is exceptional, by Lemma 1.6.3, there exists a curve $C$ contracted by $h$ such that $(E \cdot C) < 0$.

Since $C$ is an $h$-relative curve and $\rho(X/Y) = 1$, $-E$ is $h$-ample. Suppose that $E$ does not coincide with the exceptional set of $h$, then there exists a relative curve $C'$ not contained in $E$. This implies that $(E \cdot C') \geq 0$, a contradiction. Therefore, the exceptional set of $h$ is a prime divisor.

(4) Take any prime divisor $F$ of $Y$. Since $X$ is $\mathbf{Q}$-factorial and $\rho(X/Y) = 1$, there exists a rational number $r$ such that $h_*^{-1}F + rE \equiv 0$ over $Y$. There exists a positive integer $p$ such that $p(h_*^{-1}F + rE)$ is a Cartier divisor.

By the contraction theorem (Theorem 2.4.1), there exists a Cartier divisor $F'$ on $Y$ such that $p(h_*^{-1}F + rE) \sim h^*F'$. Since $h$ is birational, $pF \sim F'$, which means that $F$ is $\mathbf{Q}$-Cartier.

(5) Write $K_X + B = h^*(K_Y + B_Y) + eE$. Since $-(K_X + B)$ and $-E$ are $h$-ample, we know that $e > 0$.

(6) follows from (5). $\qquad\square$

Let $(X, B)$ be a $\mathbf{Q}$-factorial DLT pair and let $f : X \to S$ be a projective morphism. If $K_X + B$ is relatively nef, then $f : (X, B) \to S$ is already minimal. If not, then by the cone theorem, there exists an extremal ray $R$ in $\overline{\mathrm{NE}}(X/S)$ such that $K_X + B$ takes negative values on $R \setminus \{0\}$. Take $h : X \to Y$ to be the contraction morphism associated to $R$. By Theorem 2.5.1, we have the following three cases:

(1) *Divisorial contraction*: $h$ is birational and the exceptional set is a prime divisor.
(2) *Small contraction*: $h$ is birational and the exceptional set is of codimension at least 2.
(3) *Mori fiber space*: $\dim Y < \dim X$.

For a divisorial contraction, the new pair $(Y, B_Y)$ has the same property as $(X, B)$. If $K_Y + B_Y$ is not relatively nef, that is, it is not a minimal model, then we can continue to consider contraction morphisms. Moreover, since $\rho(Y/S) = \rho(X/S) - 1$, there cannot be infinitely many divisorial contractions in this procedure. So we may expect to get a minimal model by induction on the Picard number $\rho(X/S)$.

For example, for a pair where $X$ is a smooth surface and $B = 0$, a divisorial contraction is the contraction of a $(-1)$-curve (see Example 2.4.7). Then after finitely many divisorial contractions, there is no $(-1)$-curve, and we reach a minimal model in the classical sense. This model is either a minimal model in the sense of this book, or admits a further contraction morphism. By dimension

118          *2 The Minimal Model Program*

reason, this contraction morphism is not small, hence is a Mori fiber space, that is, a ruled surface or $\mathbf{P}^2$.

However, this is not the case in higher dimensions due to the existence of small contractions. In dimension 3, small contractions appear only if $X$ is singular or $B \neq 0$ (see Example 2.4.9). In dimensions 4 and higher, small contractions can appear even if $X$ is smooth and $B = 0$ (see [62]).

Although Mori fiber spaces are not birational, it is interesting to be able to handle them in the same category of contraction morphisms. For example, in the situation of algebraic surfaces, they can be constructed from ruled surfaces or the morphism from $\mathbf{P}^2$ to a point. A Mori fiber space is also called a *Fano fibration*.

In general, an algebraic variety $X$ is called a *uniruled variety* if it is covered by a family of rational curves. In other words, this condition means that there exists an algebraic variety $Z$ with $\dim Z = \dim X - 1$ and a dominant rational map $Z \times \mathbf{P}^1 \dashrightarrow X$. Uniruledness is a property invariant under birational equivalence.

As later discussed by the lengths of extremal rays (Section 2.8), each irreducible component of any fiber of a contraction morphism is always uniruled, unless it is a point. One image of the MMP is that "if you contract redundant rational curves by contraction morphisms, then you will get a minimal model." In particular, an algebraic variety with a Mori fiber space structure is a uniruled variety. Moreover, Hacon and M$^c$Kernan showed further that the fibers of contraction morphisms are always *rationally connected* ([37]).

For Mori fiber spaces we have the following result:

**Proposition 2.5.2** *Let* $h\colon X \to Y$ *be a Mori fiber space. Then* $Y$ *is* **Q**-*factorial.*

*Proof* We may assume that $\dim Y > 0$. Take any prime divisor $E$ on $Y$ and take a prime divisor $D$ on $X$ such that $h(D) = E$. Since $X$ is **Q**-factorial, there exists a positive integer $d$ such that $dD$ is Cartier. Since $\rho(X/Y) = 1$ and there exists a curve $C$ contained in a fiber of $h$ such that $D \cap C = \emptyset$, we get $D \equiv_Y 0$. Applying the basepoint-free theorem to $h$, there exists a Cartier divisor $E'$ on $Y$ and a rational function $g$ on $X$ such that $dD = h^*E' + \operatorname{div}(g)$. Since $\operatorname{div}(g)$ does not intersect the generic fiber of $h$, there exists a rational function $g'$ on $Y$ such that $g = h^*(g')$. So there exists a positive integer $e$ such that $eE = E' + \operatorname{div}(g')$, and hence $E$ is **Q**-Cartier.      $\square$

## 2.5.2 Flips

The existence of small contractions is a phenomenon appearing only in dimensions 3 and higher, which is completely different from the situation

### 2.5 Types of Contraction Morphisms and the Minimal Model Program 119

of dimension 2. If $X \to Y$ is a small contraction and we consider the pair $(Y, h_* B)$, then we are in trouble because $K_Y + h_* B$ is not **R**-Cartier. Indeed, if $K_Y + h_* B$ is **R**-Cartier, then we can consider its pullback by $h$. Since $X$ and $Y$ are isomorphic in codimension 1, we must have $h^*(K_Y + h_* B) = K_X + B$. On the other hand, take any curve $C$ contracted by $h$, then $((K_X + B) \cdot C) < 0$, which contradicts to the projection formula (before Proposition 1.4.3).

By this reason, we need to construct a new pair by an operation called a flip. The new pair obtained by a flip has the same properties as the original pair. Flips and divisorial contractions are completely different operations in geometry, but they are very similar in the point view of numerical geometry.

**Definition 2.5.3** Let $(X, B)$ be a **Q**-factorial DLT pair and let $f : X \to S$ be a projective morphism. Assume that $g : X \to Y$ is a small contraction morphism associated to a $(K_X + B)$-negative extremal ray. Then another birational projective morphism $g^+ : X^+ \to Y$ is called the *flip* of $g$ if the following conditions are satisfied:

(1) $g^+$ is isomorphic in codimension 1, that is, the exceptional set of $g^+$ does not contain any prime divisor.
(2) $K_{X^+} + B^+$ is $g^+$-ample, here $B^+$ is the strict transform of $B$.

Here note that the positivity of log canonical divisors $K_X + B$ and $K_{X^+} + B^+$ are reversed. The birational transform $(g^+)^{-1} \circ g$ is also called a flip.

When considering the existence of the flip of a small contraction, as ampleness is an open condition, it suffices to consider the case that $B$ is a **Q**-divisor without loss of generality. Indeed, the ampleness of $-(K_X + B)$ and $K_{X^+} + B^+$ is not changed after perturbing $B$ slightly. Similarly, it suffices to consider KLT pairs instead of DLT pairs.

**Example 2.5.4** Let us give examples of flips. Both examples are flips of toric varieties ([120]).

(1) Let us consider the example by Francia ([25]). Here $\dim X = 3$, $B = 0$, and $X$ is singular. We denote $X = X^-$. Originally, this example intended to claim that "the minimal model theory is impossible in dimensions 3 and higher," but later it was included into the development of the minimal model theory, and become the simplest example of flips (see Figure 2.1).
Consider the locally free sheaf $F = \mathcal{O}_{\mathbf{P}^1}(-1) \oplus \mathcal{O}_{\mathbf{P}^1}(-2)$ of rank two over $C^+ = \mathbf{P}^1$, take $X^+$ to be the total space of the corresponding vector bundle, that is,

$$X^+ = \operatorname{Spec}_{C^+}\left(\bigoplus_{m=0}^{\infty} \operatorname{Sym}^m F^*\right).$$

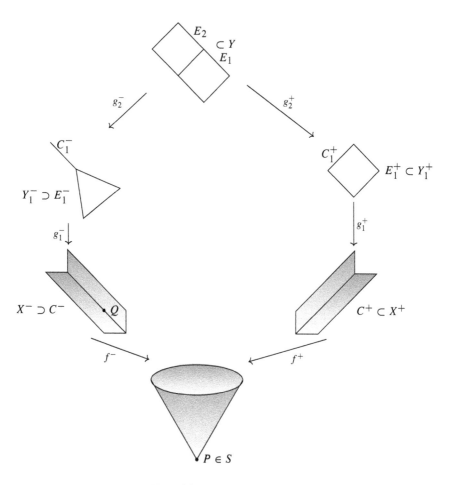

Figure 2.1 An example of flips.

$X^+$ is a smooth 3-dimensional algebraic variety which contains $C^+$ as the 0-section, and the cotangent bundle $N_{C^+/X^+}$ is isomorphic to $F$. Hence $(K_{X^+} \cdot C^+) = 1$. Set

$$S = \operatorname{Spec} H^0(X^+, \mathcal{O}_{X^+}) = \operatorname{Spec}\left(\bigoplus_{m=0}^{\infty} H^0(C^+, \operatorname{Sym}^m F^*)\right),$$

then there is a natural birational morphism $f^+ \colon X^+ \to S$. The exceptional set of $f^+$ coincides with the 0-section $C^+$, and $f^+(C^+) = P$ is a point. Hence $K_{X^+}$ is $f^+$-ample.

### 2.5 Types of Contraction Morphisms and the Minimal Model Program   121

Take $g_1^+ : Y_1^+ \to X^+$ to be the blowup of $X^+$ along $C^+$. The exceptional set $E_1^+$ of $g_1^+$ is isomorphic to the ruled surface $\mathbf{P}(F^*)$. Take $l_1^+$ to be a fiber of $g_1^+|_{E_1^+}$ and $C_1^+$ the curve with negative intersection on $E_1^+$. Note that $C_1^+$ is a section of $g_1^+|_{E_1^+}$. The cotangent bundle $N_{C_1^+/Y_1^+}$ is isomorphic to $\mathcal{O}_{\mathbf{P}^1}(-1)^{\oplus 2}$.

Take $g_2^+ : Y \to Y_1^+$ to be the blowup of $Y_1^+$ along $C_1^+$. The exceptional set $E_2$ of $g_2^+$ is isomorphic to the product $\mathbf{P}^1 \times \mathbf{P}^1$. Take $l_2^+$ to be a fiber of $g_2^+|_{E_2}$ and $l_2^-$ the fiber of the other projection of $E_2$. On $Y_1^+$ and $Y$, $g_1^+$ and $g_2^+$ are divisorial contractions. Denote $l_1 = (g_2^+)_*^{-1} l_1^+$.

Since $\dim N^1(Y/S) = 3$, we have

$$\overline{NE}(Y/S) = \langle l_1, l_2^+, l_2^- \rangle.$$

Here the symbol $\langle \ \rangle$ means the convex cone generated by the elements within there. We have $(K_Y \cdot l_1) = 0$ and $(K_Y \cdot l_2^+) = (K_Y \cdot l_2^-) = -1$. Take $R_2^+, R_2^-$ to be the extremal rays generated by $l_2^+, l_2^-$. The contraction morphism associated to $R_2^+$ is just $g_2^+$. The contraction morphism $g_2^- : Y \to Y_1^-$ associated to $R_2^-$ contracts the exceptional divisor $E_2$ of $g_2^+$ in the other direction.

Take $E_1 = (g_2^+)_*^{-1} E_1^+$. Since $((K_Y + E_1) \cdot l_1) = -2$, if we consider the pair $(Y, E_1)$, then $l_1$ also generates an extremal ray, so the corresponding contraction morphism exists and is a divisorial contraction contracting $E_1$. But we do not consider this contraction morphism here.

$E_1^- = (g_2^-)_* E_1$ is isomorphic to $\mathbf{P}^2$ and $l_1^- = (g_2^-)_* l_1$ is a line. As $\dim N^1(Y_1^-/S) = 2$, $\overline{NE}(Y_1^-/S)$ is generated by $l_1^-$ and $C_1^- = (g_2^-)_* l_2^+$. Here $(K_{Y_1^-} \cdot l_1^-) = -1$ and $(K_Y \cdot C_1^-) = 0$. Take $R_1^-$ to be the extremal ray generated by $l_1^-$, the corresponding contraction morphism $g_1^- : Y_1^- \to X^-$ contracts $E_1^-$ to a singular point $Q$ on $X^-$. As $\mathcal{O}_{E_1^-}(E_1^-) \cong \mathcal{O}_{\mathbf{P}^2}(-2)$, the singular point $Q$ is a cyclic quotient singularity of type $\frac{1}{2}(1, 1, 1)$.

Take $C^- = (g_1^-)_* C_1^-$, then $\overline{NE}(X^-/S)$ is generated by $C^-$. We have $(K_{X^-} \cdot C^-) = -1/2$. Here it might seem strange that the intersection number is a fractional, but this is because $K_{X^-}$ is not Cartier. In fact, $C^-$ passes through the singular point $Q$ and $2K_{X^-}$ becomes Cartier near $Q$.

$-K_{X^-}$ is $f^-$-ample and the morphism $f^- : X^- \to S$ is a small contraction. The morphism $f^+ : X^+ \to S$ is just the flip of $f^-$.

(2) There are examples of flips from smooth algebraic varieties in general dimensions.

Consider the locally free sheaf $F = \mathcal{O}_E(-1)^{\oplus(t+1)}$ of rank $t+1$ over the projective space $E = \mathbf{P}^s$ and take its total space $X = \operatorname{Spec}_E(\bigoplus_{m=0}^{\infty} \operatorname{Sym}^m F^*)$.

## 2 The Minimal Model Program

$X$ is a smooth $(s + t + 1)$-dimensional algebraic variety which contains $E$ as the 0-section, and the cotangent bundle $N_{E/X}$ is isomorphic to $F$. Set

$$S = \operatorname{Spec} H^0(X, \mathcal{O}_X) = \operatorname{Spec} \left( \bigoplus_{m=0}^{\infty} H^0(E, \operatorname{Sym}^m F^*) \right),$$

then there is a natural birational morphism $f : X \to S$. The exceptional set of $f$ coincides with the 0-section $E$ of $F$. View a line $C$ on $E$ as a curve on $X$, we have $(K_X \cdot C) = t - s$ and $f$ is a small contraction if $s > t$.

Take homogeneous coordinates $x_0, \ldots, x_s$ on $E$ and coordinates $y_0, \ldots, y_t$ along the direction of fibers of $F$, then

$$\bigoplus_{m=0}^{\infty} H^0(E, \operatorname{Sym}^m F^*) \cong k[x_i y_j]_{0 \le i \le s, 0 \le j \le t}$$

is symmetric with respect to $x_i, y_j$, so we can make another construction as follows.

Consider the locally free sheaf $F^+ = \mathcal{O}_{E^+}(-1)^{\oplus(s+1)}$ of rank $s+1$ over $E^+ = \mathbf{P}^t$ and take its total space $X^+ = \operatorname{Spec}_{E^+}(\bigoplus_{m=0}^{\infty} \operatorname{Sym}^m (F^+)^*)$. Then there is an isomorphism

$$S \cong \operatorname{Spec} H^0(X^+, \mathcal{O}_{X^+}) = \operatorname{Spec} \left( \bigoplus_{m=0}^{\infty} H^0(E^+, \operatorname{Sym}^m (F^+)^*) \right)$$

and a natural birational morphism $f^+ : X^+ \to S$. The exceptional set of $f^+$ coincides with the 0-section $E^+$ of $F^+$. View a line $C^+$ on $E^+$ as a curve on $X^+$, we have $(K_{X^+} \cdot C^+) = s - t$. If $s > t$, then $f^+$ is the flip of $f$.

Also, if $s = t$, then $K_X = f^* K_S$ and $K_{X^+} = (f^+)^* K_S$, which is an example of a birational transform so-called a *flop*. In particular, if $s = t = 1$, then $S$ is the same as in Example 1.1.4(2), and this flop is called *Atiyah's flop*.

The pair obtained by a flip admits the same property as the original one:

**Theorem 2.5.5** *Let $(X, B)$ be a $\mathbf{Q}$-factorial DLT pair and let $f : X \to S$ be a projective morphism. Let $R$ be an extremal ray of $\overline{\operatorname{NE}}(X/S)$ such that $K_X + B$ takes negative values on $R \setminus \{0\}$. Suppose that the contraction morphism $g : X \to Y$ associated to $R$ is small and the flip $g^+ : X^+ \to Y$ of $g$ exists. Then $X^+$ is $\mathbf{Q}$-factorial, $(X^+, B^+)$ is DLT, and $\rho(X/S) = \rho(X^+/S)$.*

*Proof* Take any prime divisor $E^+$ on $X^+$ and denote by $E$ the strict transform of $E^+$ on $X$. Since $X$ is $\mathbf{Q}$-factorial, $E$ is a $\mathbf{Q}$-Cartier divisor. Since

## 2.5 Types of Contraction Morphisms and the Minimal Model Program  123

$\rho(X/Y) = 1$, there exists a real number $r$ such that $E + r(K_X + B) \equiv_Y 0$. As $g$ is a birational morphism, by the basepoint-free theorem, $E_0 = g_*(E + r(K_X + B))$ is $\mathbf{R}$-Cartier and $g^* E_0 = E + r(K_X + B)$. Since $K_{X^+} + B^+$ is $\mathbf{R}$-Cartier, $E^+ = (g^+)^* E_0 - r(K_{X^+} + B^+)$ is $\mathbf{R}$-Cartier. Therefore, $X^+$ is $\mathbf{Q}$-factorial. Then it is easy to see that $\rho(X/S) = \rho(X^+/S)$. The fact that $(X^+, B^+)$ is DLT can be concluded from the next Theorem 2.5.6. □

### 2.5.3 Decrease of Canonical Divisors

Although flips and divisorial contractions look very different, the following theorem shows that they are similar in the sense that both are operations that make canonical divisors smaller.

**Theorem 2.5.6** ([76, Proposition 5.1.11]) *Let $(X, B)$ be a $\mathbf{Q}$-factorial DLT pair and let $f: X \to S$ be a projective morphism. Let $R$ be an extremal ray of $\overline{\mathrm{NE}}(X/S)$ such that $K_X + B$ takes negative values on $R \setminus \{0\}$, take $h: X \to Y$ to be the associated contraction morphism. Consider the following two cases:*

*(1) $h: X \to Y$ is a divisorial contraction.*
*(2) $h: X \to Y$ is a small contraction with flip $h^+: X^+ \to Y$.*

*In each case, take a normal algebraic variety $Z$ with birational projective morphisms as the following: in case (1) take $g: Z \to X$; in case (2) take $g: Z \to X$ and $g^+: Z \to X^+$ such that $h \circ g = h^+ \circ g^+$. For each case, $\mathbf{R}$-divisors $C, C'$ on $Z$ can be determined as the following:*

*(1) $g^*(K_X + B) = K_Z + C$ and $(h \circ g)^*(K_Y + h_* B) = K_Z + C'$.*
*(2) $g^*(K_X + B) = K_Z + C$ and $(g^+)^*(K_{X^+} + B^+) = K_Z + C'$.*

*Then we have $C \geq C'$. Moreover, the support of $C - C'$ coincides with the inverse image $g^{-1}(\mathrm{Exc}(h))$ of the exceptional set of $h$.*

*Proof* In case (1), take $E$ to be the exceptional divisor of $h$, then we can write $K_X + B - h^*(K_Y + h_* B) = eE$ with $e > 0$. The assertion of the theorem is clear.

Let us consider case (2). Since $K_{X^+} + B^+$ is $h^+$-ample and $\rho(X^+/Y) = 1$, there exists an $h^+$-ample Cartier divisor $D^+$ and a positive real number $d$ such that $K_{X^+} + B^+ \equiv_Y d D^+$. Take $D$ to be the strict transform of $D^+$ on $X$. Then $C - C' = d(g^* D - (g^+)^* D^+)$. Here note that this is not only a numerical equivalence but indeed an equality because the differences are supported on the exceptional locus (cf. Lemma 1.6.3).

124          *2 The Minimal Model Program*

We may replace $Y$ by an affine open subset intersecting $h(\mathrm{Exc}(h))$. In this case, since $D^+$ is ample, there exists a sufficiently large integer $m$ such that $|mD^+|$ is free. On the other hand, $D$ is negative along the fibers of $h|_{\mathrm{Exc}(h)}$, hence the base locus of $|mD|$ coincides with $\mathrm{Exc}(h)$. Since $X$ and $X^+$ are isomorphic in codimension 1, there is a 1–1 correspondence between elements in linear systems. Therefore, $m(g^*D - (g^+)^*D^+)$ is the fixed part of the linear system $|mg^*D|$, and its support coincides with $g^{-1}(\mathrm{Exc}(h))$. $\qquad\square$

### 2.5.4 The Existence and the Termination of Flips

The existence of flips can be reduced to a special case of the finite generation of canonical rings:

**Theorem 2.5.7** *Let $(X, B)$ be a $\mathbf{Q}$-factorial DLT pair where $B$ is a $\mathbf{Q}$-divisor and let $f : X \to Y$ be a small contraction. Then the following conditions are equivalent:*

*(1) The flip $f^+ : X^+ \to Y$ exists.*
*(2) The graded $\mathcal{O}_Y$-algebra*

$$R(X/Y, K_X + B) = \bigoplus_{m=0}^{\infty} f_*(\mathcal{O}_X(\llcorner m(K_X + B)\lrcorner))$$

*is finitely generated.*

*Moreover, there is an isomorphism*

$$X^+ \cong \mathrm{Proj}_Y R(X/Y, K_X + B).$$

*In particular, the flip is unique if it exists.*

*Proof* Assume (1). Since $X$ and $X^+$ are isomorphic in codimension 1, we have

$$R(X/Y, K_X + B) \cong \bigoplus_{m=0}^{\infty} f_*^+(\mathcal{O}_{X^+}(\llcorner m(K_{X^+} + B^+)\lrcorner)).$$

Since $K_{X^+} + B^+$ is a relatively ample $\mathbf{Q}$-divisor, $R(X/Y, K_X + B)$ is finitely generated and

$$X^+ \cong \mathrm{Proj}_Y R(X/Y, K_X + B).$$

Moreover, the latter part of the assertion of the theorem follows.

Assume (2). Take $X^+ = \mathrm{Proj}_Y R(X/Y, K_X + B)$ and the natural projection $f^+ : X^+ \to Y$. By construction, there exists a positive integer $r$ and a relatively ample divisor $H$ on $X^+$ such that

## 2.5 Types of Contraction Morphisms and the Minimal Model Program 125

$$f_*^+(\mathcal{O}_{X^+}(mH)) \cong f_*(\mathcal{O}_X(mr(K_X + B)))$$

for any positive integer $m$. Since $f$ is isomorphic in codimension 1, $f_*^+(\mathcal{O}_{X^+}(mH))$ is a reflexive sheaf on $Y$.

We will show that $f^+$ is isomorphic in codimension 1. Assume that $f^+$ contracts a prime divisor $E$, consider the coherent sheaf $F$ supported on $E$ satisfying the following exact sequence

$$0 \to \mathcal{O}_{X^+}(mH) \to \mathcal{O}_{X^+}(mH + E) \to F(mH) \to 0.$$

Here $E$ is not assumed to be **Q**-Cartier. Since $H$ is relatively ample, we can take $m$ sufficiently large such that $R^1 f_*^+(\mathcal{O}_{X^+}(mH)) = 0$ and $f_*^+(F(mH)) \neq 0$. However, as $f^+(E)$ is of codimension at least 2 and $f_*^+(\mathcal{O}_{X^+}(mH))$ is reflexive, $f_*^+\mathcal{O}_{X^+}(mH) \to f_*^+\mathcal{O}_{X^+}(mH + E)$ is an isomorphism. This is a contradiction, so there is no such prime divisor $E$.

So $f^+$ is isomorphic in codimension 1. By construction, $K_{X^+} + B^+$ is $f^+$-ample and therefore $f^+$ is the flip. □

The following theorem is called the conjecture on existence of flips before it was finally proved by Hacon and M$^c$Kernan ([36]).

**Theorem 2.5.8** (*Existence of flips*) *Let $(X, B)$ be a **Q**-factorial DLT pair and let $f : X \to S$ be a projective morphism. Assume that $g : X \to Y$ is a small contraction morphism associated to a $(K_X + B)$-negative extremal ray $R$. Then the flip $g^+ : X^+ \to Y$ always exists.*

The proof will be in Chapter 3. This theorem is a special case of the finite generation theorem of canonical rings, but it is also an essential part in the proof of the latter theorem.

Divisorial contractions decrease Picard numbers by 1, but in contrast, flips preserve Picard numbers. Therefore, to make the MMP work, we need the following conjecture on termination of flips.

**Conjecture 2.5.9** (*Termination of flips*) *Let $(X, B)$ be a **Q**-factorial DLT pair and let $f : X \to S$ be a projective morphism. Then there does not exist any infinite sequence of flips:*

$$(X, B) = (X_0, B_0) \dashrightarrow (X_1, B_1) \dashrightarrow \cdots$$
$$\dashrightarrow (X_n, B_n) \dashrightarrow (X_{n+1}, B_{n+1}) \dashrightarrow \cdots .$$

*Here $\alpha_n : (X_n, B_n) \dashrightarrow (X_{n+1}, B_{n+1})$ is a flip over $S$ and $B_n$ is the strict transform of $B$.*

Since $X_n$ are all isomorphic in codimension 1, their spaces of divisors can be viewed as the same. Note that $B_n$ are constant under this identification.

126 2 *The Minimal Model Program*

## 2.5.5 Minimal Models and Canonical Models

In Section 1.12, we defined when a morphism $f: X \to S$ or $f: (X, B) \to S$ is said to be minimal. In this subsection, for a morphism $f: X \to S$ or $f: (X, B) \to S$, we define its minimal model and canonical model:

**Definition 2.5.10** (1) Let $X$ be a normal algebraic variety with **Q**-factorial terminal singularities and let $f: X \to S$ be a projective morphism. Another normal algebraic variety $X'$ with **Q**-factorial terminal singularities with a projective morphism $f': X' \to S$ such that there exists a birational map $\alpha: X \dashrightarrow X'$ with $f = f' \circ \alpha$ is called a *minimal model* of $f: X \to S$ if the following conditions are satisfied. Sometimes it is also called a *terminal model*, or more accurately, a **Q**-*factorial terminal minimal model*.

(a) $\alpha$ is surjective in codimension 1. That is, any prime divisor on $X'$ is the strict transform of a prime divisor on $X$.
(b) If we take a normal algebraic variety $Z$ with birational projective morphisms $g: Z \to X$ and $g': Z \to X'$ such that $g' = \alpha \circ g$, then the difference $g^* K_X - (g')^* K_{X'}$ is effective, and its support contains all $g_*^{-1} E$ where $E$ is a prime divisor contracted by $\alpha$.
(c) $K_{X'}$ is relatively nef.

A normal algebraic variety $Y$ with a projective morphism $f'': Y \to S$ and a projective morphism $h: X' \to Y$ such that $f' = f'' \circ h$ is called a *canonical model* or an *ample model* of $f: X \to S$ if the following conditions are satisfied.

(d) $h$ is surjective with connected geometric fibers.
(e) There exists an $f''$-ample **Q**-divisor $H$ such that $h^* H \sim_{\mathbf{Q}} K_{X'}$.

(2) Let $(X, B)$ be a **Q**-factorial DLT pair and let $f: X \to S$ be a projective morphism. Another **Q**-factorial DLT pair $(X', B')$ with a projective morphism $f': X' \to S$ such that there exists a birational map $\alpha: X \dashrightarrow X'$ with $f = f' \circ \alpha$ is called a *minimal model* of $f: (X, B) \to S$ if the following conditions are satisfied. Sometimes it is also called a *log minimal model*, or more accurately, a **Q**-*factorial DLT minimal model*.

(a) $\alpha$ is surjective in codimension 1 and $B' = \alpha_* B$.
(b) If we take a normal algebraic variety $Z$ with birational projective morphisms $g: Z \to X$ and $g': Z \to X'$ such that $g' = \alpha \circ g$, then the difference $g^*(K_X + B) - (g')^*(K_{X'} + B')$ is effective, and its support contains all $g_*^{-1} E$ where $E$ is a prime divisor contracted by $\alpha$.
(c) $K_{X'} + B'$ is relatively nef.

## 2.5 Types of Contraction Morphisms and the Minimal Model Program  127

A normal algebraic variety $Y$ with a projective morphism $f'': Y \to S$ and a projective morphism $h: X' \to Y$ such that $f' = f'' \circ h$ is called a *canonical model*, a *log canonical model*, or an *ample model* of $f: (X, B) \to S$ if the following conditions are satisfied.

(d) $h$ is surjective with connected geometric fibers.

(e) There exists an $f''$-ample **R**-divisor $H$ such that $h^* H \sim_{\mathbf{R}} K_{X'} + B'$.

**Remark 2.5.11** (1) A minimal model defined as above is (log) minimal in the sense of Definition 1.12.1, so Proposition 1.12.2 can be applied. Indeed, there is no such $C_j$ in condition (b) of Proposition 1.12.2 by the assumption "sujective in codimension 1."

(2) By condition (a), prime divisors contracted by $g$ are also contracted by $g'$. Hence the support of $g^*(K_X + B) - (g')^*(K_{X'} + B')$ is contracted by $g'$.

(3) Condition (b) says that there is a reason for a prime divisor contracted by $\alpha$ to be contracted.

(4) The minimal model and canonical model defined in the former part of the definition are special cases of the log version defined in the latter part. Indeed, if $B = 0$ and $X$ is terminal in the given pair $(X, B)$, then $Y$ is also terminal by condition (b). Therefore, when considering the existence of minimal models, it suffices to consider the log version. In this book we will consider the log version in general, and usually the word "log" will be omitted.

For a given morphism $f: (X, B) \to S$, its minimal model is not necessarily unique. But its canonical model can be proved to be unique if exists:

**Theorem 2.5.12** *Let $(X, B)$ be a **Q**-factorial DLT pair and let $f: X \to S$ be a projective morphism. For $i = 1, 2$, assume that there exist minimal models $f_i': (X_i', B_i') \to S$ with birational maps $\alpha_i: X \dashrightarrow X_i'$, and canonical models $f_i'': Y_i \to S$ with projective morphisms $h_i: X_i' \to Y_i$. Then the following assertions hold:*

*(1) The induced birational map $\beta: X_1' \dashrightarrow X_2'$ is isomorphic in codimension 1 and $(X_i', B_i')$ $(i = 1, 2)$ are $K$-equivalent to each other.*

*(2) There exists an isomorphism $e: Y_1 \to Y_2$ such that $f_1'' = f_2'' \circ e$.*

*Proof* (1) We can take a smooth algebraic variety $Z$ with a birational projective morphism $g: Z \to X$ such that $g_i = \alpha_i \circ g$ is a birational morphism for $i = 1, 2$. Denote $g_1^*(K_{X_1'} + B_1') - g_2^*(K_{X_2'} + B_2') = E$. Assume, to the contrary, that $E \neq 0$. Write $E = E^+ - E^-$ into parts with positive and negative coefficients. By symmetry, we may assume that $E^+ \neq 0$.

Since $g^*(K_X + B) \geq g_1^*(K_{X_1'} + B_1') = g_2^*(K_{X_2'} + B_2') + E$, every irreducible component of $E^+$ is contracted by $g_2$. By the negativity lemma (Lemma 1.6.3), there exists a curve $C$ contracted by $g_2$ such that $(E^+ \cdot C) < 0$ and $(E^- \cdot C) \geq 0$. Hence $(g_1^*(K_{X_1'} + B_1') \cdot C) = (g_2^*(K_{X_2'} + B_2') \cdot C) + (E \cdot C) < 0$. This contradicts the fact that $K_{X_1'} + B_1'$ is relatively nef. This shows the equality.

By this equality, we know that the set of divisors contracted by $\alpha_i$ is independent of $i$, which implies that $\beta$ is isomorphic in codimension 1.

(2) For each $i = 1, 2$, there exists an $f_i''$-ample $\mathbf{R}$-divisor $H_i$ such that $h_i^* H_i = K_{X_i'} + B_i'$. Hence a curve $C$ on $Z$ is contracted by $h_i \circ g_i : Z \to Y_i$ if and only if $(g_i^* h_i^* H_i \cdot C) = 0$. Since $g_1^*(K_{X_1'} + B_1') = g_2^*(K_{X_2'} + B_2')$, this condition is independent of $i$. Hence we get the conclusion by Zariski's main theorem. $\qquad\square$

**Example 2.5.13** Consider $X_0$ to be the hypersurface defined by the equation $x_1 x_2 + x_3 x_4 = 0$ in the projective space $\mathbf{P}^4$ with homogeneous coordinates $x_0, \ldots, x_4$. $X_0$ is the projective cone over $\mathbf{P}^1 \times \mathbf{P}^1$ with vertex $P = [1 : 0 : 0 : 0 : 0]$, and $P \in X_0$ is a terminal singularity. Take $\bar{B}$ to be a general hypersurface not passing $P$ and $B_0 = \bar{B} \cap X_0$. Assume that the degree $d = \deg(\bar{B})$ is at least 3, then $K_{X_0} + B_0 = \mathcal{O}_{X_0}(d - 3)$ is nef.

Blowing up the ideal $(x_1, x_3)$ or $(x_1, x_4)$ on $X_0$, we get two small resolutions $g_i : X_i \to X_0$ ($i = 1, 2$). $g_i$ is isomorphic outside $P$ and $g_i^{-1}(P)$ is isomorphic to $\mathbf{P}^1$. Take $B_i$ to be the strict transform of $B_0$ on $X_i$.

Then $(X_i, B_i)$ is a minimal model of $(X_0, B_0)$. The induced birational map $\alpha : X_1 \dashrightarrow X_2$ is the Atiyah flop (see Example 2.5.4(2)).

If one would like to have an example without boundaries $B_i$, suppose that $d \geq 4$, then one can consider the cyclic covering $\pi_0 : X_0' \to X_0$ of degree $d$ ramified along $B_0$, and do a similar construction. Here if $B_0$ is defined by the equation $f(x) = 0$, then the covering map $\pi_0$ is given by $t^d = f(x)$. In this case, $K_{X_0'} = \pi_0^*(K_{X_0} + (d - 1)B_0/d)$ and $K_{X_0'}$ is nef.

### 2.5.6 The Minimal Model Program

We introduce the formal definition of the MMP. Starting from an arbitrary $\mathbf{Q}$-factorial DLT pair $(X, B)$ and a projective morphism $f : X \to S$, in order to get a minimal model or a Mori fiber space, we have the following *MMP* which is a process consists of a sequence of birational operations.

(1) Given a $\mathbf{Q}$-factorial DLT pair $(X, B)$ and a projective morphism $f : X \to S$.
(2) If $K_X + B$ is relatively nef, then $(X, B)$ is minimal and the MMP ends here.
(3) If $K_X + B$ is not relatively nef, then there exists a contraction morphism $h : X \to Y$ associated to an extremal ray.

## 2.5 Types of Contraction Morphisms and the Minimal Model Program  129

(a) If $h$ is a divisorial contraction, then $(Y, B_Y)$ is again a $\mathbf{Q}$-factorial DLT pair. Here $B_Y = h_* B$. The Picard number drops by 1: $\rho(Y/S) = \rho(X/S) - 1$. Replace $(X, B)$ by the new pair $(Y, B_Y)$ and go back to (1).

(b) If $h$ is a small contraction, then take the flip $h^+: X^+ \to Y$ and $(X^+, B^+)$ is again a $\mathbf{Q}$-factorial DLT pair. Here $B^+$ is the strict transform of $B$. The Picard number is not changed: $\rho(X^+/S) = \rho(X/S)$. Replace $(X, B)$ by the new pair $(X^+, B^+)$ and go back to (1).

(c) If $h$ is a Mori fiber space, then the MMP ends here.

The MMP is to repeat this operation. If the termination of flips is true, then the operations in (3b) stop after finitely many times, and eventually we get into case (2) or (3c).

**Example 2.5.14** Let $(X, B)$ be a $\mathbf{Q}$-factorial DLT pair and let $f : X \to S$ be a projective morphism. Let us consider the case $\rho(X/S) = 2$. The corresponding MMP is called a *2-ray game*. We explain as follows.

The closed cone of curves $\overline{NE}(X/S)$ must be a fan generated by two extremal rays $R_1, R_2$ in $N_1(X/S)$. If $K_X + B$ is not nef over $S$, then for at least one extremal ray, say $R_1$, $((K_X + B) \cdot R_1) < 0$.

Assume that the corresponding contraction morphism $\phi: (X, B) \to Y$ is small and $\phi': (X', B') \to Y$ is the flip. Then again we have $\rho(X'/S) = 2$ and $\overline{NE}(X'/S)$ is a fan generated by two extremal rays $R'_1, R'_2$. Suppose that $R'_2$ is the extremal ray generated by curves contracted by $\phi'$, then by the property of flips, $((K_{X'} + B') \cdot R'_2) > 0$. If $K_{X'} + B'$ is not nef over $S$, then $((K_{X'} + B') \cdot R'_1) < 0$. Therefore, the choice of the extremal ray is unique, and the same operation can be repeated in a single direction.

The 2-ray game can be easily understood using the spaces of divisors. The nef cone $\overline{Amp}(X/S)$ must be a fan generated by two extremal rays $L_1, L_2$ in $N^1(X/S)$. Take $L_1$ to be the extremal ray corresponding to $\phi$, that is, $(L_1 \cdot R_1) = 0$.

As the induced map $X \dashrightarrow X'$ by the flip is isomorphic in codimension 1, we can identify $N^1(X/S) \cong N^1(X'/S)$. Then after the flip, the nef cone $\overline{Amp}(X'/S)$ is a fan generated by two extremal rays $L'_1, L'_2$ in $N^1(X/S)$, one of them, say $L'_2$, is just $L_1$. This is because they both coincide with the pullback of $\overline{Amp}(Y/S)$.

Therefore, we can view this flip as moving from one room $\overline{Amp}(X/S)$ to another room $\overline{Amp}(X'/S)$. The next contraction corresponds to the wall $L'_1$ on the other side. This is similar to the MMP with scaling introduced in Section 2.6.

**Remark 2.5.15** In the formulation of the MMP, we can make similar arguments by just assuming that the pairs are $\overline{\text{KLT}}$ instead of DLT. Indeed, as $X$ is assumed to be $\mathbf{Q}$-factorial, if $(X, B)$ is $\overline{\text{KLT}}$, then for $\epsilon \in (0, 1)$, $(X, (1-\epsilon)B)$ is KLT. If $K_X + B$ is not nef, then $K_X + (1 - \epsilon)B$ is not nef for a sufficiently small $\epsilon$.

## 2.6 Minimal Model Program with Scaling

In each step of the MMP, when there exist more than one extremal rays, we just choose one of them. The way to choose is arbitrary. The so-called *MMP with scaling* or *directed MMP* proceeds by choosing the extremal ray in an efficient way. The MMP with scaling goes toward the minimal model straight in one direction. In this way, the termination of flips is easier to control. Except for lower dimensional cases, to prove the termination of flips is an extremely hard problem, but it is sightly hopeful if we only consider the termination for the MMP with scaling.

Originally the MMP uses convex geometry, but the MMP with scaling is particularly compatible with convex geometry. The idea of such an MMP was first seen in [122], and then was much further developed to become a basic tool in [16]. As for the termination of flips, it might not be true for the general MMP, but it is expected to be true for the MMP with scaling.

To begin with, let $(X, B)$ be a $\mathbf{Q}$-factorial $\overline{\text{KLT}}$ pair and let $f : X \to S$ be a projective morphism, a *scale* is an $\mathbf{R}$-divisor $H$ satisfying the following properties:

(1) $(X, B + H)$ is $\overline{\text{KLT}}$.
(2) $K_X + B + H$ is relatively nef.

The idea is to use $H$ to control the progress of MMP. Starting from $(X, B) = (X_0, B_0)$, we construct the MMP for $(X, B)$ with scaling of $H$ such that in the $n$th step we have a $\mathbf{Q}$-factorial $\overline{\text{KLT}}$ pair $(X_n, B_n)$ satisfying the following properties:

(1) $(X_n, B_n + t_{n-1} H_n)$ is $\overline{\text{KLT}}$.
(2) $K_{X_n} + B_n + t_{n-1} H_n$ is relatively nef.

Here $H_n$ the strict transform of $H$, and $t_n$ is defined as the following threshold:

$$t_n = \min\{t \geq 0 \mid K_{X_n} + B_n + t H_n \text{ is relatively nef}\}.$$

## 2.6 Minimal Model Program with Scaling

We set $t_{-1} = 1$. When $n = 0$, by assumption, $t_0 \leq 1$. Assume that $K_X + B$ is not relatively nef, then $t_0 > 0$. When $n > 0$, by construction, $K_{X_n} + B_n + t_{n-1} H_n$ is relatively nef, and hence $t_n \leq t_{n-1}$.

The inductive construction of this MMP is as follows. Take $n \geq 0$. Assume that we already have $(X_n, B_n)$. If $t_n = 0$, then $K_{X_n} + B_n$ is relatively nef and the MMP ends. If $t_n > 0$, then we proceed to the next step by the following lemma. We treat in this section the case where $(X, B + H)$ is KLT and $B + H$ is relatively big. We will consider the general case in Section 2.10 after further preparation.

**Lemma 2.6.1** *Suppose that $(X, B + H)$ is KLT and $B + H$ is relatively big. If $t_n > 0$, then there exists a $(K_{X_n} + B_n)$-negative extremal ray $R_n$ such that*

$$((K_{X_n} + B_n + t_n H_n) \cdot R_n) = 0.$$

*Proof* Since $B_n + t_n H_n$ is relatively big, for a sufficiently small positive real number $\epsilon$, $(X_n, B_n + (t_n - \epsilon) H_n)$ is KLT and there are only finitely many $(K_{X_n} + B_n + (t_n - \epsilon) H_n)$-negative extremal rays (Corollary 2.4.5). Since $K_{X_n} + B_n + t_n H_n$ is relatively nef, for $0 < \epsilon' < \epsilon$, a $(K_{X_n} + B_n + (t_n - \epsilon') H_n)$-negative extremal ray is also a $(K_{X_n} + B_n + (t_n - \epsilon) H_n)$-negative extremal ray. So among them there exists an extremal ray realizing the threshold $t_n$. Note that this extremal ray is also a $(K_{X_n} + B_n)$-negative extremal ray. □

We can use the extremal ray $R_n$ in the above lemma to proceed the MMP. Since $K_{X_n} + B_n + t_n H_n$ is relatively nef and numerically trivial along this extremal ray, the strict transform $K_{X_{n+1}} + B_{n+1} + t_n H_{n+1}$ is relatively nef. Also note that $(X_n, B_n + t_n H_n)$ is $\overline{\text{KLT}}$, which implies that $(X_{n+1}, B_{n+1} + t_n H_{n+1})$ is $\overline{\text{KLT}}$. In this way, we inductively constructed the MMP with scaling of $H$. Note that we get a nonincreasing sequence $1 \geq t_0 \geq t_1 \geq \cdots$.

The MMP with scaling can be visualized as the following (Figure 2.2). For simplicity, let us consider that the MMP consists of flips. In this case, the vector spaces $N^1(X_n/S)$ can be all identified with $N^1(X/S)$. Under this identification, the point corresponding to $K_{X_n} + B_n$ does not depend on $n$.

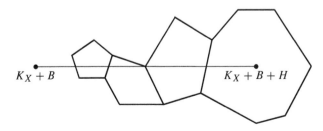

Figure 2.2 Directed MMP.

Let us track the changing of nef cones $\overline{\mathrm{Amp}}(X_n/S)$ in $N^1(X/S)$. By the cone theorem, observed from $K_X + B$, the surface of the nef cone $\overline{\mathrm{Amp}}(X_n/S)$ is locally a polyhedron (here note that the inside is invisible). Choosing an extremal ray corresponds to choosing a face, and taking the flip means passing through this face and moving from one room $\overline{\mathrm{Amp}}(X_n/S)$ to another room $\overline{\mathrm{Amp}}(X_{n+1}/S)$. Such an operation is usually called a *wall crossing*.

The condition in the beginning is that $K_X + B + H \in \overline{\mathrm{Amp}}(X/S)$. Consider the line $L$ in $N^1(X/S)$ connecting $K_X + B + H$ and $K_X + B$. In each step of the MMP with scaling, we choose a face intersecting $L$. Note that

$$K_X + B + t_n H \in \overline{\mathrm{Amp}}(X_n/S) \cap \overline{\mathrm{Amp}}(X_{n+1}/S) \cap L$$

and all rooms line up along $L$. So such an MMP moves from $K_X + B + H$ to $K_X + B$ directly along this line, and the termination is easier to consider.

**Remark 2.6.2** Without the assumptions that $(X, B + H)$ is KLT and $B + H$ is relatively big, we can still consider the MMP with scaling. In this case, the proof of Lemma 2.6.1 is a little more complicated (by Corollary 2.10.12).

Birkar et al. showed the termination of flips in the following special but very important case. The proof will be in Chapter 3.

**Theorem 2.6.3** ([16]) *Let $(X, B)$ be a $\mathbf{Q}$-factorial KLT pair and let $f : X \to S$ be a projective morphism. Assume that $B$ is relatively big. Then any MMP with scaling terminates.*

As a corollary, we can show the existence of minimal models for varieties of general type, or oppositely the existence of Mori fiber spaces for varieties with non-pseudo-effective canonical divisors (see Chapter 3 for details):

**Corollary 2.6.4** *Let $(X, B)$ be a $\mathbf{Q}$-factorial KLT pair and let $f : X \to S$ be a projective morphism.*

*(1) Assume that $K_X + B$ is not relatively pseudo-effective over $S$. Then there exists a Mori fiber space which is a birational model of $(X, B)$.*

*(2) Assume that $K_X + B$ is relatively big over $S$. Then $(X, B)$ has a minimal model. Moreover, by the basepoint-free theorem, $(X, B)$ has a canonical model.*

## 2.7 The Existence of Rational Curves

Given an algebraic variety, whether there exists a rational curve, and how many rational curves there are if they exist, are very important questions. We will give a proof of Theorem 2.7.2 which states that there are many rational curves

## 2.7 The Existence of Rational Curves

on algebraic varieties with canonical divisors satisfying certain negativity. For example, $\mathbf{P}^1$ is the only smooth projective curve with negative canonical divisor ($-K$ is ample).

In order to prove this theorem, we first take the *reduction* of the given algebraic variety to positive characteristics, and then proceed the discussion by methods specific in positive characteristics. Applying *Frobenius morphisms*, there is a method to get a morphism from $\mathbf{P}^1$ by deforming a given morphism and degenerating it by taking a limit. This method was originally discovered by Mori ([99]), and so far is the only method to prove the existence of rational curves in general situation. The existence of rational curves is also a very important problem in complex geometry, but this theorem has no analytic proof. We can say that this is a theorem unique to algebraic geometry which includes positive characteristics in its coverage.

### 2.7.1 Deformation of Morphisms

First, in order to construct the space of all deformations of a morphism, or the *moduli space* of morphisms, we introduce the definition of *Hilbert scheme* by Grothendieck ([34]). For details we refer to [85].

**Definition 2.7.1** Fix a projective morphism $f : X \to S$ between Noetherian schemes and a relatively ample sheaf $H$. For a closed subscheme $Z$ of a fiber $X_s = f^{-1}(s)$ of $f$, the polynomial

$$P_Z(m) = \chi(Z, mH) = \sum_{p \geq 0} \dim_{k(s)} H^p(Z, mH)$$

in integer $m$ is called the *Hilbert polynomial* of $Z$. Fixing a polynomial $P$, there exists a moduli space for all closed subschemes of fibers of $f$ whose Hilbert polynomials coincide with $P(m)$. It is a projective scheme $g : \mathrm{Hilb}^P(X/S) \to S$ over $S$ and is called the *Hilbert scheme*. Here a moduli space is a scheme satisfying the following universal property.

There exists a closed subscheme $\mathcal{Z}$ in the fiber product $X \times_S \mathrm{Hilb}^P(X/S)$, which is called the *universal family*, satisfying the following conditions:

(1) The first projection $p_1$ maps every fiber $p_2^{-1}(t)$ of the second projection $p_2 : \mathcal{Z} \to \mathrm{Hilb}^P(X/S)$ isomorphically to a closed subscheme of $X_{g(t)}$, whose Hilbert polynomial is $P(m)$.
(2) For any $S$-scheme $T \to S$ and any closed subscheme $Z_T$ of $X \times_S T$ such that the Hilbert polynomial of every fiber of the second projection $Z_T \to T$ is $P(m)$, there exists a unique morphism $T \to \mathrm{Hilb}^P(X/S)$ such that the pullback $\mathcal{Z} \times_{\mathrm{Hilb}^P(X/S)} T$ from the universal family $\mathcal{Z}$ coincides with $Z_T$.

134                    2 *The Minimal Model Program*

Note that a family with constant Hilbert polynomial is automatically flat. By taking disjoint union for all polynomials, we denote $\mathrm{Hilb}(X/S) = \bigsqcup_P \mathrm{Hilb}^P(X/S)$.

The moduli space of morphisms is defined to be the moduli space of graphs of morphisms. Let $X \to S$ and $Y \to S$ be projective $S$-schemes such that $X$ is flat over $S$, and take $G \subset X_s \times Y_s$ to be the graph of a morphism between fibers $g \colon X_s \to Y_s$. Fix a relatively ample sheaf $H$ on $X \times_S Y$ and take $P(m) = \chi(G, mH)$. Consider the Hilbert scheme $\mathrm{Hilb}^P(X \times_S Y/S)$, and take $\pi \colon \bar{\mathcal{G}} \to \mathrm{Hilb}^P(X \times_S Y/S)$ to be the universal family. Then the set of points in $\mathrm{Hilb}^P(X \times_S Y/S)$, whose fibers in the universal family are graphs of morphisms between fibers of $X$ and $Y$, is an open subset. Indeed, a closed subscheme $G'$ of $X_{s'} \times Y_{s'}$ is the graph of a morphism $X_{s'} \to Y_{s'}$ if and only if the first projection $p_1 \colon G' \to X_{s'}$ is an isomorphism, therefore being a graph of a morphism is an open condition. This open subset is denoted by $\mathrm{Hom}_S^P(X, Y)$ and called the *moduli space of morphisms*.

The theory of infinitesimal deformation is very useful when studying the structure of Hilbert schemes. For example, let us assume that $X$ is a smooth projective algebraic variety over a field $k$ and $Z$ is a smooth closed subvariety. Then $Z$ determines a point $[Z] \in \mathrm{Hilb}(X/k) = \mathrm{Hilb}(X)$. Then the *Zariski tangent space* $T_{\mathrm{Hilb}(X),[Z]} = (\mathfrak{m}_{[Z]}/\mathfrak{m}_{[Z]}^2)^*$ of the point $[Z]$ is isomorphic to $H^0(Z, N_{Z/X})$. Here $N_{Z/X}$ is the cotangent bundle of $Z \subset X$ and $\mathfrak{m}_{[Z]} \subset \mathcal{O}_{\mathrm{Hilb}(X),[Z]}$ is the maximal ideal of the local ring. On the other hand, the *obstruction space* is $H^1(Z, N_{Z/X})$. That is, the completion of $\mathrm{Hilb}(X)$ along $[Z]$ can be expressed by $h^1(Z, N_{Z/X})$ equations in the completion of $h^0(Z, N_{Z/X})$-dimensional affine space along the origin. Therefore, we have the inequality

$$\dim_{[Z]} \mathrm{Hilb}(X) \geq h^0(Z, N_{Z/X}) - h^1(Z, N_{Z/X}).$$

This can be also applied to moduli spaces of morphisms. Consider the deformation of a morphism between smooth projective algebraic varieties $g \colon X \to Y$, the cotangent bundle of $G$ is given by $N_{G/X \times Y} \cong p_2^* T_Y$. Here $T_Y$ is the tangent bundle of $Y$ and $p_2 \colon G \to Y$ is the second projection. Therefore, we have the inequality

$$\dim_{[g]} \mathrm{Hom}_k(X, Y) \geq h^0(X, g^* T_Y) - h^1(X, g^* T_Y).$$

### 2.7.2 The Bend-and-Break Method

**Theorem 2.7.2** ([98]) *Let $X$ be a normal projective algebraic variety of dimension $n$ over an algebraically closed field of arbitrary characteristic. Take*

## 2.7 The Existence of Rational Curves

$C$ to be a curve on $X$ which does not pass through any singular point of $X$, fix a point $P$ on $C$ and take an ample divisor $H$ on $X$. Suppose that $C$ is not a rational curve and the inequality $(K_X \cdot C) < 0$ holds. Then there exists a rational curve $L$ on $X$ passing through $P$ satisfying the inequality

$$(H \cdot L) \leq \frac{2n(H \cdot C)}{(-K_X \cdot C)}.$$

Here note that $C$ and $L$ might have singularities, and $L$ might pass through singularities of $X$.

*Proof* First, let us prove the case that the characteristic $p$ of $k$ is positive. The point of the proof is that by using Frobenius morphisms, we can make coverings with sufficiently high degrees while keeping the genus of the curve fixed.

Take the normalization $\nu: C' \to C$ and denote by $g$ the genus of $C'$. By assumption, $g > 0$. Take the Frobenius morphism $f': C'_q \to C'$ of degree $q = p^m$ which is a power of the characteristic. Here $f'$ is the morphism defined over $k$ by taking $q$th power of coordinates, which exists only in positive characteristics. The genus of $C'_q$ is again $g$. Take $f: C'_q \to X$ to be the composition morphism.

Since $(K_X \cdot C) < 0$, we can take $q = p^m$ sufficiently large such that the following inequality holds:

$$b = \lfloor \frac{q(-K_X \cdot C) - 1}{n} \rfloor + 1 - g > 0.$$

Take $b$ distinct points $P_1, \ldots, P_b$ on $C'_q$, consider the divisor $B = \sum_{i=1}^{b} P_i$. Consider the deformation of the morphism $f: C'_q \to X$ fixing $B$. As the deformation of the morphism $f$ is the deformation of its graph $G$, by fixing $B$ it means that the graphs after deformation contain all points $(P_i, f(P_i))$ for all $i$. The set of all such deformations is a closed subscheme $\mathrm{Hom}_k(C'_q, X; B)$ in the moduli space $\mathrm{Hom}_k(C'_q, X)$.

We can compute the dimension of $\mathrm{Hom}_k(C'_q, X; B)$ by infinitesimal deformation theory. The Zariski tangent space of $\mathrm{Hom}_k(C'_q, X)$ at the point $[f]$ is isomorphic to $H^0(C'_q, f^*T_X)$, and the Zariski tangent space of its closed subscheme $\mathrm{Hom}_k(C'_q, X; B)$ is isomorphic to $H^0(C'_q, f^*T_X \otimes \mathcal{O}_{C'_q}(-B))$. Also the obstruction space becomes $H^1(C'_q, f^*T_X \otimes \mathcal{O}_{C'_q}(-B))$ instead of $H^1(C'_q, f^*T_X)$. Therefore, the completion of $\mathrm{Hom}_k(C'_q, X; B)$ at the point $[f]$ can be expressed by $h^1(C'_q, f^*T_X \otimes \mathcal{O}_{C'_q}(-B))$ linear equations in the completion of the $h^0(C'_q, f^*T_X \otimes \mathcal{O}_{C'_q}(-B))$-dimensional affine space at the origin. Hence we get an estimate of the dimension

$$\dim_{[f]} \mathrm{Hom}_k(C_q', X; B) \geq \chi(C_q', f^*T_X \otimes \mathcal{O}_{C_q}(-B))$$
$$= \deg_{C_q'}(f^*T_X \otimes \mathcal{O}_{C_q}(-B)) + n(1-g)$$
$$= q(-K_X \cdot C) - nb + n(1-g) \geq 1.$$

Here the first equality is derived from the Riemann–Roch formula since $T_X$ is a locally free sheaf of rank $n$.

From the above argument, there exists a nontrivial deformation family $F: C_q' \times T \to X$ of $f$ fixing $B$, parametrized by a smooth affine algebraic curve $T$. Here $T$ has a basepoint $t_0$ such that $F(P, t_0) = f(P)$ for all $P \in C_q'$, and also $F(P_i, t) = f(P_i)$ for all $1 \leq i \leq b$ and all $t \in T$. On the other hand, since $g > 0$ and $b > 0$, the morphism $C_q' \to C$ has no deformation itself. Therefore, the image of $F$ is not contained in $C$, that is, $F(C_q' \times T) \not\subset C$. We note that the above argument does not work if $g = 0$. Indeed there are still deformations of the morphism $f: C_q' \to C$ even if $b$ is large, because $\deg_{C_q'}(f^*T_C)$ becomes large.

Compactify the affine curve $T$ into a smooth projective algebraic curve $\bar{T}$. We can extend $F$ to a birational map $C_q' \times \bar{T} \dashrightarrow X$. Resolving this birational map by a sequence of blowups on points of indeterminacy, we can get a birational morphism $\mu: Y \to C_q' \times \bar{T}$ and a morphism $h = F \circ \mu: Y \to X$. Here $\mu$ is obtained by repeatedly blowing up points on the smooth projective surface $C_q' \times \bar{T}$. In each blowup step in this procedure, if the image of the center of the blowup in $C_q' \times \bar{T}$ lies on $T_i = P_i \times \bar{T}$ $(i = 1, \ldots, b)$, we denote the exceptional divisors by $\bar{E}_{i,j}$ $(j = 1, \ldots, n_i)$. Denote the total transforms of all such exceptional divisors on $Y$ by $E_{i,j}$ $(j = 1, \ldots, n_i)$. Take $T_0 = P \times \bar{T}$ for a general point $P$ on $C_q'$ and take $T_i'$ to be the strict transform of $T_i$ by $\mu$. We may take $P$ such that $\mu$ is isomorphic over $T_0$ (so we can view $T_0$ as a curve on $Y$), and we have the linear equivalence

$$T_i' \sim T_0 - \sum_{j=1}^{n_i} \epsilon_{i,j} E_{i,j}$$

for $i = 1, \ldots, b$. Here the value of $\epsilon_{i,j}$ is given by $\epsilon_{i,j} = 1$ or $0$ depending on whether the center of the blowup corresponding to $E_{i,j}$ is on the strict transform of $T_i$ or not.

Take $C_0 = C_q' \times t_0 \subset Y$, since the morphism $C_0 \to C$ is of degree $q$,

$$(h^*H \cdot C_0) = q(H \cdot C).$$

Also $(T_0 \cdot C_0) = 1$. Since $N^1(Y)_{\mathbf{R}}$ is generated by $C_0, T_0$ and exceptional divisors of $\mu$, there exist integers $c$ and $e_{i,j}$ such that

$$h^*H \equiv cC_0 + q(H \cdot C)T_0 - \sum_{i,j} e_{i,j} E_{i,j} + E.$$

## 2.7 The Existence of Rational Curves

Here $E$ is a divisor whose support is contained in the exceptional divisors whose images are not on $T_i$. Since $h^*H$ is nef, $c \geq 0$ and $e_{i,j} \geq 0$.

Since $\dim h(Y) = 2$, $(h^*H)^2 > 0$. Note that

$$(h^*H)^2 = 2cq(H \cdot C) + \sum_{i,j} e_{i,j}^2 (E_{i,j})^2 + E^2.$$

Since $(E^2) \leq 0$,

$$2cq(H \cdot C) - \sum_{i,j} \epsilon_{i,j}^2 e_{i,j}^2 \geq 2cq(H \cdot C) - \sum_{i,j} e_{i,j}^2 > 0.$$

Also for every $i$,

$$c - \sum_{j=1}^{n_i} \epsilon_{i,j} e_{i,j} = (h^*H \cdot T_i') = 0.$$

Therefore,

$$2q(H \cdot C) \sum_{i,j} \epsilon_{i,j} e_{i,j} > b \sum_{i,j} \epsilon_{i,j}^2 e_{i,j}^2.$$

This implies that there exist indices $i_0$ and $j_0$ such that $\epsilon_{i_0,j_0} = 1$ and

$$2q(H \cdot C) > b\epsilon_{i_0,j_0} e_{i_0,j_0} > 0,$$

which means that

$$(E_{i_0,j_0} \cdot T_{i_0}') = \epsilon_{i_0,j_0} > 0$$

and

$$0 < (h^*H \cdot E_{i_0,j_0}) = e_{i_0,j_0} < \frac{2q(H \cdot C)}{b}.$$

Hence there exists an irreducible component $L'$ of $E_{i_0,j_0}$ such that $L = h(L')$ is a rational curve, $P_{i_0} \in L$, and

$$(H \cdot L) < \frac{2q(H \cdot C)}{b}.$$

Recall that $q = p^m$, and by the definition of $b$, we have

$$\lim_{m \to \infty} \frac{2q(H \cdot C)}{b} = \frac{2n(H \cdot C)}{(-K_X \cdot C)},$$

so by taking $m$ sufficiently large, we have

$$(H \cdot L) \leq \frac{2n(H \cdot C)}{(-K_X \cdot C)}.$$

Here note that the left-hand side is always an integer.

138                    2 *The Minimal Model Program*

We have shown that for the images of any $b$ points on $C'_q$, there exists a rational curve $L$ passing through one of them and $(H \cdot L)$ satisfies the required inequality. Next, we use this to show that for any point $P \in C$, there exists a rational curve $L$ such that $P \in L$ and the degree $(H \cdot L)$ satisfies the required inequality.

In the Hilbert scheme $\mathrm{Hilb}(X)$, the set of points corresponding to all rational curves on $X$ is a locally closed subset. This is because for a family of algebraic curves, genus is lower semicontinuous. Moreover, if we only consider all rational curves of degree (i.e. the intersection number with $H$) bounded from above by a constant number, then the set is a locally closed subset of finite type. From the above argument, we conclude that there exists an irreducible locally closed subset $Z \subset \mathrm{Hilb}(X)$ such that if we take $\mathcal{U}_Z \subset X \times Z$ to be the restriction of the universal family $\mathcal{U} \subset X \times \mathrm{Hilb}(X)$ on $Z$, then the fibers of the second projection $p_2 \colon \mathcal{U}_Z \to Z$ are rational curves on $X$ of degree bounded by $2n(H \cdot C)/(-K_X \cdot C)$ and the image of the first projection $p_1(\mathcal{U}_Z)$ contains a non-empty open subset of $C$. Take $\bar{Z}$ to be the closure of $Z$ in $\mathrm{Hilb}(X)$ and take $\mathcal{U}_{\bar{Z}} \subset X \times \bar{Z}$ to be the restriction of the universal family. Then all irreducible components of the fibers of the second projection $p_2 \colon \mathcal{U}_{\bar{Z}} \to \bar{Z}$ are rational curves, and the image of the first projection $p_1(\mathcal{U}_{\bar{Z}})$ contains $C$. Therefore, there exists a rational curve passing through any fixed point on $C$ with degree bounded by $2n(H \cdot C)/(-K_X \cdot C)$.

We can construct rational curves on algebraic varieties defined over a field of characteristic 0 by lifting the above result to characteristic 0. The proof essentially uses the property of Hilbert schemes again.

All given data as $X$, $H$, $C$ can be described by finitely many polynomial equations in finitely many invariables with finitely many coefficients in $k$.

By adding all coefficients of those equations to $\mathbf{Z}$ and localizing the subring of $k$ by adding the inverses of finitely many elements, we can construct a finitely generated $\mathbf{Z}$-algebra $R$ satisfying the following conditions.

(1) There exists a projective morphism $X_R \to \mathrm{Spec}\, R$ such that all the geometric fibers $X_t$ are normal and the generic geometric fiber $X_{\bar{\eta}}$ is isomorphic to $X$. Here for a geometric point $t$ of $\mathrm{Spec}\, R$, we denote by $X_t$ the fiber over $t$.
(2) There exists an ample Cartier divisor $H_R$ on $X_R$ whose restriction on $X_{\bar{\eta}}$ is $H$.
(3) There exists a closed subscheme $C_R$ of $X_R$ such that for any geometric point $t$ of $\mathrm{Spec}\, R$, the fiber $C_t$ is an irreducible algebraic curve on $X_t$ which does not pass through the singularities of $X_t$ and is not a rational curve.

## 2.8 The Lengths of Extremal Rays

Here note that all conditions on fibers are open conditions, so we can make appropriate localization to remove bad sets.

Consider the universal family on the Hilbert scheme

$$\mathcal{U} \subset X_R \times_{\text{Spec } R} \text{Hilb}(X_R/\text{Spec } R).$$

Then there exists a locally closed subset of finite type $Z_R \subset \text{Hilb}(X_R/\text{Spec } R)$ satisfying the following: For any geometric point $t$ of Spec $R$, the set of points in $\text{Hilb}(X_R/\text{Spec } R)$ corresponding to rational curves $L$ on $X_t$ such that

$$(H_t \cdot L) \le \frac{2n(H \cdot C)}{(-K_X \cdot C)}$$

coincides with $Z_t$. As the right-hand side is a constant, the degree of $L$ is bounded from above uniformly.

Take the closure $\bar{Z}_R$ in $\text{Hilb}(X_R/\text{Spec } R)$ and take the restriction of the universal family $\mathcal{U}_{\bar{Z}_R} \subset X_R \times_{\text{Spec } R} \bar{Z}_R$. Then any irreducible component of any geometric fiber of the second projection $p_2 \colon \mathcal{U}_{\bar{Z}_R} \to \bar{Z}_R$ is a rational curve with degree bounded from above.

As the residue field of a geometric point $t$ is of positive characteristic, by the argument of the first part, the image of the first projection $p_1(\mathcal{U}_{\bar{Z}_t})$ contains $C_t$. Since $\bar{Z}_R$ is a closed subscheme of finite type, it follows that $C_R \subset p_1(\mathcal{U}_{\bar{Z}_R})$. In particular, $C_{\bar{\eta}} \subset p_1(\mathcal{U}_{\bar{Z}_{\bar{\eta}}})$. This finishes the proof. $\qquad\square$

The argument in the proof uses a method to deform the curve until its limit breaks up with a piece (irreducible component) of rational curve, which is called the *bend-and-break method*.

## 2.8 The Lengths of Extremal Rays

In this section we define the "length" of an extremal ray and prove a theorem stating that it is bounded by a constant depending only on the dimension. This theorem also contains the assertion that extremal rays are generated by rational curves, which is essential for many boundedness results and termination results.

As the proof uses the existence theorem of rational curves proved in Section 2.7, it is based on algebraic geometry in positive characteristics. In addition to this, we use the vanishing theorem which is specific in characteristic 0. This theorem was also used to give an alternative proof of the discreteness of extremal rays in the cone theorem (Step 5') in Section 2.4.2.

140         *2 The Minimal Model Program*

For an extremal ray $R$ of a morphism $f: (X, B) \to S$, the minimal value of the intersection numbers $-((K_X + B) \cdot C)$ for all irreducible curves $C$ whose classes are contained in $R$, is called the *length* of $R$.

First, we begin with generalizing the vanishing theorem for complex analytic varieties.

**Theorem 2.8.1** ([115, Theorem 3.7]) *Let $f: X \to S$ be a projective surjective morphism from a complex manifold to a possibly singular complex variety, let $B$ be an $\mathbf{R}$-divisor with normal crossing support and coefficients in $(0, 1)$, and let $D$ be a Cartier divisor on $X$. Assume that $D - (K_X + B)$ is relatively nef and relatively big. Then $R^p f_*(\mathcal{O}_X(D)) = 0$ for any $p > 0$.*

The theorem is proved by generalizing the Kodaira-type vanishing theorem for compact complex manifolds to *weakly 1-complete* complex manifolds. A complex manifold is said to be *weakly 1-complete* if there exists a plurisubharmonic $C^\infty$-function $\phi$ such that the subset $X_c = \{x \in X \mid \phi(x) \le c\}$ is compact for all $c \in \mathbf{R}$. For a positive line bundle $L$ on a weakly 1-complete complex manifold $X$, $H^p(X, K_X + L) = 0$ for all $p > 0$ ([112, 113]), the same as the Kodaira vanishing theorem.

**Theorem 2.8.2** *Let $(X, B)$ be a KLT pair and let $f: X \to Y$ be a birational projective morphism to a normal algebraic variety. Assume that $-(K_X + B)$ is $f$-ample. Take $E$ to be any irreducible component of the exceptional set $\mathrm{Exc}(f)$ and denote $n = \dim E - \dim f(E)$. Then the set $\{C_t\}$ of all rational curves $C_t$ such that $C_t$ is contracted by $f$ to a point and satisfies the inequality $0 > ((K_X + B) \cdot C_t) > -2n$, covers $E$; that is, $\bigcup_t C_t = E$.*

*Proof* For a flat family whose generic fiber is a rational curve, any irreducible component of its special fiber is again a rational curve. Therefore, it suffices to show that, passing through a general point of $E$, there exists a rational curve contracted by $f$ satisfying the required inequality. Replacing $Y$ by an affine open subset intersecting $f(E)$ and cutting $Y$ by general hyperplanes, it suffices to consider the case when $f(E)$ is a point. We assume this in the following.

We need the following lemma.

**Lemma 2.8.3** *Take $v: E' \to E$ to be the normalization and take an $f$-ample divisor $H$ on $X$. Then*

$$(H^{n-1} \cdot (K_X + B) \cdot E) > ((v^*H)^{n-1} \cdot K_{E'}).$$

*Proof* We may assume that $H$ is very ample. Cutting by hyperplanes in $|H|$ for $n - 1$ times, we get $C \subset X_0$ from $E \subset X$. Since $\dim E = n$, $\dim C = 1$. Denote $B_0 = B|_{X_0}$ and $v^{-1}(C) = C'$.

Since $K_{X_0} = (K_X + (n-1)H)|_{X_0}$ and $K_{C'} = (K_{E'} + (n-1)\nu^*H)|_{C'}$, if the required inequality fails, then $((K_{X_0} + B_0) \cdot C) \leq \deg K_{C'}$. Then there exists a Cartier divisor $A_0$ on $C$ such that $((K_{X_0} + B_0) \cdot C) \leq \deg A_0$ and $H^0(C', K_{C'} - \nu^*A_0) \neq 0$. By the trace map we have $H^0(C, \omega_C(-A_0)) \neq 0$. Here $\omega_C$ is the canonical sheaf of $C$.

On the other hand, since $C$ is 1-dimensional, we can take a sufficiently small analytic neighborhood $V \subset Y$ of $f(C)$ and denote $U = f^{-1}(V) \cap X_0$ such that there exists a Cartier divisor $A$ on $U$ where $A_0 = A|_C$ and the support of $A$ does not intersect with irreducible components of $\mathrm{Exc}(f|_U)$ other than $C$. Since $((K_{X_0} + B_0) \cdot C) \leq \deg A_0$, $A - (K_{X_0} + B_0)$ is relatively nef for $f : U \to V$.

By the complex analytic version of the vanishing theorem (Theorem 2.8.1), $R^1 f_*(\mathcal{O}_U(A)) = 0$. Therefore, $H^1(C, A_0) = 0$ and $H^0(C, \omega_C(-A_0)) = 0$ by the *Serre duality theorem*, which is a contradiction. So the inequality is proved. $\square$

Go back to the proof of the theorem. If $n = 1$, then by $\deg K_{E'} < ((K_X + B) \cdot E) < 0$, it is easy to see that $E' \cong \mathbf{P}^1$ and $-2 < ((K_X + B) \cdot E)$. Moreover, by the vanishing theorem, $R^1 f_* \mathcal{O}_X = 0$, which implies that $E \cong \mathbf{P}^1$.

Suppose that $n > 1$. By taking the degree of $H$ sufficiently large, we may assume that $C$ is not a rational curve. By the lemma, $(K_{E'} \cdot C') < ((K_X + B) \cdot C) < 0$, we can apply Theorem 2.7.2 to $C' \subset E'$. Note that $M = -\nu^*(K_X + B)$ is ample on $E'$, so passing through any point on $C'$, there exists a rational curve $L'$ satisfying $(M \cdot L') \leq 2n(M \cdot C')/(-K_{E'} \cdot C') < 2n$. Then $L = \nu(L')$ is the rational curve we are looking for. $\square$

**Corollary 2.8.4** *Let $f : (X, B) \to S$ be a projective morphism from a $\mathbf{Q}$-factorial KLT pair. Take an extremal ray $R$ in $\overline{\mathrm{NE}}(X/S)$ such that $((K_X + B) \cdot R) < 0$. Take $E$ to be the exceptional set of the corresponding contraction morphism $h$ and denote $n = \dim E - \dim h(E)$. Here $E = X$ if $h$ is a Mori fiber space. Then $E$ is covered by rational curves $L$ such that $L$ is contracted by $h$ and $-((K_X + B) \cdot L) < 2n$ (respectively, $\leq 2n$) if $E \neq X$ (respectively, $E = X$).*

*Proof* If $E \neq X$, this is Theorem 2.8.2. If $E = X$, this is by Theorem 2.7.2. $\square$

## 2.9 The Divisorial Zariski Decomposition

In algebraic surface theory, the intersection theory of divisors is a very powerful tool. Since the intersection number is a symmetric bilinear form, the

# 2 The Minimal Model Program

Zariski decomposition theory can be developed in a strong form. In higher dimensional algebraic geometry, it is difficult to construct a strong Zariski decomposition, but if restricted to codimension 1, the "divisorial Zariski decomposition" can be easily constructed, and is sufficiently useful.

**Definition 2.9.1** Let $f: X \to S$ be a projective morphism from a normal $\mathbf{Q}$-factorial algebraic variety to a quasi-projective algebraic variety, let $D$ be a relatively pseudo-effective $\mathbf{R}$-divisor, and let $H$ be a relatively ample divisor. If

$$N = \lim_{t \downarrow 0} \inf\{D' \mid D + tH \equiv_S D' \geq 0\}$$

has finite coefficients as an $\mathbf{R}$-divisor, then we define the relative *divisorial Zariski decomposition* $D = P + N$ of $D$ over $S$ by taking $P = D - N$. Here $P$ is called the *numerically movable part* and $N$ is called the *numerically fixed part*.

If $D = P$, then $D$ is said to be *numerically movable*. The cone consisting of numerical equivalence classes of all numerically movable $\mathbf{R}$-divisors is denoted by $\overline{\mathrm{Mov}}(X/S) \subset N^1(X/S)$ and called the *numerically movable cone*.

Let us give a further explanation about the definition. Fixing $H$ and a positive number $t$, since $[D + tH] \in \mathrm{Big}(X/S)$, $D + tH$ is numerically equivalent to an effective divisor. Therefore, the effective $\mathbf{R}$-divisor

$$N_t = \inf\{D' \mid D + tH \equiv_S D' \geq 0\}$$

can be defined. Here the infimum of $\mathbf{R}$-divisors is defined by taking the infimum of the coefficients of each irreducible component. Since $H$ is numerically free, we know that $N_{t'} \geq N_t$ if $t' \leq t$. But here we should be careful that by taking limit $N = \lim_{t \downarrow 0} N_t$, the coefficients of $N$ may go to infinity though the number of irreducible components is bounded by $\rho(X/S)$ (cf. Lemma 2.9.3). An example given by Lesieutre ([93]) shows that this can happen. Therefore, the relative divisorial Zariski decomposition can be defined only if $N$ is an $\mathbf{R}$-divisor, that is, none of its coefficients is infinity. Nevertheless we know the existence of the relative divisorial Zariski decomposition in the following cases:

**Lemma 2.9.2** ([116, Lemma III.4.3]) *The relative divisorial Zariski decomposition of a relatively pseudo-effective $\mathbf{R}$-divisor $D$ exists if one of the following conditions holds:*

*(1) $D$ is relatively numerically equivalent to an effective $\mathbf{R}$-divisor.*
*(2) $\mathrm{codim}(f(N_i)) \leq 1$ hold for all irreducible component $N_i$ of $N$. In particular, they hold if $S$ is a point.*

## 2.9 The Divisorial Zariski Decomposition

If $\dim X = 2$, then the divisorial Zariski decomposition and the classical Zariski decomposition coincide ([61]).

**Lemma 2.9.3** *Assume that the relative divisorial Zariski decomposition $D = P + N$ exists. Then*

*(1) $N$ is uniquely determined as an effective $\mathbf{R}$-divisor and the number of irreducible components of $N$ is bounded by $\rho(X/S)$.*
*(2) $P$ is relatively pseudo-effective.*
*(3) $N$ and $P$ are independent of the choice of the relatively ample divisor $H$.*

*Proof* (1) As $H$ is numerically free, $N_{t'} \geq N_t$ for $t' \leq t$. On the other hand, the number of irreducible components of $N_t$ is bounded by the number of numerically linearly independent $\mathbf{R}$-divisors, which is $\rho(X/S)$. So

$$N = \lim_{t \to 0} N_t$$

is uniquely determined as an effective $\mathbf{R}$-divisor.

(2) $P$ is relatively pseudo-effective just because

$$P = \lim_{t \to 0}(D + tH - N_t),$$

where $D + tH - N_t$ is relatively pseudo-effective.

(3) For another relatively ample divisor $H'$, there exist positive integers $m, m'$ such that $mH - H'$ and $m'H' - H$ are both relatively ample. Hence $N$ is independent of the choice of $H$. $\qquad\qquad\square$

**Lemma 2.9.4** *(1) The numerically movable cone $\overline{\mathrm{Mov}}(X/S)$ is a closed cone, and we have the following inclusions*

$$\overline{\mathrm{Amp}}(X/S) \subset \overline{\mathrm{Mov}}(X/S) \subset \overline{\mathrm{Eff}}(X/S).$$

*(2) Let $\alpha\colon X \dashrightarrow Y$ be a birational map between normal $\mathbf{Q}$-factorial algebraic varieties projective over a quasi-projective algebraic variety $S$. Assume that $\alpha$ is isomorphic in codimension 1, then the natural isomorphism $\alpha_*\colon N^1(X/S) \to N^1(Y/S)$ induces a bijective map $\alpha_*(\overline{\mathrm{Mov}}(X/S)) = \overline{\mathrm{Mov}}(Y/S)$.*

*Proof* (1) Let $D$ be a relatively pseudo-effective $\mathbf{R}$-divisor and let $H$ be a relatively ample divisor. If $D + tH \in \overline{\mathrm{Mov}}(X/S)$, then it is easy to see that $N_{t'} = 0$ for any $t' > t$. Hence if $D + tH \in \overline{\mathrm{Mov}}(X/S)$ for any $t > 0$, then $D \in \overline{\mathrm{Mov}}(X/S)$. So the numerically movable cone is closed.

If $D$ is relatively nef, then $D + tH$ is relatively ample and hence the nef cone is contained in the numerically movable cone.

144      2 *The Minimal Model Program*

(2) Recall that $\alpha$ is well defined by Lemma 1.5.13. Take relatively ample divisors $H_X$ and $H_Y$ on $X$ and $Y$ such that $H_Y - \alpha_* H_X$ is relatively ample. If $\inf\{D' \mid D + t H_X \equiv_S D' \geq 0\} = 0$, then $\inf\{D'' \mid \alpha_* D + t H_Y \equiv_S D'' \geq 0\} = 0$, which means that the image of a numerically movable divisor is numerically movable. $\qquad\square$

**Remark 2.9.5** If $\dim X = 2$, then being numerically movable is equivalent to being nef. Hence in this case the numerically movable cone coincides with the nef cone, and the divisorial Zariski decomposition is the classical Zariski decomposition.

For a pair $(X, B)$, the divisors that should be contracted in order to get a minimal model can be determined by the divisorial Zariski decomposition of $K_X + B$:

**Theorem 2.9.6** *Let $(X, B)$ be a $\mathbf{Q}$-factorial DLT pair and let $f : X \to S$ be a projective morphism to a quasi-projective variety. Assume that there exists a minimal model $\alpha : (X, B) \dashrightarrow (Y, C)$ with induced projective morphism $g : Y \to S$. Then the divisorial Zariski decomposition $K_X + B = P + N$ over $S$ exists. Moreover, let $E$ be a prime divisor on $X$, then $E$ is contracted by $\alpha$ (that is, $\alpha_* E = 0$) if and only if $E$ is an irreducible component of $N$.*

*Proof* Note that $K_X + B$ is relatively pseudo-effective since it has a minimal model, hence we can consider the divisorial Zariski decomposition.

Take birational projective morphisms $p : Z \to X$ and $q : Z \to Y$ from a normal algebraic variety $Z$ such that $q = \alpha \circ p$. By assumption, the *discrepancy* $G = p^*(K_X + B) - q^*(K_Y + C)$ is effective and $E$ is contracted by $\alpha$ if and only if $p_*^{-1} E$ is an irreducible component of $G$.

In the following we show that $N = p_* G$.

Take a relatively ample divisor $H'$ on $Y$ and a relatively ample divisor $H$ on $X$ such that $H - p_* q^* H'$ is relatively ample. For any $t > 0$, since $K_Y + C + t H'$ is relatively ample and

$$K_X + B + t H = p_* q^*(K_Y + C + t H') + t(H - p_* q^* H') + p_* G,$$

we have

$$\inf\{D' \mid K_X + B + t H \equiv_S D' \geq 0\} \leq p_* G.$$

Therefore, $N$ is well defined and $N \leq p_* G$.

Conversely, if $K_X + B + t H \equiv_S D' \geq 0$, then $\alpha_* D' \equiv_S K_Y + C + t \alpha_* H$ and

$$p^* D' - q^* \alpha_* D' \equiv_S p^*(K_X + B + t H) - q^*(K_Y + C + t \alpha_* H)$$
$$= G + t(p^*(H) - q^*(\alpha_* H)).$$

## 2.9 The Divisorial Zariski Decomposition

Note that both sides are exceptional divisors over $Y$, so they are actually equal by the negativity lemma. Therefore,

$$p^* N_t \geq G + t(p^*(H) - q^*(\alpha_* H)).$$

Taking the limit as $t \to 0$, we can see that $N \geq p_* G$. □

**Remark 2.9.7** (1) If $\dim X = 2$, contracting all those divisors in $N$, or in other words contracting all $(-1)$-curves, will produce a minimal model. If $\dim X \geq 3$, then the situation becomes much more complicated because the geometry in codimension 2 or higher is involved.

(2) The Zariski decomposition of a divisor $D$ on an algebraic surface is discovered by Zariski ([144]) during the study of the section ring $\bigoplus_{m=0}^{\infty} H^0(X, mD)$ of the divisor $D$. It is a consequence of the intersection theory of divisors and the general theory of symmetric bilinear forms. In particular, if we consider the Zariski decomposition of the canonical divisor, then the numerically movable part coincides with the pullback of the canonical divisor on the minimal model. In this sense, we can say that the Zariski decomposition of canonical divisors is equivalent to the minimal model theory.

Generalizing this consideration, the log version of the existence of minimal models in dimension 2 can be proved as an application of the Zariski decomposition ([51]). Moreover, [30] generalized the Zariski decomposition to pseudo-effective divisors.

In dimension 2, the intersection theory of divisors is available so that we can use the general theory of symmetric bilinear forms to define the Zariski decomposition, but this is not the case in dimensions 3 and higher. So in [61], the divisorial Zariski decomposition was defined only for big divisors using limits of linear systems. [116] pushed this forward and generalized the definition to pseudo-effective divisors. In [16], the fixed part was defined using **R**-linear equivalence. Here the definition is simplified by replacing **R**-linear equivalence with numerical equivalence.

Similar to the case of dimension 2, if the numerically movable part is nef, then indeed we can get a minimal model. In order to deal with problems caused by subsets of codimensions 2 and higher, we need to replace $X$ by blowups. Although this approach to the minimal model is not successful, it might be helpful for understanding the problem. In this book, we use flips instead of blowups to deal with subsets of codimensions 2 and higher.

In addition, there is also an analytical approach to the analytical Zariski decomposition, which has played a certain role ([138]).

146            *2 The Minimal Model Program*

If the numerically movable part is not 0, then we can make many global sections by adding a little positivity:

**Theorem 2.9.8** (Nakayama [116]) *Let $D$ be a pseudo-effective $\mathbf{R}$-divisor on a normal projective $\mathbf{Q}$-factorial algebraic variety $X$. Take $D = P + N$ to be the divisorial Zariski decomposition. If $P \not\equiv 0$, then there exists an ample divisor $H$ such that the function in positive integer $m$ satisfies*

$$\lim_{m \to \infty} \dim H^0(X, \llcorner mD \lrcorner + H) = \infty.$$

*Proof* Since $N$ is effective, we may assume that $D = P$. Consider the *numerical base locus*

$$\mathrm{NBs}(D) = \lim_{t \downarrow 0} \left( \bigcap \{\mathrm{Supp}(D') \mid D + tH \equiv D' \geq 0\} \right).$$

Since $N = 0$, $\mathrm{NBs}(D)$ has no irreducible component of codimension 1. Also, since this is a limit of an increasing sequence of closed subsets, it is a union of at most countably many subvarieties of codimension at least 2. Therefore, we may take a very general smooth curve $C \subset X$ by cutting by very general hyperplanes such that $C \cap \mathrm{NBs}(D) = \emptyset$. Since $D \not\equiv 0$, $(D \cdot C) > 0$.

Fix an ample divisor $H$ and take $L_m = \llcorner mD \lrcorner + H$. We will show that if $m$ is a sufficiently large integer, then the natural map $H^0(X, L_m) \to H^0(C, L_m|_C)$ is surjective. Note that $\llcorner mD \lrcorner \cdot C = (mD \cdot C) - ((mD - \llcorner mD \lrcorner) \cdot C)$ can be arbitrarily large if $m$ is sufficiently large, since $(D \cdot C) > 0$ and $((mD - \llcorner mD \lrcorner) \cdot C)$ is bounded. Hence $H^0(C, L_m|_C)$ can be arbitrarily large and the theorem can be proved by the above surjectivity.

Note that $C$ does not pass through the singular locus of $X$, consider $g \colon Y \to X$ to be the blowup along $C$, and denote by $E$ the exceptional divisor. For any $t > 0$, there exists an effective $\mathbf{R}$-divisor $D_m \equiv mD + tH$ such that its support does not intersect $C$, and $(Y, g^*D_m)$ is KLT in a neighborhood of $E$.

Note that

$$g^*L_m - E - (K_Y + g^*D_m)$$
$$= g^*(\llcorner mD \lrcorner + H - (K_X + D_m)) - (n-1)E$$
$$\equiv g^*((1-t)H - (mD - \llcorner mD \lrcorner) - K_X) - (n-1)E.$$

Here $n = \dim X$. Note that we may take $H$ sufficiently large comparing to irreducible components of $K_X$, $E$, $D$, and $t$ sufficiently small such that the right-hand side is ample.

By the Nadal vanishing theorem,

$$H^1(Y, I(Y, g^*D_m) \otimes \mathcal{O}_Y(g^*L_m - E)) = 0.$$

By assumption,

$$E \cap \text{Supp}(\mathcal{O}_Y/I(Y, g^*D_m)) = \emptyset,$$

hence the natural map

$$H^0(Y, g^*L_m) \to H^0(E, (g^*L_m)|_E)$$

is surjective. This proves the claim. □

Conversely, if the function $\dim H^0(X, \llcorner mD \lrcorner + H)$ of positive integer $m$ is bounded, then we say that the *numerical Iitaka–Kodaira dimension* of $D$ is 0, which is denoted by $\nu(X, D) = 0$. In general, we define the numerical Iitaka–Kodaira dimension as the following:

**Definition 2.9.9** The *numerical Iitaka–Kodaira dimension* $\nu(X, D)$ of an **R**-divisor $D$ is defined to be the minimal integer $\nu$ satisfying the following property ([116]): For any fixed $H$, there exists a positive real number $c$, such that for any positive integer $m$, the inequality

$$\dim H^0(X, \llcorner mD \lrcorner + H) \leq cm^{\nu}$$

holds. If $D$ is not pseudo-effective, then we denote $\nu(X, D) = -\infty$.

This definition corresponds to the definition of the Iitaka–Kodaira dimension $\kappa(X, D)$, which is just the minimal integer $\kappa$ satisfying that there exists a positive real number $c$ such that for any positive integer $m$, the inequality

$$\dim H^0(X, \llcorner mD \lrcorner) \leq cm^{\kappa}$$

holds.

## 2.10 Polyhedral Decompositions of a Cone of Divisors

A *polytope* in a real vector space is the convex hull of finitely many points. It is called a *rational polytope* if all its vertices are points with rational numbers as coordinates (rational points). In this section, we consider polyhedral decompositions of a cone of divisors with respect to minimal models or canonical models and their applications. A line segment is an important example of a polytope, and the MMP with scaling is related to the decomposition of this polytope.

When changing the coefficients $b_i$ in the log canonical divisor $K_X + \sum_i b_i B_i$, the corresponding canonical model changes. This phenomenon is similar to that quotient spaces change according to polarizations in the *geometric invariant theory (GIT)*.

148            *2 The Minimal Model Program*

### 2.10.1 Rationality of Sections of Nef Cones

Applying the lengths of extremal rays, we can show that sections of nef cones are rational polytopes:

**Theorem 2.10.1** (Shokurov [129]) *Let $X$ be a normal $\mathbf{Q}$-factorial algebraic variety, let $f : X \to S$ be a projective morphism, and let $B_1, \ldots, B_t$ be effective $\mathbf{Q}$-divisors. Assume that $(X, B_i)$ is $\overline{KLT}$ for all $i$. Take $P$ to be the smallest convex closed subset containing all $B_i$ in the real vector space of $\mathbf{R}$-divisors on $X$ and denote $N = \{B' \in P \mid K_X + B' \text{ is relatively nef}\}$. Take $\{R_j\}$ to be the set of all extremal rays $R$ such that there exists a point $B' \in P$ with $((K_X + B') \cdot R) < 0$. Take $H_j = \{B'' \in P \mid ((K_X + B'') \cdot R_j) = 0\}$ to be the rational hyperplane section of $P$ determined by $R_j$. Then the following assertions hold:*

*(1) For any interior point $x$ in $P$, take $U$ to be a sufficiently small neighborhood of $x$, then it intersects only finitely many rational hyperplanes $H_j$.*
*(2) $N$ is a rational polytope.*

*Proof* (1) Assume that any neighborhood $U$ of $x$ intersects infinitely many distinct $H_j$. Then there exists a rational line in the smallest real linear space containing $P$ passing through a sufficiently near neighborhood of $x$ with the following property: $L \cap U$ is an open subset of the rational closed interval $L \cap P = [B, C]$ intersecting infinitely many $H_j$ at distinct points. Denote $L \cap H_j = (1 - t_j)B + t_j C$ and take $t_0 \in (0, 1)$ to be a limit point of the set $\{t_j\}$.

By construction, either $((K_X + B) \cdot R_j) < 0$ or $((K_X + C) \cdot R_j) < 0$ holds. By the lengths of extremal rays, we can take a rational curve $l_j$ generating $R_j$ such that either

$$0 < (-(K_X + B) \cdot l_j) \le 2b$$

or

$$0 < (-(K_X + C) \cdot l_j) \le 2b.$$

Here $b$ is the maximal dimension of fibers of $f$.

There exists a positive integer $m$ such that $m(K_X + B)$ and $m(K_X + C)$ are both Cartier. Then by definition, there are only finitely many values of

$$t_j = \frac{-m((K_X + B) \cdot l_j)}{m((K_X + C) \cdot l_j) - m((K_X + B) \cdot l_j)}.$$

Therefore, there is no accumulation of $\{t_j\}$, a contradiction.

## 2.10 Polyhedral Decompositions of a Cone of Divisors 149

(2) By the cone theorem, the nef set $N$ is the intersection of the inner sides of the hyperplanes $H_j$. Therefore, by (1), $N$ is a rational polytope in the interior of $P$. We only need to investigate the neighborhood of the boundary of $P$.

Take $L$ to be any rational linear subspace contained in the smallest linear space containing $P$, we will prove that $N \cap L$ is a rational polytope by induction on $\dim L$. If $P \subset L$, then this is the assertion of the theorem. Take $P_L$ to be the smallest face of $P$ containing $L \cap P$. We may replace $P$ by $P_L$ and assume that $P = P_L$, that is, $L$ contains an interior point of $P$.

If $\dim L = 1$, then $N \cap L$ is a point or a closed interval. Every endpoint is a rational point: This is clear if the point is on the boundary of $P$, and this follows from (1) if the point is an interior point of $P$.

Now assume that $\dim L > 1$. For any face $P'$ of $P$, $N \cap P' \cap L$ is a rational polytope by the inductive hypothesis. Since $N$ is locally a rational polytope near interior points of $P$, it suffices to show that $N \cap L$ is locally a rational polytope near every vertex $B$ of $N \cap P' \cap L$.

Take any rational line $L' \subset L$ passing through $B$ and containing an interior point of $P$ and write $P \cap L' = [B, C]$. Then $N \cap L' = [B, (1 - t_0)B + t_0 C]$ for some $t_0 \in [0, 1]$. Here $t_0$ is a rational number by (1). If $t_0 \neq 0, 1$, then $(1 - t_0)B + t_0 C$ is an interior point of $P$, and there exists an index $j$ such that $L' \cap H_j = \{(1 - t_0)B + t_0 C\}$. Since $(m(K_X + B) \cdot l_j) > 0$ and after multiplying by $m$ it becomes an integer, by the argument of (1), there exists a constant $c > 0$ independent of the choice of $L'$ such that $t_0 \geq c$. Therefore, there exists a sufficiently small neighborhood $U$ of $B$ such that $N \cap L \cap U$ is a cone with vertex $B$.

Take a general rational hyperplane $M$ sufficiently near to $B$, then $N \cap L \cap M$ is a rational polytope by the inductive hypothesis, hence $N \cap L \cap U$ is a cone over this rational polytope. This finishes the proof that $N \cap L$ is a rational polytope. Hence $N$ is a rational polytope. $\qquad \square$

**Remark 2.10.2** In this theorem, since we fix finitely many divisors in the beginning, the section of the nef cone is a rational polytope. In general, such a statement is not true for $N^1(X/S)$ since there are infinitely many divisors. For example, the surface of the nef cone of an Abelian variety is defined by the equation $(D^n) = 0$, which is not linear.

### 2.10.2 Polyhedral Decomposition according to Canonical Models

For a given pair $(X, B)$, its minimal model is not unique in general, but its canonical model is unique if it exists. Therefore, we first consider the decomposition according to canonical models:

150  2 *The Minimal Model Program*

**Theorem 2.10.3** (Polyhedral decomposition I [73, 129]) *Let $f : X \to S$ be a projective morphism from a normal $\mathbf{Q}$-factorial algebraic variety to a quasi-projective variety and let $B_1, \ldots, B_t$ be effective $\mathbf{R}$-divisors such that $(X, B_i)$ is $\overline{KLT}$ for all $i$. Take $V$ to be the affine subspace generated by all $B_i$ in the real vector space of divisors. Take $P'$ to be the polytope generated by all $B_i$. Consider the following convex closed subset of $P'$:*

$$P = \left\{ B = \sum_i b_i B_i \in P' \mid [K_X + B] \in \overline{\mathrm{Eff}}(X/S) \right\}.$$

*Assume the following conditions:*

- *For each point $B \in P$, there exists a minimal model $\alpha : (X, B) \dashrightarrow (Y, C)$ and a canonical model $g : Y \to Z$ of $f : (X, B) \to S$.*
- *For each point $B \in P$, there exists a polytope $P'_B \subset V$ containing $B$ as an interior point in the topology of $V$ such that if we denote*

$$P_B = \{ B' \in P'_B \cap P' \mid [K_Y + \alpha_* B'] \in \overline{\mathrm{Eff}}(Y/Z) \},$$

*then for any $B' \in P_B$, the morphism $g : (Y, \alpha_* B') \to Z$ admits a minimal model and a canonical model.*

*Then there exists a decomposition into finitely many disjoint subsets*

$$P = \coprod_{j=1}^{s} P_j$$

*and rational maps $\beta_j : X \dashrightarrow Z_j$ satisfying the following properties:*

*(1) $B \in P_j$ if and only if $\beta_j$ gives the canonical model of $f : (X, B) \to S$.*
*(2) The closures $\bar{P}_j$ of $P_j$ are unions of polytopes. In particular, $P$ is a polytope.*
*(3) If $P_j \cap \bar{P}_{j'} \neq \emptyset$, then there exists a morphism $f_{jj'} : Z_{j'} \to Z_j$ such that $\beta_j = f_{jj'} \circ \beta_{j'}$.*

Here note that $P_j$ is not necessarily connected.

*Proof* First, note that $(X, B)$ is $\overline{KLT}$ for any $B \in P'$, therefore we can use the framework of the minimal model theory. We prove the theorem by induction on $\dim V$. If $\dim V = 0$, then the assertion is trivial. Assume that $\dim V \geq 1$. Fix any point $B \in P$. It is an arbitrary point which is not necessarily a rational point.

Take the minimal model $\alpha : (X, B) \dashrightarrow (Y, C)$ and the canonical model $g : (Y, C) \to Z$. There exists an $\mathbf{R}$-Cartier divisor $H$ on $Z$ relatively ample over $S$ such that $K_Y + C = g^* H$. Take the polytope $P'_B$, for any $B' \in P_B$,

## 2.10 Polyhedral Decompositions of a Cone of Divisors
151

take the minimal model $\alpha' \colon (Y, \alpha_* B') \dashrightarrow (Y', C')$, and the canonical model $g' \colon (Y', C') \to Z'$ of $g \colon (Y, \alpha_* B') \to Z$. Take $h \colon Z' \to Z$ to be the natural morphism. There exists an $\mathbf{R}$-Cartier divisor $H'$ on $Z'$ relatively ample over $Z$ such that $K_{Y'} + C' = (g')^* H'$. Take a sufficiently small real number $\delta$ such that $(1 - \delta) h^* H + \delta H'$ is relatively ample over $S$. Take $B'' = (1 - \delta) B + \delta B'$ and $C'' = (1 - \delta) \alpha_* C + \delta C'$, then the negativity still holds, so $\alpha' \circ \alpha \colon (X, B'') \dashrightarrow (Y', C'')$ is a minimal model of $f \colon (X, B'') \to S$ and $g' \colon (Y', C'') \to Z'$ is the canonical model.

We can take such $\delta$ independent of $B'$ but depending only on $B$. First, we take $P_B'$ sufficiently small such that for any $B' \in P_B' \cap P$, $K_X + B'$ is negative with respect to $\alpha$, so we do not need to worry about the negativity. Since $H$ is ample over $Z$, we may take a sufficiently small $\epsilon > 0$ such that $(H \cdot \Gamma_Z) > \epsilon$ for any relative curve $\Gamma_Z$ on $Z$. Then we may take $\delta = \epsilon / (2\epsilon + 4 \dim X)$. Indeed, we will show that $K_{Y'} + (1 - 2\delta) \alpha_* C + 2\delta C'$ is relatively nef over $S$, which implies that $(1 - 2\delta) h^* H + 2\delta H'$ is relatively nef over $S$, and therefore $(1 - \delta) h^* H + \delta H'$ is relatively ample over $S$. Assume, to the contrary, that $K_{Y'} + (1 - 2\delta) \alpha_* C + 2\delta C'$ is not relatively nef over $S$, then there exists a negative extremal ray $R$, which is also a $(K_{Y'} + C')$-negative extremal ray since $K_{Y'} + \alpha_* C$ is relatively nef over $S$. (Here recall that $K_Y + C$ is relatively numerically trivial over $Z$, and hence $K_{Y'} + \alpha_* C$ is crepant to $K_Y + C$, which is nef over $S$.) By the lengths of extremal rays, $R$ is generated by a rational curve $\Gamma$ such that $((K_{Y'} + C') \cdot \Gamma) \geq -2 \dim X$. Note that $\Gamma$ is not contacted over $Z$ as $K_{Y'} + C'$ is nef over $Z$, therefore $(h^* H \cdot \Gamma) \geq \epsilon$ and then

$$((K_{Y'} + (1 - 2\delta) \alpha_* C + 2\delta C') \cdot \Gamma) \geq 0,$$

a contradiction.

Therefore, to summarize, if we take $P_B'$ sufficiently small, then for any $B' \in P_B$, $(Y', C'')$ and $Z'$ are minimal and canonical models for both $f \colon (X, B') \to S$ and $g \colon (Y, \alpha_* B') \to Z$. In particular, $P_B = P_B' \cap P$. Also we can see that they are minimal and canonical models for $f \colon (X, (1 - t) B + t B') \to S$ for any $0 < t \leq 1$.

The boundary $\partial (P_B' \cap P')$ of $P_B' \cap P'$ (as a subset of $V$) is a finite union of $(\dim V - 1)$-dimensional polytopes $(\partial (P_B' \cap P'))_k$. By the above argument, $P_B$ is a cone over $P_B \cap (\partial (P_B' \cap P'))$ with vertex at $B$. We can apply the inductive hypothesis to $(\partial (P_B' \cap P'))_k$ and $(Y, C) \to Z$. Here to check the second condition, we use the second condition on $X$ and the fact that $X$ and $Y$ have the same minimal model and canonical model for any divisor in $P_B$. Then $P_B \cap (\partial (P_B' \cap P'))$ can be decomposed into a disjoint union of finitely many subsets whose closures are polytopes, and these subsets correspond to canonical models of $(Y, \alpha_* B') \to Z$.

152                           2 *The Minimal Model Program*

Cones over these polytopes with vertex at $B$ gives the decomposition of $P_B$ into finitely many (not necessarily rational) polytopes. Since $P'$ is compact, it can be covered by finitely many such $P'_B$, and the first two assertions are proved. For the third assertion, just take $B \in P_j \cap \bar{P}_{j'}$.                           □

### 2.10.3 Polyhedral Decomposition according to Minimal Models

Next, we consider the decomposition according to minimal models:

**Theorem 2.10.4** (Polyhedral decomposition II [73, 129]) *Keep the assumption in Theorem 2.10.3. Then for each $P_j$, there is a finite disjoint decomposition*

$$P_j = \coprod_{k=1}^{t_j} Q_{j,k}$$

*satisfying the following properties: fix an arbitrary birational map $\alpha \colon X \dashrightarrow Y$ such that*

$$Q = \{B \in P \mid \alpha \text{ is a minimal model of } f \colon (X, B) \to S\}$$

*is non-empty, then*

*(1) $Q$ is locally closed, whose closure is a polytope.*
*(2) There exists an index $j$ such that $Q \subset \bar{P}_j$.*
*(3) If $Q \cap P_j \neq \emptyset$ for some $j$, then there exists $k$ such that $Q \cap P_j = Q_{j,k}$.*
*(4) The closure of $\bar{Q}_{j,k}$ is a polytope for any $j, k$.*

**Remark 2.10.5** For any fixed $j, k$, it is possible that there are infinitely many $\alpha$ such that $Q \cap P_j = Q_{j,k}$. For example, for a pair $(X, B)$ satisfying $K_X + B \equiv_S 0$, there might be infinitely many birational maps $\alpha$ inducing minimal models (Example 2.10.7).

*Proof* (1) Taking the intersection of $P$ with the pre-image of the nef cone $\overline{\mathrm{Amp}}(Y/S)$ by the push-forward map $\alpha_*$ induced by the birational map and cutting by finitely many linear inequalities given by negativity of log canonical divisors, we can obtain $Q$. The former is a closed polytope by Theorem 2.10.1, and the latter is an open condition, hence we get the conclusion.

(2) It is easy to see that if $B, B' \in Q$, then $tB + (1 - t)B' \in Q$ for any $t \in [0, 1]$. Hence $Q$ is a convex set. Take a relative interior point $B \in Q$, that is, an interior point of $Q$ in the affine subspace generated by $Q$, take $g \colon Y \to Z$ to be the canonical model of $(Y, \alpha_* B)$. Then $[\alpha_*(K_X + B)] \in g^* \mathrm{Amp}(Z/S)$ and $g^* \overline{\mathrm{Amp}}(Z/S)$ is a face of $\overline{\mathrm{Amp}}(Y/S)$.

## 2.10 Polyhedral Decompositions of a Cone of Divisors 153

For any $B' \in Q$, since $[\alpha_*(K_X + B')] \in \overline{\mathrm{Amp}}(Y/S)$ and $Q$ is convex, we have $[\alpha_*(K_X + B')] \in g^*\overline{\mathrm{Amp}}(Z/S)$. Moreover, if $B'$ is another relative interior point, then $[\alpha_*(K_X + B')] \in g^*\mathrm{Amp}(Z/S)$. Hence if we take $P_j$ to be the subset corresponding to the canonical model $g \circ \alpha$, then $Q \subset \bar{P}_j$.

(3) Given two birational maps $\alpha_i : X \dashrightarrow Y_i$ $(i = 1, 2)$ with corresponding subsets $\emptyset \neq Q_i \subset P$. Assume that there exist morphisms $g_i : Y_i \rightarrow Z$ such that $\beta = g_1 \circ \alpha_1 = g_2 \circ \alpha_2$ corresponds to some $P_j$. Consider the birational map $\gamma : Y_1 \dashrightarrow Y_2$ determined by $\alpha_2 = \gamma \circ \alpha_1$. We claim that if $\gamma$ is isomorphic in codimension 1, then $Q_1 \cap P_j = Q_2 \cap P_j$.

Indeed, take a point $B \in Q_1 \cap P_j$, we can write $K_{Y_1} + \alpha_{1*}B = g_1^*H$ for a relatively ample $\mathbf{R}$-divisor $H$ on $Z$. Since $\gamma$ is isomorphic in codimension 1, $K_{Y_2} + \alpha_{2*}B = g_2^*H$. Therefore, $B \in Q_2 \cap P_j$.

In particular, if $Q_1 \cap Q_2 \cap P_j \neq \emptyset$, then the minimal models corresponding to a point $B \in Q_1 \cap Q_2 \cap P_j$ are isomorphic in codimension 1, and therefore $Q_1 \cap P_j = Q_2 \cap P_j$.

In summary, by the above argument, we get a disjoint decomposition

$$P_j = \coprod_\alpha (Q \cap P_j),$$

where $\alpha$ runs over all birational contractions $\alpha : X \dashrightarrow Y$, and $Q \cap P_j$ depends only on the set of divisors contracted by $\alpha$.

Take $B_{j,l}$ to be vertices of $\bar{P}_j$ and take $\{E_m\}$ to be the set of prime divisors appearing in the numerically fixed part of some $K_X + B_{j,l}$. Note that $\{E_m\}$ is a finite set and contains all prime divisors appearing in the numerically fixed part of $K_X + B$ for any $B \in P_j$. So by Theorem 2.9.6, there are finitely many possibilities for the set of prime divisors contracted by $\alpha$, and hence the decomposition of $P_j$ is finite.

(4) Since $\bar{P}_j$ is a union of polytopes and $\bar{Q}$ is a polytope, $\bar{Q}_{j,k}$ is a polytope. Here we remark that $Q_{j,k}$ and $\bar{Q}_{j,k}$ are convex. $\quad\square$

**Corollary 2.10.6** *In Theorems 2.10.3 and 2.10.4, if all $B_i$ are $\mathbf{Q}$-divisors, then $P$, $P'$, $\bar{Q}_{j,k}$ are all rational polytopes and $\bar{P}_j$ is a union of rational polytopes.*

*Proof* $\bar{Q}$ is determined by cutting the pullback of the nef cone of the minimal model by finitely many linear inequalities with rational coefficients. As the section of the nef cone is a rational polytope by Theoren 2.10.1, $\bar{Q}$ is also a rational polytope.

If $P_j$ contains an interior point as a subset of $V$, then a general point of $P_j$ is contained in some $Q \subset \bar{P}_j$. So the closure of interior points of $P_j$ is the

154     *2 The Minimal Model Program*

union of such $\bar{Q}$, which is a union of rational polytopes. Therefore, $P$ is also a rational polytope.

If $P_j$ does not contain an interior point as a subset of $V$, then $P_j$ is contained in a union of faces of rational polytopes. In this case, we may just replace $P$ by those faces in the beginning to get the same conclusion.

$\bar{Q}_{j,k}$ is the intersection of a rational polytope and a union of rational polytopes, hence is a rational polytope. $\qquad\square$

**Example 2.10.7** Consider a general hypersurface $X$ in $\mathbf{P}^2 \times \mathbf{P}^1 \times \mathbf{P}^1$ of type $(3, 2, 2)$. $X$ is a smooth projective 3-dimensional algebraic variety with $K_X \sim 0$. We consider the polyhedral decomposition of the pseudo-effective cone of this example, in which there are infinitely many rational polytopes. This is also an example such that the quotient group of the birational automorphism group by the biregular automorphism group $\mathrm{Bir}(X)/\mathrm{Aut}(X)$ is an infinite group.

Denote by $P_1, P_2, P_3$ the projective spaces in the fiber product, take $L_i$ to be the pullback of hyperplanes $H_i$ in $P_i$ by the projection $p_i \colon X \to P_i$ ($i = 1, 2, 3$). $L_1, L_2, L_3$ form a basis of the real linear space $N_1(X)$. The nef cone $\overline{\mathrm{Amp}}(X)$ is the simplicial cone generated by $L_1, L_2, L_3$.

The projection $p_i \colon X \to P_i$ corresponds to the extremal ray $\langle L_i \rangle$. Here $\langle \rangle$ means the generated cone. The generic fiber of $p_1$ is an elliptic curve and the generic fibers of $p_2, p_3$ are K3 surfaces. The projection $p_{ij} \colon X \to P_i \times P_j$ corresponds to the face $\langle L_i, L_j \rangle$. The generic fiber of $p_{23}$ is an elliptic curve and the generic fibers of $p_{12}, p_{13}$ are sets of two points. Taking a Stein factorization, they give small contractions $q_{12}, q_{13}$.

Express the equation of $X$ by $f(x, y)z_0^2 + g(x, y)z_0 z_1 + h(x, y)z_1^2 = 0$. Here $[x_0 : x_1 : x_2]$, $[y_0 : y_1]$, $[z_0 : z_1]$ are homogeneous coordinates of $P_1, P_2, P_3$ and $f, g, h$ are homogeneous polynomials of degree 3 for $x_0, x_1, x_2$ and of degree 2 for $y_0, y_1$. The exceptional set of $q_{12} \colon X \to Y_{12}$ is defined by $f = g = h = 0$, which consists of 54 copies of $\mathbf{P}^1$, since $(3\bar{H}_1 + 2\bar{H}_2)^3 = 54$ on $P_1 \times P_2$ where $\bar{H}_1, \bar{H}_2$ are pullbacks of $H_1, H_2$.

As $p_{12} \colon X \to P_1 \times P_2$ gives a degree 2 extension of function fields, $X$ admits a birational automorphism induced by the Galois group $\mathbf{Z}/(2)$, which is a birational map $\alpha \colon X \dashrightarrow X$ exchanging two points in the generic fiber of $p_{12}$, and given by $(x, y, [z_0 : z_1]) \mapsto (x, y, [hz_1 : fz_0])$. Indeed, this birational map is nontrivial but preserves the equation of $X$ and $q_{12}$. By the form of this transformation,

$$\alpha^* L_1 = L_1,$$
$$\alpha^* L_2 = L_2,$$
$$\alpha^* L_3 = 3L_1 + 2L_2 - L_3.$$

## 2.10 Polyhedral Decompositions of a Cone of Divisors

Here to distinguish from the same $X$, we denote $\alpha\colon X_0 \dashrightarrow X_1$. We consider $(X_1, 0)$ as a nontrivial minimal model of $(X_0, 0)$.

For $p_{13}\colon X \to P_1 \times P_3$, we can similarly define the birational map $\beta\colon X_0 \dashrightarrow X_{-1}$ exchanging two points in the generic fiber. We have $\beta^* L_1 = L_1$, $\beta^* L_2 = 3L_1 - L_2 + 2L_3$, and $\beta^* L_3 = L_3$.

Note that $\alpha^2$ and $\beta^2$ are the identity map, but $\alpha$ and $\beta$ are not commutative. For each $n \in \mathbf{Z}$, we inductively define birational maps $\alpha_n\colon X_0 \dashrightarrow X_n$ by $\alpha \circ \alpha_n = \alpha_{-n+1}$, $\beta \circ \alpha_n = \alpha_{-n-1}$. If we take $M_k = \frac{3}{2}(k^2 + k)L_1 + (k + 1)L_2 - kL_3$, then

$$\alpha_n^* L_1 = L_1,$$

$$\alpha_n^* L_2 = \begin{cases} M_{2m} & n = 2m; \\ M_{2m} & n = 2m + 1, \end{cases}$$

$$\alpha_n^* L_3 = \begin{cases} M_{2m-1} & n = 2m; \\ M_{2m+1} & n = 2m + 1. \end{cases}$$

So the image of the nef cone $\alpha_n^* \overline{\mathrm{Amp}}(X_n)$ is generated by $L_1, M_{n-1}, M_n$, which is different from each other for each $n$. So we get a subgroup $\mathbf{Z}/(2) * \mathbf{Z}/(2) \subset \mathrm{Bir}(X)$ of the birational automorphism group.

The pseudo-effective cone is decomposed into nef cones:

$$\overline{\mathrm{Eff}}(X) = \bigcup_{n \in \mathbf{Z}} \alpha_n^* \overline{\mathrm{Amp}}(X_n).$$

Indeed, the right-hand side is generated by $L_1$ and $M_k$; all these vertices correspond to morphisms to lower dimensional varieties, and the limit of the rays generated by $M_k$ is $L_1$ as $|k| \to \infty$, so divisors outside this cone cannot be effective. Moreover, since $L_1$ and $M_k$ are all effective, the pseudo-effective cone coincides with the effective cone $\mathrm{Eff}(X)$.

This cone is decomposed into infinitely many rational polyhedral cones, and each of them corresponds to a birational map to a minimal model of $X$. The reason that infinitely many cones appear is because the finite-dimensional space of divisor classes is the projection of the space of all divisors, which is of infinite dimension.

### 2.10.4 Applications of Polyhedral Decompositions

The polyhedral decomposition theorem plays an important role in the proof of the existence of minimal models in Chapter 3. Here we introduce other applications as the finiteness of crepant blowups, the termination of MMP with

156                    2 The Minimal Model Program

scaling, the fact that birational minimal models are connected by flops, and the
generalization of MMP with scaling under weaker conditions.

For a KLT pair $(X, B)$, a *crepant blowup* of $(X, B)$ is a birational projective
morphism $g: (Y, C) \rightarrow (X, B)$ from a $\mathbf{Q}$-factorial KLT pair such that $g^*(K_X + B) = K_Y + C$. In particular, if $(Y, C)$ admits no crepant blowup other than
automorphisms, then it is called a *maximal crepant blowup*.

As an application of [16], we can get the following corollary by the argument
in [61]:

**Corollary 2.10.8** (*Crepant blowups*) *For a KLT pair $(X, B)$, there exists a
maximal crepant blowup for $(X, B)$. Moreover, the set of crepant blowups of
$(X, B)$ is finite up to isomorphisms.*

*Proof* Take a very log resolution $f: \tilde{Y} \rightarrow (X, B)$ of $(X, B)$ and define $\tilde{C}$
by the equation $f^*(K_X + B) = K_{\tilde{Y}} + \tilde{C}$. Write $\tilde{C} = \tilde{C}^+ - \tilde{C}^-$ into the
positive part and the negative part. Take a minimal model $g: (Y, C) \rightarrow (X, B)$
of $f: (\tilde{Y}, \tilde{C}^+) \rightarrow X$ (the existence of such minimal model is by [16]). Since
$[K_{\tilde{Y}} + \tilde{C}^+] = [\tilde{C}^-] \in N^1(Y/X)$ and all irreducible components of $\tilde{C}^-$ are
contracted by $f$, the set of divisors contracted by the rational map $\alpha: \tilde{Y} \dashrightarrow Y$ induced by the minimal model coincides with the support of $\tilde{C}^-$ by the
negativity lemma. That is, the set of exceptional divisors of $g$ coincides with
the set of exceptional divisors of $f$ with nonnegative coefficients in $\tilde{C}$. As $f$ is
a very log resolution, any blowup of $\tilde{Y}$ does not create new prime divisors in
the latter set. In particular, suppose that there exists a nonisomorphic projective
birational map $g_0: Y_0 \rightarrow Y$ from another normal variety and write $g_0^*(K_Y + C) = K_{Y_0} + C_0$, then there exists a $g_0$-exceptional prime divisor $E$ on $Y_0$
and its coefficient in $C_0$ is negative by the construction of $\tilde{C}$. Therefore, $g$ is a
maximal crepant blowup.

Since $g$ is birational, for any divisor $D$ on $Y$, there exists an effective divisor
$D'$ on $Y$ such that $D \equiv_X D'$. For any sufficiently small $\epsilon > 0$, $(Y, C + \epsilon D')$
is KLT. By [16], there exists a minimal model over $X$ and the canonical model
exists by the basepoint-free theorem. Hence by the polyhedral decomposition
theorem, there exists a polyhedral decomposition in a neighborhood of the
origin of $N^1(Y/X)$ corresponding to the canonical models. Taking cones of
those polytopes, we get a decomposition of $N^1(Y/X)$ into polyhedral cones.

For any maximal crepant blowup $g': (Y', C') \rightarrow X$, the set of exceptional
divisors of $g'$ coincides with the set of exceptional divisors of $f$ with
nonnegative coefficients in $\tilde{C}$ as discrete valuations on $k(X)$. Indeed, if this is
not the case, we can take a common very log resolution and a minimal model

## 2.10 Polyhedral Decompositions of a Cone of Divisors 157

over $Y'$ as above to create a nontrivial crepant blowup of $Y'$. Therefore, $Y$ and $Y'$ are isomorphic in codimension 1. The image of $\overline{\mathrm{Amp}}(Y'/X)$ under the natural homomorphism $N^1(Y'/X) \to N^1(Y/X)$ coincides with one of the above polyhedral cones. Hence there are only finitely many maximal crepant blowups. Note that here we use the fact that $\overline{\mathrm{Amp}}(Y'/X)$ determines $Y' \to X$ up to isomorphisms by Lemma 1.5.13.

For a crepant blowup $g'' : (Y'', C'') \to X$, we can take a maximal crepant blowup $(Y', C')$ of $(Y'', C'')$, which is also a maximal crepant blowup of $(X, B)$, and the nef cone $\overline{\mathrm{Amp}}(Y''/X)$ corresponds to a face of $\overline{\mathrm{Amp}}(Y'/X)$. Hence there are only finitely many such things. $\qquad \square$

Assuming the existence of minimal models and canonical models, we can show the termination of flips in MMP with scaling. Note that if there exists a sequence of flips that terminates, then it implies the existence of minimal models, but be aware that this is different with that any sequence of flips terminates.

**Corollary 2.10.9** (*Termination of MMP with scaling*) *Let* $f : (X, B) \to S$ *be a projective morphism from a* **Q**-*factorial KLT pair. Consider the MMP with scaling of H. Here* $(X, B + H)$ *is KLT,* $[K_X + B] \in \overline{\mathrm{Eff}}(X/S)$ *and* $[K_X + B + H] \in \mathrm{Big}(X/S) \cap \overline{\mathrm{Amp}}(X/S)$. *Assume that there exists a minimal model and canonical model for* $(X, B)$. *Then this MMP terminates.*

*Proof* Take $X = X_0$ and denote each step of the MMP by $\alpha_i : X_i \dashrightarrow X_{i+1}$. Since there are only finitely many divisorial contractions, after removing finitely many steps, we may assume that $\alpha_i$ are all flips. From now on we use the same notation for strict transforms of a divisor.

Since $K_X + B + H$ is relatively big and $K_X + B$ is relatively pseudo-effective, for any $t > 0$, $K_{X_i} + B + tH$ is relatively big, hence its minimal model exists, by the existence of minimal models in [16]. Moreover, by the basepoint-free theorem, its canonical model exists. By assumption, the minimal model and canonical model exist if $t = 0$. We may apply the polyhedral decomposition theorem to the segment $[B, B + H]$, and get a decomposition into finitely many interval $P_j$. To simplify the notation, we denote $B + tH$ by $t$ and consider the decomposition on $[0, 1]$. Take

$$t_i = \min\{t \in \mathbf{R} \mid K_{X_i} + B + tH \text{ is relatively nef}\},$$
$$t'_i = \max\{t \in \mathbf{R} \mid K_{X_i} + B + tH \text{ is relatively nef}\}.$$

In other words, the interval $Q_i$ in which $X \dashrightarrow X_i$ gives a minimal model is just $[t_i, t'_i]$. Recall that for the extremal ray corresponding to $\alpha_i$, we have

$((K_{X_i} + B) \cdot R) < 0$, $((K_{X_{i+1}} + B) \cdot R) > 0$, $((K_{X_i} + B + t_i H) \cdot R) = ((K_{X_{i+1}} + B + t_i H) \cdot R) = 0$. Hence $t_i = t'_{i+1}$.

We consider the case $t_i = t_{i+1} > 0$. In this case, take $(Y, C)$ to be the common canonical model of $(X_i, B + t_i H)$ and $(X_{i+1}, B + t_i H)$. Since $K_{X_i} + B + t_i H$ is relatively nef and relatively big over $S$, $g_{i+1} \colon (X_{i+1}, B + t_{i+1} H) \to (Y, C)$ is a crepant blowup. Then by Corollary 2.10.8, there are only finitely many such $g_{i+1}$, that is, there does not exist any infinite sequence

$$0 < t_i = t_{i+1} = t_{i+2} = \cdots .$$

As a consequence, there are infinitely many distinct nontrivial intervals $Q_i$ if the MMP does not terminate.

Suppose that there exists an interval $P_j$ in which $t_i$ is an interior point, take $\beta \colon X \dashrightarrow Y$ to be the corresponding canonical model. Possibly changing the index $i$, we can take $Q_i$ such that there exists $t > t_i$ in $P_j \cap Q_i$. Also, we can take $Q_{i'}$ ($i' > i$) such that there exists $t' < t_i$ in $P_j \cap Q_{i'}$. In this case, there exist morphisms $g_i \colon X_i \to Y$ and $g_{i'} \colon X_{i'} \to Y$. By construction, there exists an **R**-divisor $H_Y$ on $Y$ such that $K_{X_i} + B + tH = g_i^* H_Y$ and $K_{X_{i'}} + B + tH = g_{i'}^* H_Y$, but the former is relatively nef while the latter is not, a contradiction. Therefore, if the interval $Q_i$ contains an interior point, then its closure coincides with the closure of the interval $P_j$, and there are only finitely many such $Q_i$.

From the above discussion, the MMP terminates. $\qquad\square$

For a given pair, minimal models, if they exist, are not unique in general. However, we can show that minimal models are connected by elementary birational maps, so-called "flops."

A birational map $\alpha \colon (X, B) \dashrightarrow (Y, C)$ between two **Q**-factorial DLT pairs is called a *flop* if there exist birational projective morphisms $f \colon (X, B) \to (Z, D)$ and $g \colon (Y, C) \to (Z, D)$ to a third pair satisfying the following conditions:

(1) $\alpha = g^{-1} \circ f$.
(2) $f, g$ are isomorphic in codimension 1.
(3) $\rho(X/Z) = \rho(Y/Z) = 1$.
(4) $f^*(K_Z + D) = K_X + B$ and $g^*(K_Z + D) = K_Y + C$.

The definition is the same as flips except for condition (4). Different from a flip, we require that the levels of canonical divisors are preserved.

**Corollary 2.10.10** (*Flop decomposition*) *Let $f \colon (X, B) \to S$ be a projective morphism from a KLT pair. Assume that it admits a minimal model and a*

## 2.10 Polyhedral Decompositions of a Cone of Divisors 159

*canonical model. Then any two minimal models $\alpha_i : (X, B) \dashrightarrow (Y_i, C_i)$
$(i = 1, 2)$ are connected by a sequence of flops.*

*Proof* By Lemma 2.5.12, $Y_i$ are isomorphic in codimension 1, and have the same canonical model. Take $g_i : Y_i \to Z$ to be the morphism to the canonical model. Take a general relatively ample effective **Q**-divisor $H_2$ on $Y_2$. After replacing $H_2$ by $\epsilon H_2$ for some sufficiently small $\epsilon > 0$, we may assume that $(Y_1, C_1 + H_2)$ is KLT. Here we use the same notation for strict transforms of a divisor.

We can run a $(K_{Y_1} + C_1 + H_2)$-MMP over $Z$ with scaling of a general relatively ample divisor, and reach a minimal model $Y'$ such that $K_{Y'} + C_1 + H_2$ is nef over $Z$. Since $K_{Y_i} + C_i$ is numerically trivial over $Z$, $K_{Y_2} + C_2 + H_2$ is ample over $Z$ and $Y' \to Y_2$ is the canonical model. As $Y_i$ are isomorphic in codimension 1 and $Y_2$ is **Q**-factorial, $Y_2 \simeq Y'$ and the MMP is a sequence of flips, which is also a sequence of flops with respect to $(Y_1, C_1)$. $\square$

**Remark 2.10.11** In [72], the same result is proved without assuming the existence of canonical models.

Applying the polyhedral decomposition theorem, we can generalize the framework of MMP with scaling to the case when the boundary is not big:

**Corollary 2.10.12** ([12]) *Let $f : X \to S$ be a projective morphism from a normal **Q**-factorial algebraic variety and let $B, C$ be **R**-divisors. Assume that $(X, B)$ and $(X, B + C)$ are $\overline{KLT}$, $K_X + B + C$ is relatively nef, and $K_X + B$ is not relatively nef. Take*

$$t_0 = \min\{t \mid K_X + B + tC \text{ is relatively nef}\}.$$

*Then there exists an extremal ray $R$ in $\overline{\mathrm{NE}}(X/S)$ such that $((K_X + B + t_0 C) \cdot R) = 0$ and $((K_X + B) \cdot R) < 0$.*

*Proof* Take effective **Q**-divisors $B_1, \dots, B_s$ such that $(X, B_i)$ is $\overline{KLT}$ and the spanned rational polytope $P$ contains $B, B + C$. By Theorem 2.10.1, $N = \{B' \in P \mid K_X + B' \text{ is relatively nef}\}$ is a rational polytope.

Consider all $(K_X + B)$-negative extremal rays $R_k$, and take $l_k$ to be a curve generating $R_k$ with $((K_X + B) \cdot l_k) \geq -2 \dim X$. Take the real number $t_k$ determined by $((K_X + B + t_k C) \cdot R_k) = 0$, then $\sup_k t_k = t_0$. Assume, to the contrary, that $t_k < t_0$ for all $k$, we claim that there is no infinite sequence $\{t_k\}$ converging to $t_0$.

Note that we may take rational points $B'_i$ in $P$ and real numbers $b_i > 0$ $(i = 1, \dots, u)$ such that $\mathrm{Supp}(B_i) = \mathrm{Supp}(B)$, $\sum b_i = 1$, and $B = \sum b_i B'_i$. Moreover, $(X, B'_i)$ is $\overline{KLT}$ and $((K_X + B'_i) \cdot l_k) \geq -2 \dim X$ for all $i, k$. Since

160    *2 The Minimal Model Program*

$N$ is a rational polytope, there exist rational points $C_j$ in $N$ and real numbers $c_j > 0$ $(j = 1, \ldots, v)$ such that $\sum c_j = 1$ and $B + t_0 C = \sum c_j C_j$.

Take a positive integer $m$ such that $mK_X$, $mB'_i$, and $mC_j$ are all Cartier. Then we have integers $m_{ik}$, $n_{jk}$ as the following:

$$m_{ik} = (m(K_X + B'_i) \cdot l_k) \geq -2m \dim X;$$
$$n_{jk} = (m(K_X + C_j) \cdot l_k) \geq 0.$$

Moreover, since $\sum_i m_{ik} b_i < 0$, there are only finitely many possible values of $m_{ik}$.

Since $K_X + B + t_k C = (1 - t_k/t_0)(K_X + B) + t_k/t_0(K_X + B + t_0 C)$, we have $(1 - t_k/t_0) \sum_i b_i m_{ik} + t_k/t_0 \sum_j c_j n_{jk} = 0$. Therefore,

$$1 - t_0/t_k = \frac{\sum_j c_j n_{jk}}{\sum_i b_i m_{ik}},$$

which is in a discrete subset of $\mathbf{R}$, and the claim is proved.    $\square$

## 2.11 Multiplier Ideal Sheaves

The goal of this section is to give an algebraic definition of a multiplier ideal sheaf and introduce the Nadel vanishing theorem. The theory of multiplier ideal sheaves is a basic tool in the $L^2$-theory in complex analysis and multiplier ideal sheaves are defined for line bundles with metrics. Here we only consider the case when metrics are defined algebraically. Also we consider a so-called adjoint ideal sheaf which is the log version of the multiplier ideal sheaf.

### 2.11.1 Multiplier Ideal Sheaves

It is classical in complex analysis to investigate functions which are not $L^2$ by multiplying functions to make them $L^2$, but it has been found in recent years that the multiplier ideal sheaf consisting of all multiplier functions is very useful in algebraic geometry.

**Definition 2.11.1** For a pair consisting of a normal algebraic variety $X$ and an effective $\mathbf{R}$-divisor $B$ such that $K_X + B$ is $\mathbf{R}$-Cartier, the *multiplier ideal sheaf* $I(X, B)$ is defined as the following. Take a log resolution $f : Y \to (X, B)$ of the pair $(X, B)$, write $f^*(K_X + B) = K_Y + C$, then

$$I(X, B) = f_*(\mathcal{O}_Y(\ulcorner -C \urcorner)).$$

## 2.11 Multiplier Ideal Sheaves
161

**Proposition 2.11.2** *(1) The multiplier ideal sheaf $I(X, B)$ is a nonzero coherent ideal sheaf, and it does not depend on the choice of log resolutions.*

*(2) $R^p f_*(\mathcal{O}_Y(\ulcorner -C \urcorner)) = 0$ for any $p > 0$.*

*(3) The cosupport of the multiplier ideal sheaf, or the support of $\mathcal{O}_X / I(X, B)$, coincides with the non-KLT locus of $(X, B)$. In particular, $I(X, B) = \mathcal{O}_X$ if and only if $(X, B)$ is KLT.*

*Proof* (1) Since the irreducible components of $C$ with negative coefficients are contracted by $f$, $I(X, B)$ is a coherent subsheaf of $\mathcal{O}_X$.

Take $f_1 \colon Y_1 \to X$ to be another log resolution. By the desingularization theorem, there exists a log resolution dominating both $f$ and $f_1$. So we only need to consider the case that $f_1$ dominates $f$, that is, there exists a morphism $g \colon Y_1 \to Y$ such that $f_1 = f \circ g$. Write $f_1^*(K_X + B) = K_{Y_1} + C_1$. It suffices to show that

$$g_* \mathcal{O}_{Y_1}(\ulcorner -C_1 \urcorner) = \mathcal{O}_Y(\ulcorner -C \urcorner).$$

The left-hand side is naturally contained in the right-hand side. We prove the inverse inclusion.

Denote by $F$ the normal crossing divisor which is the union of the exceptional divisors of $f$ and the strict transform of the support of $B$. If $g$ is a permissible blowup with respect to $(Y, F)$, then the equality of both sides can be directly checked. In general, $g$ is dominated by a sequence of such blowups, so the inverse inclusion can be proved.

(2) As $-C - K_Y$ is relatively numerically trivial over $X$ and $f$ is birational, $-C - K_Y$ is relatively nef and relatively big over $X$. Then we can apply the vanishing theorem to get the conclusion.

(3) Write $C = C^+ - C^-$ where $C^+, C^-$ are effective $\mathbf{R}$-divisors with no common irreducible component. Then as in the proof of Lemma 1.11.9, by (2) we know that the natural map

$$\mathcal{O}_X \simeq f_* \mathcal{O}_Y \to f_*(\mathcal{O}_{\llcorner C^+ \lrcorner}(\ulcorner C^- \urcorner))$$

is surjective. Hence $f_* \mathcal{O}_{\llcorner C^+ \lrcorner} \simeq f_*(\mathcal{O}_{\llcorner C^+ \lrcorner}(\ulcorner C^- \urcorner))$. On the other hand, $\mathcal{O}_X / I(X, B) \simeq f_*(\mathcal{O}_{\llcorner C^+ \lrcorner}(\ulcorner C^- \urcorner))$ and the support of $f_* \mathcal{O}_{\llcorner C^+ \lrcorner}$ is exactly the non-KLT locus of $(X, B)$. $\qquad\square$

The fact (2) in the above proposition seems to be a reason why multiplier ideal sheaves are useful.

**Example 2.11.3** If $X$ is smooth and the support of $B$ is normal crossing, then $I(X, B) = \mathcal{O}_X(\ulcorner -B \urcorner)$.

162                    2 The Minimal Model Program

**Exercise 2.11.4** For a birational projective morphism $f : Y \to X$ between smooth algebraic varieties, show that $R^p f_* \mathcal{O}_Y = 0$ for $p > 0$ by the same argument as in Proposition 2.11.2.

We will need the following lemma in Section 2.12:

**Lemma 2.11.5** *Let $(X, B)$ be a KLT pair, let $B'$ be an effective **R**-Cartier divisor, let $L$ be a line bundle, and let $s$ be a global section of $L$. Assume that $K_X$ is **Q**-Cartier and*

$$B' - B \leq \mathrm{div}(s).$$

*Then*

$$s \in H^0(X, L \otimes I(X, B')).$$

*Proof* Take a log resolution $f : Y \to (X, B + B')$, and write $f^*(K_X + B) = K_Y + C$ and $f^*(K_X + B') = K_Y + C'$. Note that

$$-f^* \mathrm{div}(s) \leq -f^* B' + f^* B = -C' + C.$$

Since $\ulcorner -C \urcorner \geq 0$ and $\mathrm{div}(s)$ is Cartier,

$$-f^* \mathrm{div}(s) \leq \ulcorner -C' \urcorner.$$

Therefore, $\mathcal{O}_X(-\mathrm{div}(s)) \subset I(X, B')$. $\qquad\square$

The Nadel vanishing theorem is a basic tool in the proof of the extension theorem in Section 2.12. Here, if we only consider algebraic multiplier ideal sheaves, then the Nadel vanishing theorem is an easy consequence of the Kawamata–Viehweg vanishing theorem:

**Theorem 2.11.6** (*Nadel vanishing theorem*) *Let $X$ be a normal algebraic variety and let $B$ be an effective **R**-divisor such that $K_X + B$ is **R**-Cartier. Let $f : X \to S$ be a projective morphism and let $D$ be a Cartier divisor. Assume that $D - (K_X + B)$ is relatively nef and relatively big over $S$. Then*

$$R^p f_*(\mathcal{O}_X(D) \otimes I(X, B)) = 0$$

*for any $p > 0$.*

*Proof* Take a log resolution $g : (Y, C) \to (X, B)$, then $g^* D - C - K_Y$ is relatively nef and relatively big over $X$ and over $S$. Therefore,

$$R^p g_*(\mathcal{O}_Y(g^* D + \ulcorner -C \urcorner)) = 0 \text{ and}$$
$$R^p (f \circ g)_*(\mathcal{O}_Y(g^* D + \ulcorner -C \urcorner)) = 0$$

## 2.11 Multiplier Ideal Sheaves

for any $p > 0$. The conclusion follows from the spectral sequence

$$E_2^{p,q} = R^p f_* R^q g_*(\mathcal{O}_Y(g^*D + \ulcorner -C \urcorner)) \Rightarrow R^{p+q}(f \circ g)_*(\mathcal{O}_Y(g^*D + \ulcorner -C \urcorner))$$

and

$$g_*(\mathcal{O}_Y(g^*D + \ulcorner -C \urcorner)) = \mathcal{O}_X(D) \otimes I(X, B). \qquad \square$$

For reference, we define analytic multiplier ideal sheaves. Let $X$ be a smooth complex manifold and let $L$ be a line bundle on $X$. A *singular Hermitian metric* $h$ on $L$ is a Hermitian metric allowing infinity values of the form $h = h_0 e^{-\phi}$, where $\phi$ is a locally $L^1$ function and $h_0$ is a $C^\infty$ Hermitian metric. The curvature of $h$ can be defined similarly as the curvature of a usual Hermitian metric and it is a real current of type $(1, 1)$. Then the *multiplier ideal sheaf* $I = I(L, h)$ is defined by

$$\Gamma(U, I) = \{p \in \Gamma(U, \mathcal{O}_X) \mid pe^{-\phi} \text{ is locally } L^2\}.$$

As $h$ is singular, regular functions are not necessarily $L^2$-integrable. The name "multiplier" is clear from the definition. It can be shown that $I$ is an analytic coherent ideal sheaf.

**Example 2.11.7** Let $g_i$ ($i = 1, \ldots, r$) be regular functions on a complex manifold $X$ and take divisors $B_i = \mathrm{div}(g_i)$ to be the zero divisors. Take an **R**-divisor $B = \sum_i b_i B_i$ where $b_i$ are positive real numbers. Define a singular Hermitian metric $h$ on the trivial line bundle $\mathcal{O}_X$ as

$$h = \sum_i |g_i|^{-2b_i}.$$

In this case, the algebraic multiplier ideal sheaf coincides with the analytic multiplier ideal sheaf: $I(X, B) = I(\mathcal{O}_X, h)$.

Of course, it is not always the case as in this example, so analytic multiplier ideal sheaves are more general than the algebraic multiplier ideal sheaves considered in this book. For example, singular Hermitian metrics that appear in the (algebraic) Hodge theory are known to be different from the algebraic ones as they have logarithmic growth.

The following theorem is the original form of the Nadel vanishing theorem. As the metric $h$ is not necessarily induced by a divisor, it is more general than the algebro-geometric version.

**Theorem 2.11.8** ([111]) *Let $X$ be a compact complex smooth manifold and let $L$ be a line bundle admitting a singular Hermitian metric $h$. Denote by $I$ the corresponding multiplier ideal sheaf. Assume that the curvature of $h$ is*

164              2 *The Minimal Model Program*

*semipositive and strictly positive at some point of X. Then* $H^p(X, \mathcal{O}_X(K_X + L) \otimes I) = 0$ *for any* $p > 0$.

## 2.11.2 Adjoint Ideal Sheaves

Next, we define adjoint ideal sheaves as a variant of multiplier ideal sheaves. Adjoint ideal sheaves are defined in algebraic geometry, and there is no natural analogue in complex analysis. The reason is that the logarithmic differential form $dz/z$ is not $L^2$. This definition is natural when considering residue map and doing induction on dimensions.

**Definition 2.11.9** Let $X$ be a normal algebraic variety and let $B$ be an effective **R**-divisor. Assume that $K_X + B$ is **R**-Cartier. Assume that there exists an irreducible component $Z$ in $B$ with coefficient 1. Then the *adjoint ideal sheaf* $I_Z(X, B)$ is defined as follows. Take a log resolution $f : Y \to (X, B)$ of the pair $(X, B)$, write $f^*(K_X + B) = K_Y + C$ and $W = f_*^{-1}Z$, then

$$I_Z(X, B) = f_*(\mathcal{O}_Y(\ulcorner -C \urcorner + W)).$$

The adjoint ideal sheaf measures how far the pair $(X, B)$ is from being PLT (purely log terminal). Fix an irreducible component $Z$ in $B$ with coefficient 1, then the set of points on $Z$, in a neighborhood of which $(X, B)$ is not PLT, is a closed subset of $Z$. It is called the *non-PLT locus* of $(X, B)$ with respect to $Z$.

**Proposition 2.11.10** *(1) The adjoint ideal sheaf $I_Z(X, B)$ is a nonzero coherent ideal sheaf, and it does not depend on the choice of log resolutions.*
*(2)* $R^p f_*(\mathcal{O}_Y(\ulcorner -C \urcorner + W)) = 0$ *for any* $p > 0$.
*(3) The intersection of $Z$ and the support of $\mathcal{O}_X/I_Z(X, B)$ coincides with the non-PLT locus of $(X, B)$ with respect to $Z$. In particular, $I_Z(X, B) = \mathcal{O}_X$ in a neighborhood $Z$ if and only if the pair $(X, B)$ is PLT in a neighborhood $Z$.*

*Proof* The proof is the same as that of Proposition 2.11.2.
(1) Given another log resolution $f_1 : Y_1 \to X$, we may assume that there exists a morphism $g : Y_1 \to Y$ such that $f_1 = f \circ g$. Write $f_1^*(K_X + B) = K_{Y_1} + C_1$ and $W_1 = f_{1*}^{-1}Z$. It suffices to show that

$$g_* \mathcal{O}_{Y_1}(\ulcorner -C_1 \urcorner + W_1) = \mathcal{O}_Y(\ulcorner -C \urcorner + W).$$

Then the proof is the same as that of Proposition 2.11.2.
(2) Note that $-C + W - (K_Y + W)$ is relatively nef and relatively big over $X$, and its restriction to $W$ is again relatively nef and relatively big over $Z$.

## 2.11 Multiplier Ideal Sheaves

(3) Note that, all coefficients of $C - W$ are strictly smaller than 1 if and only if $\ulcorner -C \urcorner + W \geq 0$. $\qquad\square$

The relation of multiplier ideal sheaves and adjoint ideal sheaves is as the following:

**Lemma 2.11.11** *Let $X$ be a normal algebraic variety and let $B$ be an effective $\mathbf{R}$-divisor. Assume that $K_X + B$ is $\mathbf{R}$-Cartier. Assume that there exists an irreducible component $Z$ in $B$ with coefficient 1. Assume that $Z$ is normal and write $(K_X + B)|_Z = K_Z + B_Z$. Then there is a short exact sequence:*

$$0 \to I(X, B) \to I_Z(X, B) \to I(Z, B_Z) \to 0.$$

*Therefore, $I_Z(X, B)\mathcal{O}_Z = I(Z, B_Z)$.*

*Proof* Write $(K_Y + C)|_W = K_W + C_W$ where $C_W = (C - W)|_W$. Denote $f_Z = f|_Z$, then $f_Z^*(K_Z + B_Z) = K_W + C_W$. We get the desired short exact sequence from the exact sequence

$$0 \to \mathcal{O}_Y(\ulcorner -C \urcorner) \to \mathcal{O}_Y(\ulcorner -C \urcorner + W) \to \mathcal{O}_W(\ulcorner -C_W \urcorner) \to 0$$

and $R^1 f_* \mathcal{O}_Y(\ulcorner -C \urcorner) = 0$. The last assertion follows from $I(X, B) \subset \mathcal{O}_X(-Z)$. $\qquad\square$

We can extend the Nadel vanishing theorem to adjoint ideal sheaves:

**Theorem 2.11.12** *Let $X$ be a normal algebraic variety and let $B$ be an effective $\mathbf{R}$-divisor. Assume that $K_X + B$ is $\mathbf{R}$-Cartier. Assume that there exists an irreducible component $Z$ in $B$ with coefficient 1. Let $f : X \to S$ be a projective morphism and let $D$ be a Cartier divisor. Assume that $D - (K_X + B)$ is relatively nef and relatively big over $S$ and $(D - (K_X + B))|_Z$ is relatively nef and relatively big over $f(Z)$. Then*

$$R^p f_*(\mathcal{O}_X(D) \otimes I_Z(X, B)) = 0$$

*for any $p > 0$.*

*Proof* The proof is similar to that of Theorem 2.11.6. If $Z$ is normal, then this is a consequence of Theorem 2.11.6 by using the exact sequence in Lemma 2.11.11. $\qquad\square$

Let us define a special case of logarithmic multiplier ideal sheaf, which is a general version of adjoint ideal sheaf:

**Definition 2.11.13** Let $(X, B)$ be a DLT pair consisting of a normal algebraic variety $X$ and an $\mathbf{R}$-divisor $B$ on $X$. Let $L$ be a linear system of divisors and let $m$ be a positive integer. Take $Z = \llcorner B \lrcorner$, which is not necessarily irreducible.

Take a general element $G \in L$, assume that it does not contain LC centers of the pair $(X, B)$. Then the *logarithmic multiplier ideal sheaf* $I_Z(X, B + L/m)$ is defined as the following. Take a log resolution $f : Y \to X$ of $(X, B + G)$ in strong sense, which is isomorphic over the generic point of each LC center of $(X, B)$ and resolves the base locus of $L$. Write $f^*(K_X + B) = K_Y + C$, $f^*G = P + N$, and $W = f_*^{-1}Z$. Here $P$ is a general element of the movable part of $f^*L$ and $N$ is the fixed part. By construction, $P$ is free. Then we define

$$I_Z(X, B + L/m) = f_*(\mathcal{O}_Y(\ulcorner -C - N/m \urcorner + W)).$$

**Lemma 2.11.14** *(1) The logarithmic adjoint ideal sheaf $I_Z(X, B + L/m)$ is a nonzero coherent ideal sheaf, and it does not depend on the choice of log resolutions.*
*(2) $R^p f_*(\mathcal{O}_Y(\ulcorner -C - N/m \urcorner + W)) = 0$ for any $p > 0$.*

*Proof* (1) Given another log resolution $f_1 : Y_1 \to X$, we may assume that there exists a morphism $g : Y_1 \to Y$ such that $f_1 = f \circ g$. Write $f_1^*(K_X + B) = K_{Y_1} + C_1$, $f_1^*D = P_1 + N_1$, and $W_1 = f_{1*}^{-1}Z$. It suffices to show that

$$g_*(\mathcal{O}_{Y_1}(\ulcorner -C_1 - N_1/m \urcorner + W_1)) = \mathcal{O}_Y(\ulcorner -C - N/m \urcorner + W).$$

We can reduce this to the case of permissible blowups as Proposition 2.11.2. In the case of permissible blowups, we can prove the assertion by comparing coefficients explicitly.

(2) Note that $-C - N/m + W - (K_Y + W) \equiv_X P/m$ is relatively nef and relatively big over $X$, also its restriction on each LC center of $(Y, W)$ is again relatively nef and relatively big. The conclusion follows from applying the vanishing theorem. $\qquad\square$

We can prove the Nadel vanishing theorem for logarithmic adjoint ideal sheaves:

**Theorem 2.11.15** *Let $(X, B)$ be a DLT pair, let $L$ be a linear system of divisors, let $m$ be a positive integer, let $D$ be a Cartier divisor, and let $f : X \to S$ be a projective morphism to an affine variety. Take $Z = \llcorner B \lrcorner$. Assume the following conditions:*

*(1) A general element $G \in L$ does not contain LC centers of the pair $(X, B)$.*
*(2) $D - (K_X + B + G/m)$ and its restriction to each LC center are relatively nef and relatively big over $S$ or the image of the center in $S$, respectively.*

*Then*

$$H^p(X, I_Z(X, B + L/m) \otimes \mathcal{O}_X(D)) = 0$$

*for any $p > 0$.*

*Proof* As $P$ is relatively nef over $S$, the proof is similar to that of Theorem 2.11.6. $\qquad\square$

In order to simultaneously investigate linear systems induced by multiples of a divisor, we define asymptotic multiplier ideal sheaves. They play important roles in the proof of extension theorems.

**Definition 2.11.16** Let $(X, B)$ be a DLT pair. Let $L_m$ ($m \in \mathbf{Z}_{>0}$) be a sequence of linear systems of divisors satisfying $L_m + L_{m'} \subset L_{m+m'}$, that is, $D + D' \in L_{m+m'}$ if $D \in L_m$, $D' \in L_{m'}$. Take $Z = \llcorner B \lrcorner$. Assume that there exists $m$ such that a general element $D \in L_m$ does not contain LC centers of the pair $(X, B)$. Then define the *asymptotic multiplier ideal sheaf* to be

$$I_Z(X, B + \{L_m/m\}) = \bigcup_{m>0} I_Z(X, B + L_m/m).$$

**Remark 2.11.17** By assumption, $I_Z(X, B + L_m/m) \subset I_Z(X, B + L_{m'}/m')$ if $m|m'$. By the Noetherian property, the right-hand side which is a union of infinitely many ideals is actually obtained by a sufficiently large and sufficiently divisible $m$. However, such $m$ cannot be determined priorly. This is the trick of asymptotic multiplier ideal sheaves.

The following lemma is a result on global generation of sheaves derived from the vanishing theorem, which will be used in Section 2.12. For ample sheaves the same assertion is difficult to prove, but for very ample sheaves it is easy. We use the so-called *Castelnuovo–Mumford regularity* method:

**Lemma 2.11.18** *Let $X$ be an n-dimensional quasi-projective algebraic variety, let $\mathcal{O}_X(1)$ be a very ample invertible sheaf, and let $\mathcal{F}$ be a coherent sheaf. Assume that*

$$H^p(X, \mathcal{F} \otimes \mathcal{O}_X(m)) = 0$$

*for any $m \in \mathbf{Z}_{\geq 0}$ and any $p \in \mathbf{Z}_{>0}$. Then $\mathcal{F} \otimes \mathcal{O}_X(n)$ is generated by global sections.*

*Proof* The proof is by induction on $n$. We may assume that $n > 0$. Fix any point $x \in X$. Take $\mathcal{F}_0 = H^0_{\{x\}}(\mathcal{F})$ to be the subsheaf of $\mathcal{F}$ containing all local sections whose supports are $x$, then the quotient sheaf $\mathcal{F}_1 = \mathcal{F}/\mathcal{F}_0$ has no local section whose support is $x$. Consider the exact sequence

$$0 \to \mathcal{F}_0 \to \mathcal{F} \to \mathcal{F}_1 \to 0.$$

Since $H^1(\mathcal{F}_0) = 0$ by dimension reason, $H^0(\mathcal{F}) \to H^0(\mathcal{F}_1)$ is surjective. Therefore, if $\mathcal{F}_1 \otimes \mathcal{O}_X(n)$ is generated by global sections at $x$, then so is $\mathcal{F} \otimes$

168          *2 The Minimal Model Program*

$\mathcal{O}_X(n)$. So we may assume in the beginning that $\mathcal{F}$ has no local section whose support is $x$.

Take a general global section $s$ of $\mathcal{O}_X(1)$ that vanishes at $x$. Take $X'$ to be the corresponding hyperplane passing through $x$. Take $\mathcal{O}_{X'}(1) = \mathcal{O}_X(1) \otimes \mathcal{O}_{X'}$ and $\mathcal{F}' = \mathcal{F} \otimes \mathcal{O}_{X'}(1)$. Since 0 is the only section of $\mathcal{F}$ that becomes 0 after multiplying $s$, we get an exact sequence

$$0 \to \mathcal{F} \to \mathcal{F} \otimes \mathcal{O}_X(1) \to \mathcal{F}' \to 0.$$

Hence

$$H^p(X', \mathcal{F}' \otimes \mathcal{O}_{X'}(m)) = 0$$

for any $m \geq 0$ and any $p > 0$. By the inductive hypothesis, $\mathcal{F}' \otimes \mathcal{O}_{X'}(n-1)$ is generated by global sections. Since $H^1(X, \mathcal{F} \otimes \mathcal{O}_X(n-1)) = 0$, $H^0(X, \mathcal{F} \otimes \mathcal{O}_X(n)) \to H^0(X, \mathcal{F}' \otimes \mathcal{O}_{X'}(n-1))$ is surjective, and hence $\mathcal{F} \otimes \mathcal{O}_X(n)$ is generated by global sections at $x$. $\qquad\square$

**Corollary 2.11.19** *Keep the assumptions in Theorem 2.11.15. Take a very ample divisor H on X and denote* $\dim X = n$. *Then*

$$I_Z(X, B + L/m) \otimes \mathcal{O}_X(D + nH)$$

*is generated by global sections.*

*Proof* This follows directly from Theorem 2.11.15 and Lemma 2.11.18. $\quad\square$

## 2.12 Extension Theorems

In this section, we prove extension theorems for pluri-log-canonical forms.

### 2.12.1 Extension Theorems I

There are many versions of extension theorems. The following form due to Hacon and M$^c$Kernan and Takayama is a key point in the proof of the existence of flips.

**Theorem 2.12.1** (*Extension theorem I* [35, 137]) *Let* $(X, B)$ *be a PLT pair where X is a smooth algebraic variety and B is a* **Q**-*divisor with normal crossing support. Let* $f : X \to S$ *be a projective morphism to an affine variety. Fix a positive integer* $m_0$ *such that* $D = m_0(K_X + B)$ *is an integral divisor. Assume that* $Y = \llcorner B \lrcorner$ *is irreducible. Assume the following conditions.*

## 2.12 Extension Theorems

*(1) There exists an ample **Q**-divisor A and an effective **Q**-divisor E such that $(X, Y + E)$ is PLT and*

$$B = A + E + Y.$$

*(2) There exists a positive integer $m_1$ such that the support of a general element $G \in |m_1 D|$ does not contain any LC center of $(X, \ulcorner B \urcorner)$, that is, it does not contain any irreducible component of intersections of irreducible components of B.*

*Then the natural homomorphism*

$$H^0(X, mD) \to H^0(Y, mD|_Y)$$

*is surjective for any positive integer m.*

**Remark 2.12.2** (1) The proof of the extension theorem discussed below is extremely technical, which is not just something that can be reached by calculating carefully.

(2) Trying to relax the assumptions of this theorem is an important question in applications.

*Proof* The proof follows from the following Propositions 2.12.3 and 2.12.7. $\square$

First, we use the usual multiplier ideal sheaves to reduce the problem to the extension problem for a sequence of slightly bigger divisors:

**Proposition 2.12.3** *Under condition (1) of Theorem 2.12.1, assume further that there exists an effective divisor F whose support does not contain Y such that for any sufficiently large positive integer l, the image of the natural homomorphism*

$$H^0(X, lD + F) \to H^0(Y, (lD + F)|_Y)$$

*contains the image of*

$$H^0(Y, lD|_Y) \to H^0(Y, (lD + F)|_Y).$$

*Then the restriction map*

$$H^0(X, D) \to H^0(Y, D|_Y)$$

*is surjective.*

*Proof* Take any $s \in H^0(Y, D|_Y)$ and take $D' = \mathrm{div}(s)$. By assumption, for a sufficiently large and sufficiently divisible positive integer $l$, there exists $G_l \in |lD + F|$ such that

$$G_l|_Y = lD' + F|_Y.$$

# 2 The Minimal Model Program

Here note that this is an equality of divisors, not just a linear equivalence. Take

$$B' = \frac{m_0 - 1}{lm_0} G_l + Y + E,$$

and consider the multiplier ideal sheaf $I = I(X, B')$. Note that

$$D - K_X - B'$$
$$= m_0(K_X + B) - K_X - B'$$
$$\sim_{\mathbb{Q}} (m_0 - 1)(K_X + B) + B - \frac{l(m_0 - 1)}{lm_0} D - \frac{m_0 - 1}{lm_0} F - Y - E$$
$$= A - \frac{m_0 - 1}{lm_0} F$$

is ample if $l$ is sufficiently large. Therefore, by the Nadel vanishing theorem,

$$H^1(X, I(X, B') \otimes \mathcal{O}_X(D)) = 0.$$

Take

$$C' = (B' - Y)|_Y = \frac{m_0 - 1}{m_0} D' + \frac{m_0 - 1}{lm_0} F|_Y + E|_Y,$$

then we have the following exact sequence

$$0 \to I(X, B') \to I_Y(X, B') \to I(Y, C') \to 0.$$

Hence the restriction map

$$H^0(X, I_Y(X, B') \otimes \mathcal{O}_X(D)) \to H^0(Y, I(Y, C') \otimes \mathcal{O}_Y(D|_Y))$$

is surjective. On the other hand, $(X, Y + E)$ is PLT, hence

$$\left(Y, \frac{m_0 - 1}{lm_0} F|_Y + E|_Y\right)$$

is KLT if $l$ is sufficiently large. Note that

$$C' - \frac{m_0 - 1}{lm_0} F|_Y - E|_Y \le D'.$$

Hence by Lemma 2.11.5,

$$s \in H^0(Y, I(Y, C') \otimes \mathcal{O}_Y(D|_Y)).$$

Therefore, $s$ can be extend to a global section of $H^0(X, D)$. $\qquad\square$

Let us forget the situation of the theorem for a moment and use the following notation in the following two lemmas. Let $X$ be a smooth algebraic variety, let $B$ be a normal crossing divisor, let $Y$ be an irreducible component of $B$, let $D$ be another divisor, and let $f : X \to S$ be a projective morphism to an

## 2.12 Extension Theorems

affine variety. Here all coefficients of $B$ are taken to be 1 (in the situation of the theorem $\ulcorner B \urcorner$ corresponds to $B$ here, this is the only difference). Take $C = (B - Y)|_Y$. Assume that there exists a positive integer $m_1$ such that the support of a general element $G \in |m_1 D|$ does not contain any LC center of $(X, B)$. Consider the following two sequences of linear systems on $Y$:

$$L_m^0 = |H^0(Y, mD|_Y)|,$$
$$L_m^1 = |\mathrm{Im}(H^0(X, mD) \to H^0(Y, mD|_Y))|.$$

Here $|\quad|$ denotes the corresponding linear system of the linear space. Then we can define the corresponding asymptotic multiplier ideal sheaves

$$J_C^0(Y, D|_Y) = I_C(Y, C + \{L_m^0/m\}),$$
$$J_C^1(Y, D|_Y) = I_C(Y, C + \{L_m^1/m\}).$$

As $L_m^1 \subset L_m^0$, $J_C^1(Y, D|_Y) \subset J_C^0(Y, D|_Y)$. In the case $C = 0$, we simply write $J^0(Y, D|_Y)$, $J^1(Y, D|_Y)$.

We compare the set of global sections and the set of extendable global sections as $m$ goes to infinity. The next two lemmas prove inclusion relations in two directions.

**Lemma 2.12.4**

$$H^0(Y, D|_Y) = H^0(Y, J_C^0(Y, D|_Y) \otimes \mathcal{O}_Y(D|_Y));$$
$$\mathrm{Im}(H^0(X, D) \to H^0(Y, D|_Y)) \subset H^0(Y, J_C^1(Y, D|_Y) \otimes \mathcal{O}_Y(D|_Y)).$$

*Proof* We will only show the second one. The proof of the first one is similar but easier. Take $m_1$ such that the support of a general element $G \in |m_1 D|$ does not contain any LC center of $(X, B)$. Take a log resolution $g: X' \to X$ of $(X, B + G)$, write $g^*(K_X + B) = K_{X'} + B'$, $Y' = g_*^{-1} Y$, $(K_{X'} + B')|_{Y'} = K_{Y'} + C'$, and $g^*G = P + N$. Here we may assume that $P$ is free and $N$ is the fixed part, and $(B')^+ = g_*^{-1} B$. Then

$$\mathrm{Im}(H^0(X, D) \to H^0(Y, D|_Y))$$
$$\subset H^0(Y', \mathcal{O}_{Y'}(g^*D|_Y + \llcorner -N/m_1|_{Y'} \lrcorner))$$
$$\subset H^0(Y', \mathcal{O}_{Y'}(g^*D|_Y + \ulcorner -C' - N/m_1|_{Y'} \urcorner + (C')^+))$$
$$\subset H^0(Y, J_C^1(Y, D|_Y) \otimes \mathcal{O}_Y(D|_Y)).$$

Here the first inclusion is by the fact that $\mathrm{Fix}(g^*D) \geq \ulcorner N/m_1 \urcorner$. $\qquad\square$

**Lemma 2.12.5** *Assume the following conditions:*

*(1) There exists an ample $\mathbf{Q}$-divisor $A'$ and an effective $\mathbf{Q}$-divisor $E'$ such that*
*$D = A' + E'$.*

172          *2 The Minimal Model Program*

(2) *There exists a positive integer $m'_1$ such that the support of a general element $G' \in |m'_1 E'|$ does not contain any LC center of $(X, B)$.*

*Then the following inclusion relation holds:*

$$H^0(Y, J^1_{\mathcal{C}}(Y, D|_Y) \otimes \mathcal{O}_Y(D|_Y + K_Y + C))$$
$$\subset \mathrm{Im}(H^0(X, D + K_X + B) \to H^0(Y, D|_Y + K_Y + C)).$$

*Proof* Take a sufficiently large and sufficiently divisible $m$ which realizes $J^1_{\mathcal{C}}(Y, D|_Y)$. For a general element $D_m \in |mD|$, take a log resolution $g: X' \to X$ of $(X, B + D_m + G')$ in strong sense, write $g^*(K_X + B) = K_{X'} + B'$, $Y' = g_*^{-1} Y$, $(K_{X'} + B')|_{Y'} = K_{Y'} + C'$, and $g^* D_m = P + N$. Here we may assume that $P$ is free and $N$ is the fixed part, and $(B')^+ = g_*^{-1} B$. Then $(B')^+$ has no common irreducible component with the exceptional divisors of $g$, $N$, and $g^* E'$.

Take an effective $\mathbf{Q}$-divisor $F$ supported on the exceptional divisors of $g$ such that $g^* A' - F$ is ample. Then for any sufficiently small positive number $\epsilon$,

$$g^* D - (1 - \epsilon)N/m - \epsilon(g^* E' + F) \sim_{\mathbf{Q}} (1 - \epsilon)P/m + \epsilon(g^* A' - F)$$

is ample. Moreover, since $A'$ is ample, $N/m \le g^* E'$ for any sufficiently divisible $m$. Therefore, for a sufficiently small $\epsilon$,

$$\ulcorner g^* D - (1 - \epsilon)N/m - \epsilon(g^* E' + F) \urcorner = g^* D - \llcorner N/m \lrcorner.$$

Then by the vanishing theorem,

$$H^1(X', K_{X'} + (B')^+ - Y' + g^* D - \llcorner N/m \lrcorner) = 0.$$

Hence

$$H^0(X', K_{X'} + (B')^+ + g^* D - \llcorner N/m \lrcorner)$$
$$\to H^0(Y', K_{Y'} + (C')^+ + g^* D|_{Y'} - \llcorner N/m|_{Y'} \lrcorner)$$

is surjective. On the other hand,

$$H^0(X', K_{X'} + (B')^+ + g^* D - \llcorner N/m \lrcorner) \subset H^0(X, D + K_X + B)$$

and

$$H^0(Y', K_{Y'} + (C')^+ + g^* D|_{Y'} - \llcorner N/m|_{Y'} \lrcorner)$$
$$= H^0(Y, J^1_{\mathcal{C}}(Y, D|_Y) \otimes \mathcal{O}_Y(K_Y + C + D|_Y)),$$

which proves the conclusion.                                               □

The following lemma is the core of the proof of the extension theorem:

## 2.12 Extension Theorems

**Lemma 2.12.6** *Let $(X, B)$ be a PLT pair where $X$ is a smooth algebraic variety of dimension $n$ and $B$ is an effective $\mathbf{Q}$-divisor with normal crossing support. Let $f : X \to S$ be a projective morphism to an affine variety. Fix a positive integer $m_0$ such that $D = m_0(K_X + B)$ has integral coefficients. Fix a very ample divisor $H$ on $X$ and take $M = nH$. Assume the following conditions:*

*(1) $H$ is sufficiently ample comparing to $B$ and $D$ (this condition will be clarified in the proof).*

*(2) There exists a positive integer $m_1$ such that the support of a general element $G \in |m_1 D|$ does not contain any LC center of the pair $(X, \ulcorner B \urcorner)$.*

*Then the following assertions hold:*

*(1) The inclusion relation*

$$J^0(Y, (mD + H)|_Y) \subset J^1_{\ulcorner C \urcorner}(Y, (mD + H + M)|_Y)$$

*holds for any nonnegative integer $m$.*

*(2) The inclusion relation*

$$H^0(Y, J^0(Y, (mD + H)|_Y) \otimes \mathcal{O}_Y((mD + H + M)|_Y))$$
$$\subset \mathrm{Im}(H^0(X, mD + H + M) \to H^0(Y, (mD + H + M)|_Y))$$

*holds for any nonnegative integer $m$.*

*Proof* (1) We will prove by induction on $m$. If $m = 0$, then both sides are $\mathcal{O}_Y$. Let us prove the conclusion for the case $m + 1$ assuming the case $m$.

Define an increasing sequence of integral divisors

$$Y \le B^{[1]} \le \cdots \le B^{[m_0]} = \ulcorner B \urcorner$$

by

$$\sum_{k=1}^{m_0} B^{[k]} = m_0 B.$$

Take $D_k = K_X + B^{[k]}$, $D_{\le k} = \sum_{s=1}^{k} D_s$, and $C^{[k]} = (B^{[k]} - Y)|_Y$. Also denote $D_{\le 0} = 0$ and $B^{[m_0+1]} = \ulcorner B \urcorner$. Note that $D = D_{\le m_0}$.

Here we clarify the assumption on $H$: for any $0 \le k \le m_0$,

(1) $D_{\le k} + H + M$ is free and
(2) $D_{\le k} + H - K_X - Y$ is ample.

Note that such condition does not depend on $m$.

174    2 *The Minimal Model Program*

We will prove the claim that

$$J^0(Y, (mD + H)|_Y) \subset J^1_{C^{[k+1]}}(Y, (mD + D_{\leq k} + H + M)|_Y)$$

by induction on $0 \leq k \leq m_0$. Note that the right-hand side is well defined by assumption (a) on $H$.

Once the claim is proved, take $k = m_0$, then

$$J^0(Y, ((m + 1)D + H)|_Y) \subset J^0(Y, (mD + H)|_Y)$$
$$\subset J^1_{C^{\neg}}(Y, ((m + 1)D + H + M)|_Y),$$

which proves the conclusion for the case $m + 1$ and finishes the proof of (1).

If $k = 0$, then by the inductive hypothesis,

$$J^0(Y, (mD+H)|_Y) \subset J^1_{C^{\neg}}(Y, (mD+H+M)|_Y) \subset J^1_{C^{[1]}}(Y, (mD+H+M)|_Y).$$

Assume that the claim holds up to $k - 1$, to proceed to the next step, we have the following three inclusion relations:

$$H^0(Y, J^0(Y, (mD + H)|_Y) \otimes \mathcal{O}_Y((mD + D_{\leq k} + H + M)|_Y))$$
$$\subset H^0(Y, J^1_{C^{[k]}}(Y, (mD + D_{\leq k-1} + H + M)|_Y) \otimes \mathcal{O}_Y((mD + D_{\leq k} + H + M)|_Y))$$
$$\subset \mathrm{Im}(H^0(X, mD + D_{\leq k} + H + M) \to H^0(Y, (mD + D_{\leq k} + H + M)|_Y))$$
$$\subset H^0(Y, J^1_{C^{[k+1]}}(Y, (mD + D_{\leq k} + H + M)|_Y) \otimes \mathcal{O}_Y((mD + D_{\leq k} + H + M)|_Y)).$$

Here the first inclusion relation is by the inductive hypothesis, the second is by Lemma 2.12.5, and the third is by Lemma 2.12.4. Note that

$$mD+D_{\leq k}+H+M-(K_X+Y+mD+H)-(n-1)H = D_{\leq k}+H-K_X-Y$$

is ample by assumption (b) on $H$. Hence by Corollary 2.11.19,

$$J^0(Y, (mD + H)|_Y) \otimes \mathcal{O}_Y((mD + D_{\leq k} + H + M)|_Y)$$

is generated by global sections. To summarize, we showed that

$$J^0(Y, (mD + H)|_Y) \subset J^1_{C^{[k+1]}}(Y, (mD + D_{\leq k} + H + M)|_Y).$$

(2) When $m = 0$ this is clear. For $m > 0$, using the first two inclusions above for $m - 1$ and $k = m_0$, we have

$$H^0(Y, J^0(Y, ((m - 1)D + H)|_Y) \otimes \mathcal{O}_Y((mD + H + M)|_Y))$$
$$\subset \mathrm{Im}(H^0(X, mD + H + M) \to H^0(Y, (mD + H + M)|_Y)).$$

On the other hand,

$$H^0(Y, J^0(Y, (mD + H)|_Y) \otimes \mathcal{O}_Y((mD + H + M)|_Y))$$
$$\subset H^0(Y, J^0(Y, ((m - 1)D + H)|_Y) \otimes \mathcal{O}_Y((mD + H + M)|_Y)),$$

so we conclude the proof. $\qquad\square$

## 2.12 Extension Theorems

**Proposition 2.12.7** *Under condition (2) of Theorem 2.12.1, there exists a very ample divisor F, such that for any sufficiently large positive integer m, the image of the restriction map*

$$H^0(X, mD + F) \to H^0(Y, (mD + F)|_Y)$$

*contains the image of $H^0(Y, mD|_Y)$.*

*Proof* By Lemma 2.12.6,

$$\begin{aligned}
H^0(Y, mD|_Y) &\subset H^0(Y, (mD + H)|_Y) \\
&= H^0(J^0(Y, (mD + H)|_Y) \otimes \mathcal{O}_Y((mD + H)|_Y)) \\
&\subset H^0(J^0(Y, (mD + H)|_Y) \otimes \mathcal{O}_Y((mD + H + M)|_Y)) \\
&\subset \operatorname{Im}(H^0(X, mD + H + M) \to H^0(Y, (mD + H + M)|_Y)).
\end{aligned}$$

So we may just take $F = H + M$. $\square$

### 2.12.2 Extension Theorems II

There are various versions of the extension theorem. The following theorem is close to the original form of the extension theorem. This theorem has many important corollaries such as the deformation invariance of plurigenera and the preservation of canonical singularities under deformations, but will not be used in subsequent sections.

**Theorem 2.12.8** *(Extension theorem II) Let $(X, B)$ be a PLT pair where X is a smooth algebraic variety and B is a $\mathbf{Q}$-divisor with normal crossing support. Let $f: X \to S$ be a projective morphism to an affine variety. Fix a positive integer $m_0$ such that $D = m_0(K_X + B)$ is an integral divisor. Assume that $Y = \lfloor B \rfloor$ is irreducible. Assume the following conditions.*

*(1) There exists an ample $\mathbf{Q}$-divisor A and an effective $\mathbf{Q}$-divisor E whose support does not contain Y such that*

$$K_X + B = A + E.$$

*(2) There exists a positive integer $m_1$ such that the support of a general element $G \in |m_1 D|$ does not contain any LC center of $(X, \ulcorner B \urcorner)$, that is, it does not contain any irreducible component of intersections of irreducible components of B.*

*(3) Either $m_0 \geq 2$ or $f(Y) \neq f(X)$.*

*Then the natural homomorphism*

$$H^0(X, mD) \to H^0(Y, mD|_Y)$$

*is surjective for any positive integer m.*

176              2 *The Minimal Model Program*

**Remark 2.12.9** (1) If taking $B = Y$ in Theorem 2.12.8, then it is a theorem in [71]. Theorem 2.12.1 is a generalization of this theorem.

(2) For a sufficiently large and sufficiently divisible positive integer $m$ and a general element $G \in |m(K_X + B)|$, replacing $B$ by $B' = B + \epsilon G$ for a sufficiently small $\epsilon$ and taking a log resolution, we are in a similar situation as Theorem 2.12.1. But the pair $(X, B')$ may not satisfy the conditions of Theorem 2.12.1. In fact $\ulcorner B' \urcorner$ has more irreducible components than $\ulcorner B \urcorner$. So Theorem 2.12.8 is not a corollary of Theorem 2.12.1.

*Proof* The proof is basically the same as that of Theorem 2.12.1 and we omit the details. We just slightly modify Proposition 2.12.3.

If $m_0 \geq 2$, we modify Proposition 2.12.3 by taking

$$B' = \frac{m_0 - 1 - \epsilon}{l m_0} G_l + B + \epsilon E$$

for some sufficiently small positive rational number $\epsilon$.

If $m_0 = 1$ and $f(Y) \neq f(X)$, then the statement says that

$$H^0(X, K_X + Y) \to H^0(Y, K_Y)$$

is surjective, which is a consequence of a Kollár-type vanishing theorem saying that

$$H^1(X, K_X) \to H^1(X, K_X + Y)$$

is injective (see [71, Theorem 2.8]). Here the condition $f(Y) \neq f(X)$ is used to guarantee that $Y$ is contained in the pullback of a Cartier divisor on $S$. $\square$

An important corollary of Theorem 2.12.8 is the following theorem on deformation invariance of plurigenera:

**Corollary 2.12.10** (Siu [133]) *Let $f : X \to S$ be a smooth projective morphism between smooth algebraic varieties. Assume that the fiber $X_\eta = f^{-1}(\eta)$ over the generic point $\eta \in S$ is of general type. Then for any positive integer $m$, the plurigenus $\dim H^0(X_s, m K_{X_s})$ of a fiber $X_s = f^{-1}(s)$ is independent of the choice of $s \in S$.*

*Proof* We may assume that $S$ is a smooth affine curve. Fix any point $s \in S$. Fix an effective ample divisor $A$ on $X$ and take $A_\eta = A|_{X_\eta}$. Since $X_\eta$ is of general type, there exists a sufficiently large positive integer $m_1$ such that we may write $m_1 K_{X_\eta} \sim A_\eta + E_\eta$ for some effective divisor $E_\eta$. Taking the closure, we may write $m_1 K_X \sim A + E$ for some effective divisor $E$ which does not contain $X_s$.

## 2.12 Extension Theorems

177

Then $B = Y = X_s$ satisfies conditions in Theorem 2.12.8. By Theorem 2.12.8, for any positive integer $m$,

$$H^0(X, m(K_X + X_s)) \to H^0(X_s, mK_{X_s})$$

is surjective. Then the conclusion follows from the upper semicontinuity. □

The following corollary, stating that flat deformations of canonical singularities are again canonical singularities, is important in the study of moduli spaces of algebraic varieties:

**Corollary 2.12.11** ([70]) *Let $f : X \to S$ be a flat morphism from an algebraic variety $X$ to a smooth affine curve $S$. Fix $x \in X$ and $s = f(x) \in S$. Assume that the fiber $X_s = f^{-1}(s)$ over $s$ has at worst canonical singularities at $x$. Then the ambient space $X$ has at worst canonical singularities at $x$. In particular, there exists a neighborhood $U \subset X$ of $x$ such that for any $s' \in S$, $X_{s'} \cap U$ has at worst canonical singularities.*

*Proof* Replacing $X$ by a sufficiently small affine neighborhood of $x$, we may assume that $X_s$ has at worst canonical singularities. Take a log resolution $g : X' \to X$ of the pair $(X, X_s)$, denote by $B = Y$ the strict transform of $X_s$. Since $X_s$ is normal, we may assume that $X$ is also normal if $X$ is sufficiently small. Since $X_s$ has at worst canonical singularities, there exists a positive integer $m$ such that $mK_{X_s}$ is Cartier and the natural map

$$H^0(Y, mK_Y) \to H^0(X_s, mK_{X_s})$$

is isomorphic. Applying Theorem 2.12.1 or 2.12.8 to the birational morphism $g$, we have that

$$H^0(X', m(K_{X'} + Y)) \to H^0(Y, mK_Y)$$

is surjective. Therefore, the composition

$$H^0(X', m(K_{X'} + Y)) \to H^0(X, m(K_X + X_s)) \to H^0(X_s, mK_{X_s})$$

is surjective. So if $X$ is sufficiently small, a nowhere vanishing section of $mK_{X_s}$ extends to a nowhere vanishing section of $m(K_X + X_s)$ and a global section of $m(K_{X'} + Y)$. This implies that $m(K_X + X_s)$ is also Cartier and $m(K_{X'} + Y) \geq g^*(m(K_X + X_s))$. Since $Y \leq g^* X_s$, $X$ has at worst canonical singularities.

The latter half is clear from the former half. □

**Remark 2.12.12** The technique in the proof of the extension theorem was originally developed by Siu in the proof of the deformation invariance of plurigenera ([133]). Later [70] proved the deformation preservation of canonical singularities by an algebraic interpretation of Siu's argument (see also [71, 116]). Here instead of considering limits of metrics in complex analysis, asymptotic multiplier ideal sheaves are introduced. By the Noetherian property, an asymptotic multiplier ideal sheaf is actually obtained at a finite stage without knowing that at which stage it will be obtained, so it is helpful for proving certain finiteness theorem. However, this method also has its limitation as it cannot reflect infinite limits as analytic multiplier ideal sheaves. The extension theorem introduced in this section was proved by the log version of this method ([35, 137]). After this, Siu proved the deformation invariance of plurigenera without assuming bigness of canonical divisors ([134]). An important open problem is to find the algebraic interpretation of the proof of this result. It seems that an algebraic interpretation of infinite limits described above is necessary.

# 3

# The Finite Generation Theorem

In this chapter, we prove the finite generation of canonical rings. First, for algebraic varieties of general type, we prove the existence of minimal models by induction on dimensions, then we use the semipositivity theorem for algebraic fiber spaces to reduce the problem to algebraic varieties of general type.

## 3.1 Setting of the Inductive Proof

In Birkar et al. (BCHM for short) ([16]), it turns out that for MMP (minimal model program) with scaling, induction on dimensions goes well under the assumption that the boundary contains an ample divisor. To be more accurate, we should put the following conditions on the pair $(X, B)$ and the morphism $f$. We will simply call it the *BCHM condition* in this book.

(1) $X$ is an $n$-dimensional normal $\mathbf{Q}$-factorial algebraic variety, $B$ is an effective $\mathbf{R}$-divisor on $X$, and $f : X \to S$ is a projective morphism to a quasi-projective variety.
(2) $(X, B)$ is DLT (divisorially log terminal).
(3) There exists a relatively ample $\mathbf{R}$-divisor $A$ over $S$ and an effective $\mathbf{R}$-divisor $E$ such that $B = A + E + \llcorner B \lrcorner$ and $(X, E + \llcorner B \lrcorner)$ is DLT with $\llcorner E \lrcorner = 0$.

For a projective morphism $f : (X, B) \to S$ from a $\mathbf{Q}$-factorial DLT pair, we will show the following theorems:

- (Existence of flips) For any small contraction of $K_X + B$, the flip exists.
- (Existence of PL [pre-limiting] flips) For any small contraction $X \to S$ of $K_X + B$, if there exists an irreducible component $P$ of $\llcorner B \lrcorner$ such that $-P$ is relatively ample over $S$, then the flip exists.

179

- (Existence of minimal models) If $f\colon (X, B) \to S$ satisfies the BCHM condition and $K_X + B$ is relatively pseudo-effective, then there exists a minimal model of $f\colon (X, B) \to S$.
- (Finiteness of minimal models) Suppose that $P$ is a polytope spanned by effective $\mathbf{R}$-divisors such that for any $B' \in P$, $f\colon (X, B') \to S$ satisfies the BCHM condition. Then there exist finitely many rational maps $g_k\colon X \dashrightarrow Y_k$ such that for any $B' \in P$ with $K_X + B'$ relatively pseudo-effective, there exists a minimal model of $f\colon (X, B') \to S$, and any minimal model of $f\colon (X, B') \to S$ coincides with one of $g_k$.
- (Termination of MMP with scaling) Suppose that $f\colon (X, B) \to S$ and $f\colon (X, B') \to S$ satisfy the BCHM condition. Moreover, assume that $K_X + B'$ is relatively nef. Then any MMP on $f\colon (X, B) \to S$ with scaling of $B' - B$ terminates after finitely many steps.
- (Special termination of MMP with scaling) Suppose that $f\colon (X, B) \to S$ and $f\colon (X, B') \to S$ satisfy the BCHM condition. Moreover, assume that $\llcorner B' \lrcorner \geq \llcorner B \lrcorner$ and $K_X + B'$ is relatively nef. Then any MMP on $f\colon (X, B) \to S$ with scaling of $B' - B$ is isomorphic in a neighborhood of $\llcorner B \lrcorner$ after finitely many steps.
- (Nonvanishing theorem) If $f\colon (X, B) \to S$ satisfies the BCHM condition and $K_X + B$ is relatively pseudo-effective, then there exists an effective $\mathbf{R}$-divisor $D$ such that $D \equiv_S K_X + B$.

**Remark 3.1.1** (1) The existence of PL flips and the special termination of MMP with scaling are special cases of the existence of flips and the termination of MMP with scaling.

(2) The existence of flips is a special case of the existence of minimal models. In fact, a flip is the relative canonical model of a relative minimal model.

(3) The statement of the finiteness of minimal models includes the existence of minimal models.

**Remark 3.1.2** The finite generation theorem, which is the main purpose of this book, is obtained by showing the existence of minimal models for KLT (Kawamata log terminal) pairs with $K_X + B$ big. However, the bigness of $K_X + B$ is not preserved if restricted on the boundary. On the other hand, the BCHM condition, considering DLT pairs with boundaries containing ample divisors, is preserved if restricted on the boundary. If $(X, B)$ is KLT and $K_X + B$ is big, then after replacing $B$, we may assume that $B$ contains an ample divisor. This condition is preserved in the process of MMP, so induction on dimensions works well in this situation (see Remark 3.1.4).

## 3.1 Setting of the Inductive Proof

First, we modify the BCHM condition for KLT pairs, and show that the DLT version and the KLT version can be used appropriately according to the situations. The KLT version BCHM condition is the following:

(1) $X$ is an $n$-dimensional normal $\mathbf{Q}$-factorial algebraic variety, $B$ is an effective $\mathbf{R}$-divisor on $X$, and $f : X \to S$ is a projective morphism to a quasi-projective variety.
(2) $(X, B)$ is KLT.
(3) $B$ is relatively big over $S$.

**Lemma 3.1.3** *For each statement of the existence of flips, the existence of minimal models, the finiteness of minimal models, the termination of MMP with scaling, and the nonvanishing theorem, the DLT version holds if and only if the KLT version holds.*

*Proof* Let us explain how to replace the DLT and KLT versions by each other.

Let $f : (X, B) \to S$ be a morphism satisfying the DLT version BCHM condition. Then there exists a sufficiently small positive real number $t$ such that $A + t \llcorner B \lrcorner$ is relatively ample. Take a general effective relatively ample $\mathbf{R}$-divisor $A_1 \equiv A + t \llcorner B \lrcorner$, denote $B' = A_1 + E + (1 - t) \llcorner B \lrcorner$, then $(X, B')$ is KLT. So $B \equiv B'$ and it satisfies the KLT version BCHM condition.

Conversely, let $f : (X, B) \to S$ be a morphism satisfying the KLT version BCHM condition. As $B$ is relatively big, there exists an effective relatively ample $\mathbf{R}$-divisor $A$ and an effective $\mathbf{R}$-divisor $E$ such that $B \equiv A + E$. For a sufficiently small positive real number $t$, denote $B' = (1 - t)B + tA + tE$, then $(X, (1 - t)B + tA + tE)$ is KLT. So $B \equiv B'$ and it satisfies the DLT version BCHM condition.

Here we should be a little careful about the existence of minimal models. The problem is the following: If $(X, B)$ is DLT and $(X, B')$ is KLT with $B \equiv_S B'$ and $\alpha : X \dashrightarrow X'$ is a minimal model of $(X, B') \to S$, then it is not automatically a minimal model of $(X, B) \to S$ because $(X', \alpha_* B)$ might be only LC (log canonical) but not necessarily DLT. The solution is the following: Once we show the existence of minimal models of pairs with the KLT version BCHM condition, we can show the termination of MMP with scaling, and then we may assume that $\alpha : X \dashrightarrow X'$ is obtained by a $(K_X + B')$-MMP which is also a $(K_X + B)$-MMP. Then in this case $(X', \alpha_* B)$ is DLT and hence is a minimal model of $(X, B) \to S$. $\square$

**Remark 3.1.4** Another advantage of the BCHM condition is that it is preserved by MMP. For the KLT version, this is simply because the KLT condition and the bigness of $B$ are both preserved by MMP.

182          *3 The Finite Generation Theorem*

For the DLT version, consider a morphism $(X, B) \rightarrow S$ satisfying the BCHM condition, suppose that $\alpha \colon X \dashrightarrow X'$ is obtained by several steps of MMP and $(X', B')$ is the induced pair. Write $B = A + E + \llcorner B \lrcorner$ as in the BCHM condition. Then $B' = \alpha_* A + \alpha_* E + \llcorner \alpha_* B \lrcorner$ but $\alpha_* A$ is no longer relatively ample. Pick a general relatively ample effective $\mathbf{R}$-divisor $A'$ on $X'$ and take a sufficiently small positive real number $\epsilon$ such that $A - \epsilon \alpha_*^{-1} A'$ is relatively ample. Take a general effective $\mathbf{R}$-divisor $E_1 \sim_{\mathbf{R}} A - \epsilon \alpha_*^{-1} A'$ such that $(X, E_1 + E + \llcorner B \lrcorner)$ is DLT. By taking $\epsilon$ sufficiently small, we may assume that $\alpha \colon X \dashrightarrow X'$ is also obtained by a $(K_X + E_1 + E + \llcorner B \lrcorner)$-MMP, which implies that $(X', \alpha_* E_1 + \alpha_* E + \llcorner \alpha_* B \lrcorner)$ is DLT. Note that

$$B' - (\alpha_* E_1 + \alpha_* E + \llcorner \alpha_* B \lrcorner) = \alpha_* A - \alpha_* E_1 \equiv_S \epsilon A'$$

is relatively ample. Moreover, $\llcorner \alpha_* E_1 + \alpha_* E \lrcorner = 0$ by the construction of $E_1$. So $(X', B') \rightarrow S$ satisfies the BCHM condition.

**Remark 3.1.5** For the existence of PL flips and the special termination of MMP with scaling, the KLT version makes no sense. It is natural to run MMP within the KLT category, but the point in extending to the DLT category is that, for DLT pairs, we can consider the restriction on the integral part $\llcorner B \lrcorner$, which forges the path towards inductive arguments on dimensions.

From now on, our goal is to prove the following assertions under the DLT version BCHM condition. Combining all these assertions, all theorems are proved by induction on dimensions.

(1) (Theorem 3.3.1) The existence of flips in dimension $n - 1$ and the termination of MMP with scaling in dimension $n - 1$ imply the special termination of MMP with scaling in dimension $n$.
(2) (Theorem 3.4.1) The existence of PL flips in dimension $n$, the special termination of MMP with scaling in dimension $n$, and the nonvanishing theorem for a morphism $(X, B) \rightarrow S$ satisfying the BCHM condition in dimension $n$ imply the existence of minimal models for the morphism $(X, B) \rightarrow S$.
(3) (Theorem 3.4.1) The existence of PL flips in dimension $n$ and the special termination of MMP with scaling in dimension $n$ imply the existence of flips in dimension $n$.
(4) (Theorem 3.2.1) The existence and finiteness of minimal models in dimension $n - 1$ imply the existence of PL flips in dimension $n$.
(5) (Theorem 3.4.6) The existence of minimal models in dimension $n$ implies the finiteness of minimal models in dimension $n$.

## 3.2 PL Flips

(6) (Corollary 2.10.9) The finiteness of minimal models in dimension $n$ implies the termination of MMP with scaling in dimension $n$.

(7) (Theorem 3.5.1) The existence of PL flips in dimension $n$, the special termination of MMP with scaling in dimension $n$, and the existence and finiteness of minimal models for pairs $(X, B)$ with $K_X + B$ relatively big in dimension $n$ imply the nonvanishing theorem in dimension $n$.

**Remark 3.1.6** If $K_X + B$ is relatively big, then the nonvanishing theorem automatically holds. Therefore, (3) is a special case of (2). (3) is originally proved by Shokurov, which is the origin of the induction method on dimensions.

In (4), Hacon and M$^c$Kernan showed the existence of flips in all dimensions by induction on dimensions. Before that, the only proof of the existence of flips used the classification of singularities, which is only available in dimension 3.

The proof of (4) needs the extension theorem of pluricanonical forms. The basepoint-free theorem is also an extension theorem for pluricanonical forms, but here we need to use a much more powerful extension theorem.

## 3.2 PL Flips

In this section, we introduce the proof of the existence of PL flips due to Hacon and M$^c$Kernan ([38]). Recall that a *PL contraction* $f : (X, B) \to S$ is a small contraction from a **Q**-factorial DLT pair such that $-P$ is $f$-ample for some irreducible component $P$ of $\llcorner B \lrcorner$.

**Theorem 3.2.1** (*Existence of PL flips*) *Let* $f : (X, B) \to S$ *be an $n$-dimensional PL contraction. Suppose that, under the BCHM condition, the existence and finiteness of minimal models in dimension $n - 1$ hold, then the flip of $f$ exists.*

This theorem is a pillar of new developments in the minimal model theory, it triggered the recent major redevelopment of the minimal model theory.

The existence of flips is a special case of the finite generation of canonical rings, but we need to prove it first, then we are able to launch the MMP, and then the finite generation theorem can be proved.

Let us describe the sketch of the proof. A PL flip is the flip in the special situation that $f : (X, B) \to S$ is a PL contraction. We may assume that $S$ is affine, $(X, B)$ is PLT (purely log terminal), and $Y = \llcorner B \lrcorner$ is irreducible. We may assume that $B$ is a **Q**-divisor after perturbing the coefficients.

Since $f$ is a small birational morphism, the restriction $f|_Y$ to the divisor is also a birational morphism and the BCHM condition holds on $Y$.

184                    3 The Finite Generation Theorem

In order to show the existence of the flip, it suffices to show that $R(X/S, K_X + B)$ is finitely generated. The subadjunction formula $(K_X + B)|_Y = K_Y + B_Y$ defines $B_Y$ and $(Y, B_Y)$ is KLT. Since dim $Y = n - 1$, $R(Y/S, K_Y + B_Y)$ is finitely generated by the inductive hypothesis.

If the natural ring homomorphism $R(X/S, K_X + B) \to R(Y/S, K_Y + B_Y)$ is surjective, then we can finish the proof. However, as $K_X + B$ is negative with respect to $f$, we cannot establish the vanishing of higher cohomologies, so this is in general not surjective. In other words, pluricanonical forms on $Y$ are not necessarily extendable to $X$. For this reason, a key point in the proof is to determine the set of pluricanonical forms on $Y$ that are extendable to $X$. In order to do this, we will make full use of $(n - 1)$-dimensional MMP and the vanishing theorem for multiplier ideal sheaves.

Fix a positive integer $m_0$ such that $m_0 B$ is an integral divisor. Denote $(K_X + B)|_Y = K_Y + B_Y$. For a positive integer $m$, the restriction map

$$H^0(X, mm_0(K_X + B)) \to H^0(Y, mm_0(K_Y + B_Y))$$

is not surjective in general.

Applying the extension theorem, we can determine the image of this map and express it as the space of pluricanonical forms $H^0(Y', m(K_{Y'} + B_{Y',m}))$ of another pair $(Y', B_{Y',m})$ different from $(Y, B_Y)$. Here the point is that the new variety $Y'$ can be chosen independent of $m$. For this purpose, we use the extension theorem for pluricanonical forms.

By blowing up $X$, we can resolve the base locus of the pluricanonical linear system $|mm_0(K_X + B)|$. However, this cannot be done simultaneously for all $m$, so as $m$ increases, we get an infinite tower of blowups over $X$, that is, an inverse system of algebraic varieties. This inverse system is equivalent to considering Shokurov's *b-divisor* ([130]) in the divisor level, which is hard to handle. This concept is similar to *Zariski's Riemann space*.

However, by applying the extension theorem, *b*-divisors are not needed, instead it suffices to consider an infinite sequence of $\mathbf{Q}$-divisors $B_{Y',m}$ on a fixed variety $Y'$. The blowups $X'_m \to X$ depend on $m$, but the restriction $Y' \to Y$ to $Y$ is independent of $m$. So we may consider an infinite sequence of boundaries $B_{Y',m}$ instead of the infinite tower of algebraic varieties. In general the limit $B_{Y'}$ of $B_{Y',m}$ is an $\mathbf{R}$-divisor. This is one reason that we must formulate the MMP for $\mathbf{R}$-divisors.

The minimal model theory in dimension up to $n - 1$ can be applied to $Y'$ and divisors on it, then by the existence and finiteness of minimal models, this limit can be obtained within finite steps, and there exists a positive integer $m$ such that $B_{Y'} = B_{Y',m}$.

## 3.2 PL Flips

### 3.2.1 Restriction of Canonical Rings to Divisors

First, we show the following lemma.

**Lemma 3.2.2** *Let* $R = \bigoplus_{m=0}^{\infty} R_m$ *be a sheaf of graded $\mathcal{O}_S$-algebras such that* $R_0 = \mathcal{O}_S$.

*(1) $R$ is a finitely generated graded $\mathcal{O}_S$-algebra if and only if the ideal $R_+ = \bigoplus_{m>0} R_m$ is a finitely generated $R$-module.*

*(2) If $R$ is a finitely generated graded $\mathcal{O}_S$-algebra, then the subalgebra*

$$R^{(m_1)} = \bigoplus_{m=0}^{\infty} R_{mm_1}$$

*is a finitely generated graded $\mathcal{O}_S$-algebra. Here $m_1$ is any fixed positive integer. Moreover, the converse holds if $R$ is a domain.*

*Proof* (1) The homogeneous generators of the graded $\mathcal{O}_S$-algebra $R$ are the same as the homogeneous generators of the ideal $R_+$.

(2) Suppose that $R$ is a finitely generated algebra, and take $x_1, \ldots, x_t$ to be homogeneous generators. Then $R$ is generated by $\prod_{i=1}^{t} x_i^{d_i}$ ($0 \le d_i < m_1$) as an $R^{(m_1)}$-module. Therefore, $R$ is a finitely generated $R^{(m_1)}$-module. Since $R_+$ is a direct sum of $R^{(m_1)}$-modules $M_j = \bigoplus_{m=0}^{\infty} R_{j+mm_1}$ ($1 \le j \le m_1$) and is a finitely generated $R$-module, it is also a finitely generated $R^{(m_1)}$-module. In particular, $R_+^{(m_1)} = M_{m_1}$ is a finitely generated $R^{(m_1)}$-module, and hence $R^{(m_1)}$ is a finitely generated graded $\mathcal{O}_S$-algebra.

Conversely, suppose that $R^{(m_1)}$ is a finitely generated algebra and $R$ is a domain. Then $R_+^{(m_1)}$ is a finitely generated $R^{(m_1)}$-module where $R^{(m_1)}$ is a Noetherian ring. If $M_j \ne 0$ for some $j$, take $0 \ne x \in M_j$, then the multiplication map $x^{m_1-1} \colon M_j \to R_+^{(m_1)}$ is injective. This implies that $M_j$ is a finitely generated $R^{(m_1)}$-module. Therefore, $R_+$ is a finitely generated $R^{(m_1)}$-module, which is also a finitely generated $R$-module. $\qquad \square$

For a PL contraction $f \colon (X, B) \to S$, the existence of the flip is equivalent to that the sheaf of rings

$$R = \bigoplus_{m=0}^{\infty} f_* \mathcal{O}_X(\llcorner m(K_X + B) \lrcorner)$$

is finitely generated over $\mathcal{O}_S$. Fix a positive integer $m_0$ such that $m_0(K_X + B)$ is an integral divisor, the latter is equivalent to the finite generation of

$$R^{(m_0)} = \bigoplus_{m=0}^{\infty} f_* \mathcal{O}_X(mm_0(K_X + B)).$$

186            *3 The Finite Generation Theorem*

Since $-Y$ is $f$-ample, it is proportional to $-(K_X + B)$, hence this is equivalent to the finite generation of

$$R' = \bigoplus_{m=0}^{\infty} f_* \mathcal{O}_X(mY).$$

**Lemma 3.2.3** *For any positive integer $m$, the reflexive sheaves $\mathcal{O}_X(mY)$ and $\mathcal{O}_Y(mY)$ on $X$ and $Y$ are well defined, and satisfy the exact sequence*

$$0 \to \mathcal{O}_X((m-1)Y) \to \mathcal{O}_X(mY) \to \mathcal{O}_Y(mY) \to 0.$$

*Proof* Note that $Y$ is just $\mathbf{Q}$-Cartier, so these sheaves are not invertible in general. For any point $x \in X$, take a sufficiently small neighborhood $X_x$ and take the index 1 cover $\pi_x \colon \tilde{X}_x \to X_x$ of $Y$. Then $\tilde{Y}_x = \pi_x^{-1}(Y)$ is Cartier and we can define invertible sheaves $\mathcal{O}_{\tilde{X}_x}(m\tilde{Y}_x)$ and $\mathcal{O}_{\tilde{Y}_x}(m\tilde{Y}_x)$ satisfying the exact sequence

$$0 \to \mathcal{O}_{\tilde{X}_x}((m-1)\tilde{Y}_x) \to \mathcal{O}_{\tilde{X}_x}(m\tilde{Y}_x) \to \mathcal{O}_{\tilde{Y}_x}(m\tilde{Y}_x) \to 0.$$

Here the first homomorphism is the multiplication map of the canonical section $\pi_x^* s \in \Gamma(\tilde{X}_x, \mathcal{O}_{\tilde{X}_x}(\tilde{Y}_x))$. Take the invariant parts with respect to the Galois group $\mathrm{Gal}(\tilde{X}_x / X_x)$, we get the required exact sequence. $\qquad\square$

Since $-Y$ is $f$-ample, in the exact sequence

$$0 \to f_* \mathcal{O}_X((m-1)Y) \to f_* \mathcal{O}_X(mY) \to f_* \mathcal{O}_Y(mY),$$

the last homomorphism is not surjective in general.

**Lemma 3.2.4** *If the restricted algebra*

$$R'_Y = \bigoplus_{m=0}^{\infty} \mathrm{Im}(f_* \mathcal{O}_X(mY) \to f_* \mathcal{O}_Y(mY))$$

*is finitely generated, then $R'$ is finitely generated.*

*Proof* Take $s_1, \ldots, s_k \in R'$ to be the generators of $R'_Y$ on $X$, then $R'$ is generated by $s, s_1, \ldots, s_k$. $\qquad\square$

Applying Lemma 3.2.2 again, we can reduce the problem to the finite generation of

$$R_Y = \bigoplus_{m=0}^{\infty} \mathrm{Im}(f_* \mathcal{O}_X(mm_0(K_X + B)) \to f_* \mathcal{O}_Y(mm_0(K_Y + B_Y))).$$

Here $K_Y + B_Y = (K_X + B)|_Y$. The restriction map

$$H^0(X, mm_0(K_X + B)) \to H^0(Y, mm_0(K_Y + B_Y))$$

## 3.2 PL Flips

187

is not surjective. The idea is to replace this image by the space of pluricanonical forms with respect to a new boundary smaller than $B_Y$.

First, we construct a tower of log resolutions:

**Proposition 3.2.5** ([38]) *Let $f : (X, B) \to S$ be a small contraction from a* **Q**-*factorial PLT pair. Assume that $S$ is affine, $Y = \llcorner B \lrcorner$ is irreducible, $-Y$ is $f$-ample, and $B$ is a* **Q**-*divisor. Take a positive integer $m_0$ such that the coefficients of $m_0(K_X + B)$ are integers. Then for any positive integer $m$, there exists a log resolution $\mu_m : X_m \to X$ and a* **Q**-*divisor $B'_m$ on $X_m$ satisfying the following conditions:*

*(1) Write $\mu_m^*(K_X + B) = K_{X_m} + B_m$ and $Y_m = \mu_{m*}^{-1}Y$, then the irreducible components of $B_m^+ - Y_m$ are disjoint. Here $B_m^+$ is the positive part of $B_m$, that is, we can write $B_m = B_m^+ - B_m^-$, where $B_m^+, B_m^-$ are effective divisors without common irreducible component.*

*(2) $m m_0 B'_m$ is an integral divisor such that the inequality $Y_m \le B'_m \le B_m^+$ hold.*

*(3) A general element in $|mm_0(K_{X_m} + B'_m)|$ does not contain any LC center of $(X_m, \ulcorner B'_m \urcorner)$.*

*(4) The natural map*

$$H^0(X_m, mm_0(K_{X_m} + B'_m)) \to H^0(X_m, mm_0(K_{X_m} + B_m^+))$$

$$\cong H^0(X, mm_0(K_X + B))$$

*is bijective.*

*(5) $Y_m$ is isomorphic to a fixed variety $Y'$ and $\mu_m$ induces a fixed morphism $\mu_Y : Y' \to Y$.*

*(6) Write $(K_{X_m} + B'_m)|_{Y_m} = K_{Y'} + B_{Y',m}$, then $B_{Y',m}$ satisfy the following convexity on $m$:*

$$m_1 B_{Y',m_1} + m_2 B_{Y',m_2} \le (m_1 + m_2) B_{Y',m_1+m_2}.$$

*(7) The limit $B_{Y'} = \lim_{m\to\infty} B_{Y',m}$ exists as an* **R**-*divisor, and $(Y', B_{Y'})$ is KLT.*

*Proof* Take a log resolution $\mu : (X', B') \to (X, B)$, write $\mu^*(K_X + B) = K_{X'} + B'$ and $Y' = \mu_*^{-1}Y$, we may assume that the irreducible components of $(B')^+ - Y'$ are disjoint from each other by the argument of Proposition 1.10.7. We will construct $\mu_m$ by blowing up $X'$. To make the notation simpler, we use the same notation as $X'$ to denote the variety after blowing up.

Fix $m$ and take a general element $D \in |mm_0(K_X + B)|$. As $f|_Y$ is birational, $D$ does not contain $Y$. $D$ induces an element $D' \in |mm_0(K_{X'} + (B')^+)|$. If $D'$ and $(B')^+$ have a common irreducible component, we replace them by

188                                   3 The Finite Generation Theorem

$D' - \min\{D', mm_0(B')^+\}$ and $(B')^+ - \min\{D'/mm_0, (B')^+\}$. Here note that $\min\{D', mm_0(B')^+\}$ is contained in the fixed part of the linear system, hence the relation $D' \in |mm_0(K_{X'} + (B')^+)|$ is preserved. The new $D'$ and $(B')^+$ have no common irreducible component.

Next we show that, after replacing $X'$ by blowups and replace $D'$, we may assume that $D'$ contains no LC centers of $(X', \ulcorner(B')^+\urcorner)$. By the construction of $\mu$, LC centers of $(X', \ulcorner(B')^+\urcorner)$ are irreducible components of $(B')^+$ and irreducible components of $Y' \cap ((B')^+ - Y')$. The former are already handled, but now consider the case that $D'$ contains an irreducible component $Z$ of $Y' \cap ((B')^+ - Y')$. In this case, we blow up $X'$ along $Z$, and keep blowing up $X'$ until the irreducible components of $(B')^+ - Y'$ are disjoint. Subtract certain multiples of exceptional divisors which are the common irreducible components of the pullback of $D'$ and the new $(B')^+$. If $D'$ still contains some irreducible component of $Y' \cap ((B')^+ - Y')$, repeat this process. Note that in each step of this process the blowup center is a prime divisor on $Y'$, hence $Y'$ is not changed. On the other hand, in each step of this process, at least one coefficient of $((B')^+ - Y')|_{Y'}$ is decreased by at least $\frac{1}{mm_0}$. Hence this process stops after finitely many times, and eventually $D'$ contains no irreducible component of $Y' \cap ((B')^+ - Y')$.

In this way we constructed a log resolution $\mu_m : X_m \to X$. Take $B'_m = (B')^+$. Here note that $(B')^+$ is obtained by subtracting redundant irreducible components. (1), (2), and (3) directly follow from the construction. As we only subtract the fixed part of $D'$ from $B_m^+$ and $B_m^-$ is $\mu_m$-exceptional, we have (4). (5) is from the fact that, in the beginning $\mu : X' \to X$ does not depend on $m$, and the blowups after this are along divisors on $Y'$. (6) is from the natural homomorphism

$$H^0(X, m_0 m_1(K_X + B)) \otimes H^0(X, m_0 m_2(K_X + B))$$
$$\to H^0(X, m_0(m_1 + m_2)(K_X + B)).$$

For (7), the limit exists by the inequality in (6), and the last assertion follows since $B_{Y',m} \le (B_m^+ - Y_m)|_{Y_m}$ and $(Y_m, (B_m^+ - Y_m)|_{Y_m})$ is KLT as $(X, B)$ is PLT, where $(B_m^+ - Y_m)|_{Y_m}$ is a divisor on $Y'$ independent of $m$ by the construction. $\qquad\square$

Next we apply the extension theorem:

**Theorem 3.2.6** *Under the setting of Proposition 3.2.5, the restriction map*

$$H^0(X_m, lmm_0(K_{X_m} + B'_m)) \to H^0(Y', lmm_0(K_{Y'} + B_{Y',m}))$$

*is surjective for any positive integer $l$.*

*Proof* Let us check the conditions of the extension theorem. First, $\lfloor B'_m \rfloor = Y'$ and $(X_m, B'_m)$ is PLT.

(1) Since $f \circ \mu$ is an isomorphism on the generic points of $X_m$ and $Y'$, we can write $B'_m - Y'$ as a sum of an ample **Q**-divisor and an effective **Q**-divisor whose support does not contain $Y'$.

(2) A general element in $|mm_0(K_{X_m} + B'_m)|$ does not contain any LC center of the pair $(X_m, \lceil B'_m \rceil)$. $\qquad\square$

### 3.2.2 The Existence of PL Flips

In this subsection, we prove the existence of PL flips.

Let us recall some symbols. For a divisor $D$ on a normal algebraic variety $X$, its *fixed part* $\mathrm{Fix}(D)$ and *movable part* $\mathrm{Mov}(D)$ are defined as the following:

$$|D| = \{D' \mid D \sim D' \geq 0\},$$
$$\mathrm{Fix}(D) = \inf|D|,$$
$$\mathrm{Mov}(D) = D - \mathrm{Fix}(D).$$

Here the infimum of divisors is defined by taking the infimum of each coefficient. Note that if $X$ is not projective, then the complete linear system $|D|$ is not necessarily finite-dimensional, but the fixed part is well defined as a divisor.

*Proof of Theorem 3.2.1* By the finiteness of minimal models, for sufficiently large $m$, the pairs $(Y', B_{Y',m})$ have the same canonical model $Z$. Here note that $f|_{Y'} : Y' \to S$ is birational to its image, so the KLT version BCHM condition automatically holds.

Taking a sufficiently high log resolution $Y'' \to Y'$ of $(Y', B_{Y'})$, we may assume that the induced morphism $Y'' \to Z$ is a morphism over $S$. In the following, we may replace $Y'$ by $Y''$. This is equivalent to taking "simultaneous resolution of singularities."

For positive integers $m, l$, denote

$$\tilde{P}_m = \mathrm{Mov}(mm_0(K_{X_m} + B_m^+)),$$
$$P_m = \tilde{P}_m|_{Y'},$$
$$\tilde{P}_{l,m} = \mathrm{Mov}(lmm_0(K_{X_m} + B'_m)),$$
$$P_{l,m} = \mathrm{Mov}(lmm_0(K_{Y'} + B_{Y',m})).$$

Since $Y'$ dominates $Z$ which is the canonical model of $(Y', B_{Y'_m})$, there exists a positive integer $l_m$ such that $P_{l_m,m}$ is free. After further blowups, we may assume that $\mathrm{Mov}(lmm_0(K_{X_m} + B'_m))$ is free for $1 \leq l \leq l_m$. According to Theorem 3.2.6, for $1 \leq l \leq l_m$,

$$P_{l,m} = \mathrm{Mov}(lmm_0(K_{Y'} + B_{Y',m})) = \tilde{P}_{l,m}|_{Y'}.$$

Since $\tilde{P}_{1,m} = \tilde{P}_m$, $P_{1,m} = P_m$. We have

$$l_m P_m = l_m P_{1,m} \le P_{l_m,m} \le P_{l_m m}.$$

Since

$$(K_{X_m} + B_m)|_{Y'} = K_{Y'} + \tilde{B}_{Y'} = \mu_Y^*(K_Y + B_Y),$$

we get

$$P_{m_1} + P_{m_2} \le P_{m_1+m_2} \le (m_1 + m_2)m_0(K_{Y'} + \tilde{B}_{Y'}),$$

hence the limit $P = \lim_{m \to \infty} P_m/m$ defines an **R**-divisor on $Y'$. Note that $P_{l_m,m}/l_m m m_0$ is the pullback of the log canonical divisor of the canonical model of $(Y', B_{Y'_m})$ and

$$\lim_{m \to \infty} \frac{P_m}{m} = \lim_{m \to \infty} \frac{P_{l_m,m}}{l_m m},$$

hence $P/m_0$ is the pullback of the log canonical divisor of the canonical model of $(Y', B_{Y'})$. Therefore $P$ is semi-ample, that is, it is a linear combination of free divisors with positive real coefficients.

In the following, we will show that there exists a positive integer $m_1$ such that $P = P_{m_1}/m_1$. If this is proved, then for any positive integer $l$, $l P_{m_1} = P_{lm_1}$, and hence

$$\bigoplus_{l \ge 0} H^0(Y', l P_{m_1})$$

$$\cong \bigoplus_{l \ge 0} \mathrm{Im}(H^0(X, lm_0m_1(K_X + B)) \to H^0(Y, lm_0m_1(K_Y + B_Y)))$$

is finitely generated, and the proof of the existence of PL flips is finished.

**Lemma 3.2.7** *For any positive integers $m, m'$,*

$$\mathrm{Mov}\left(\left\lceil \frac{m' P_{l_m,m}}{l_m m} - \tilde{B}_{Y'} \right\rceil\right) \le P_{m'}.$$

*Therefore, after taking the limit,*

$$\mathrm{Mov}(\lceil m' P - \tilde{B}_{Y'} \rceil) \le P_{m'}.$$

*Proof* Take a general effective **Q**-divisor $D \equiv m' \tilde{P}_{l_m,m}/l_m m$. That is, take a sufficiently large and sufficiently divisible positive integer $N$, take $D$ to be

## 3.2 PL Flips

a general element in $|Nm'\tilde{P}_{l_m,m}/l_m m|$ dividing by $N$. Also take a general effective $\mathbf{Q}$-divisor

$$D' \equiv \left\lceil \frac{m'\tilde{P}_{l_m,m}}{l_m m} - B_m \right\rceil - \left(\frac{m'\tilde{P}_{l_m,m}}{l_m m} - B_m\right).$$

Take $J$ to be the multiplier ideal sheaf of $(X_m, D+D')$. Since $P_{l_m,m} = \tilde{P}_{l_m,m}|_{Y_m}$ is free, $Y_m$ does not intersect the support of $\mathcal{O}_{X_m}/J$. As

$$\left\lceil \frac{m'\tilde{P}_{l_m,m}}{l_m m} - B_m \right\rceil - (K_{X_m} + D + D') \equiv -\mu^*(K_X + B)$$

is relatively nef and relatively big,

$$H^1\left(X_m, J\left(\left\lceil \frac{m'\tilde{P}_{l_m,m}}{l_m m} - B_m \right\rceil\right)\right) = 0.$$

Hence

$$H^0\left(X_m, \left\lceil \frac{m'\tilde{P}_{l_m,m}}{l_m m} - (B_m - Y_m)\right\rceil\right) \to H^0\left(Y', \left\lceil \frac{m'P_{l_m,m}}{l_m m} - \tilde{B}_{Y'}\right\rceil\right)$$

is surjective. On the other hand, since

$$\mu_{m*}\left(\left\lceil \frac{m'\tilde{P}_{l_m,m}}{l_m m} - (B_m - Y_m)\right\rceil\right) \le m'm_0(K_X + B),$$

we get

$$\mathrm{Mov}\left(p^*\left\lceil \frac{m'\tilde{P}_{l_m,m}}{l_m m} - (B_m - Y_m)\right\rceil\right) \le q^*\tilde{P}_{m'}.$$

Here if $m > m'$, then $p = \mathrm{id}$ and $q$ is the induced morphism $X_m \to X_{m'}$; if $m \le m'$, then $p$ is the induced morphism $X_{m'} \to X_m$ and $q = \mathrm{id}$. This proves the former assertion by restricting on $Y'$. For the latter, just take the limit. $\square$

Now, go back to the proof of the existence of PL flips. First, suppose that $P$ is a $\mathbf{Q}$-divisor. In this case, there exists a positive integer $m_1$ such that $m_1 P$ is Cartier and free. Then

$$m_1 P \le \mathrm{Mov}(\lceil m_1 P - \tilde{B}_{Y'}\rceil) \le P_{m_1} \le m_1 P.$$

Hence $m_1 P = P_{m_1}$.

Next, consider the case that $P$ is not a $\mathbf{Q}$-divisor. In this case, we can use positive real numbers $p_j$ and free Cartier divisors $L_j$ to express $P = \sum_j p_j L_j$. We may also assume that the set of $p_j$ is $\mathbf{Q}$-linearly independent. Take an effective divisor $M$ containing supports of all $L_j$, and take a sufficiently small real number $\epsilon > 0$ such that $\lfloor \tilde{B}_{Y'} + \epsilon M \rfloor \le 0$. Suppose that at least one $p_j$ is

192　　　　　　　　　　*3 The Finite Generation Theorem*

not rational, then there exists a positive integer $m$ and a free Cartier divisor $L$ such that

$$mP - \epsilon M < L < mP + \epsilon M$$

and $L \not\leq mP$. Then

$$L \leq \lceil mP + \epsilon M - \tilde{B}_{Y'} - \epsilon M \rceil,$$

which implies that

$$L \leq \text{Mov}(\lceil mP - \tilde{B}_{Y'} \rceil) \leq P_m \leq mP,$$

a contradiction. □

**Remark 3.2.8** (1) In this way, the existence of flips can be proved in any dimension. It deeply impressed me when recalling that the proof of the existence of flips in dimension 3 was very difficult ([102]). So this is a big success of the formulation of the problem by using log pairs. In the above argument, the basepoint-free theorem plays an important role in the background.

(2) $P$ is equivalent to the numerically movable part of the Zariski decomposition. In [60], assuming the existence of Zariski decomposition in the sense that the numerically movable part is nef, even if the numerically movable part might be an **R**-divisor, it can be shown that the canonical ring is finitely generated and the numerically movable part is indeed a **Q**-divisor. The technique in that proof might be applied here, but the proof introduced here uses the idea of "*saturation*" due to Shokurov ([130]).

## 3.3 The Special Termination

The special termination is an essential idea for applying induction on dimensions.

The definition of DLT pairs is suitable for the inductive argument on dimensions, as an irreducible component in the boundary with coefficient 1 determines a DLT pair of one dimension lower. The special termination theorem by Shokurov is essential in the discussion of the termination of MMP by induction on dimensions. In this section, we show the special termination of MMP with scaling.

The log version, as the generalization of the nonlog version, should be more complicated originally. For example, log terminal singularities are more complicated than terminal singularities. On the other hand, the log version has

## 3.3 The Special Termination

more freedom, because we can perturb the coefficients of the boundary. Also if there is an irreducible component in the boundary with coefficient 1, we can use the subadjunction formula to get a DLT pair of one dimension lower. In addition, for a fixed algebraic variety $X$, if the coefficients of $B$ increase, then the condition that $(X, B)$ is log terminal gets stronger, and the singularities of $X$ gets better.

First, let us recall the statement of the special termination of MMP with scaling: Suppose that $f : (X, B) \to S$ and $f : (X, B') \to S$ satisfy the BCHM condition. Assume that $\llcorner B' \lrcorner \geq \llcorner B \lrcorner$ and $K_X + B'$ is relatively nef. For an MMP on $f : (X, B) \to S$ with scaling of $B' - B$, starting from $(X_0, B_0) = (X, B)$, we get an infinite sequence of flips

$$\alpha_m : (X_m, B_m) \dashrightarrow (X_{m+1}, B_{m+1}), \qquad m = 0, 1, 2, \dots.$$

Here $B_{m+1} = \alpha_{m*} B_m$. Then there exists a positive integer $m_0$ such that for any $m \geq m_0$, $\alpha_m$ is isomorphic in a neighborhood of the integral part $\llcorner B_m \lrcorner$ of the boundary. That is,

$$\mathrm{Exc}(\alpha_m) \cap \mathrm{Supp}\, \llcorner B_m \lrcorner = \emptyset.$$

**Theorem 3.3.1** (*Special termination theorem*) *Under the BCHM condition, suppose that the existence of flips in dimension $n - 1$ and the termination of MMP with scaling in dimension $n - 1$ hold, then the special termination of MMP with scaling in dimension $n$ holds.*

**Remark 3.3.2** (1) The special termination theorem was originally proved by Shokurov ([128]). It forged the path of proving the existence of minimal models by induction on dimensions.

(2) As we will discuss in Section 3.4, the assumption above implies the existence of $\mathbf{Q}$-factorialization of KLT pairs in dimension $n-1$ (see Lemma 3.4.5 or Corollary 3.6.9). Indeed, a $\mathbf{Q}$-factorialization is a minimal model of certain birational morphism. Assuming the existence of flips in dimension $n - 1$ and the termination of MMP with scaling in dimension $n - 1$, the existence of minimal models in dimension $n - 1$ follows.

*Proof* Fix any irreducible component $Z_0$ of $Z = \llcorner B \lrcorner$. Given an MMP with scaling consisting of flips, it suffices to show that the MMP is isomorphic in a neighborhood of $Z_0$ after finitely many steps.

Recall that we can write $B = A + E + Z$ and $B' = A' + E' + Z'$, where $Z' = \llcorner B' \lrcorner$. Take a sufficiently small real number $t$ such that $A + t(Z - Z_0)$ and $A' + t(Z' - Z_0)$ are relatively ample. Take sufficiently general relatively ample effective $\mathbf{R}$-divisors $A_0, A'_0$ such that $A_0 \equiv A + t(Z - Z_0)$ and $A'_0 \equiv A' + t(Z' - Z_0)$. Then $(X, A_0 + E + (1-t)(Z - Z_0) + Z_0)$ and $(X, A'_0 + E' +$

$(1-t)(Z' - Z_0) + Z_0)$ are PLT. As $B \equiv A_0 + E + (1-t)(Z - Z_0) + Z_0$, $B' \equiv A'_0 + E' + (1-t)(Z' - Z_0) + Z_0$, and $\llcorner A_0 + E + (1-t)(Z - Z_0) + Z_0 \lrcorner = \llcorner A'_0 + E' + (1-t)(Z' - Z_0) + Z_0 \lrcorner = Z_0$, after replacing $A, A'$ by $A_0, A'_0$ and replacing $E, E'$ by $E + (1-t)(Z - Z_0), E' + (1-t)(Z' - Z_0)$, we may assume that the pairs $(X, B)$ and $(X, B')$ are PLT in the beginning. Denote $Z_m = \llcorner B_m \lrcorner$.

The sequence of flips of $(X, B)$ induces birational maps $\alpha'_m : Z_m \dashrightarrow Z_{m+1}$. Note that $\alpha'_m$ might contract divisors on $Z_m$ and might also extract new divisors on $Z_{m+1}$.

The set of the coefficients of $B$ in the pair $(X, B)$ is a finite set $\mathrm{Coeff}(B) \subset [0, 1]$. So $\mathrm{Coeff}(B_m)$ is a fixed set. However, the subadjunction formula $(K_{X_m} + B_m)|_{Z_m} = K_{Z_m} + B_{Z_m}$ defines an $\mathbf{R}$-divisor $B_{Z_m}$ on $Z_m$, while the set $\mathrm{Coeff}(B_{Z_m})$ might vary as $m$ varies.

Define the set $\Sigma \subset [0, 1]$ as the following:

$$\Sigma = \left\{ x \in [0, 1] \mid x = 1 - \frac{1}{r} + \sum_i \frac{r_i b_i}{r}, b_i \in \mathrm{Coeff}(B), r \in \mathbf{Z}_{>0}, r_i \in \mathbf{Z}_{\geq 0} \right\}.$$

Then by the subadjunction formula, $\mathrm{Coeff}(B_{Z_m}) \subset \Sigma$ for any $m$.

Here recall the following definition:

**Definition 3.3.3** A set $T \subset \mathbf{R}$ satisfies the *ascending chain condition = ACC* (respectively, *descending chain condition = DCC*) if for any sequence $\{x_n\} \subset T$ satisfying $x_n \leq x_{n+1}$ (respectively, $x_n \geq x_{n+1}$) for all $n$, there exists $n_0$ such that $x_n = x_{n+1}$ for $n \geq n_0$.

**Remark 3.3.4** The ACC and DCC are finiteness conditions. For example, finite sets satisfy both the ACC and the DCC.

In higher dimensional algebraic geometry, it is expected that certain naturally defined sets satisfy the ACC or the DCC. For example, the set of "minimal log discrepancies" (MLDs) is expected to satisfy the ACC, which is considered as a solution for the termination of flips. This reflects hidden patterns in the world of algebraic varieties, and there are many unsolved problems. We refer to [15].

**Lemma 3.3.5** *The set $\Sigma$ satisfies the DCC. Moreover, for any $\epsilon > 0$, the set $\Sigma \cap [0, 1 - \epsilon]$ is finite.*

*Proof* Consider $x = 1 - \frac{1}{r} + \sum_i \frac{r_i b_i}{r} \in \Sigma$. Since $b_i$ are in a finite set, if $x \leq 1 - \epsilon$, then it is easy to see that there are only finitely many possible values for $r$ and $r_i$. Therefore, the only accumulation point of $\Sigma$ is 1, which concludes the DCC. $\qquad\square$

## 3.3 The Special Termination

For any positive number $m$, take a common log resolution $g: Y \to X_m$ and $g': Y \to X_{m+1}$ of $(X_m, B_m)$ and $(X_{m+1}, B_{m+1})$ in strong sense. Write $K_Y + C = g^*(K_{X_m} + B_m)$ and $K_Y + C' = (g')^*(K_{X_{m+1}} + B_{m+1})$. Write $C = \sum c_i C_i$ and $C' = \sum c_i' C_i$ into prime divisors, then $c_i \geq c_i'$ for all $C_i$, and $c_i > c_i'$ for divisors supported over the exceptional set of $X_m \dashrightarrow X_{m+1}$ (cf. Theorem 2.5.6).

Take $Z' \subset Y$ to be the common strict transform of $Z_m$ and $Z_{m+1}$ and take $\bar{C}_i = C_i \cap Z'$, then $K_{Z_m} + B_{Z_m} = (g|_{Z'})_*(K_{Z'} + \sum' c_i \bar{C}_i)$ and $K_{Z_{m+1}} + B_{Z_{m+1}} = (g'|_{Z'})_*(K_{Z'} + \sum' c_i' \bar{C}_i)$. Here the sum $\sum'$ runs over all $C_i \neq Z'$.

Since $G := g^*(K_{X_m} + B_m) - (g')^*(K_{X_{m+1}} + B_{m+1}) > 0$ with $\mathrm{Supp}(G) = g^{-1}(\mathrm{Exc}(\alpha_m))$, we have $G_Z := G|_{Z'} = (g|_{Z'})^*(K_{Z_m} + B_{Z_m}) - (g'|_{Z'})^*(K_{Z_{m+1}} + B_{Z_{m+1}}) > 0$ with $\mathrm{Supp}(G_Z) = g^{-1}(\mathrm{Exc}(\alpha_m)) \cap Z'$.

As $(Z_m, B_{Z_m})$ and $(Z_{m+1}, B_{Z_{m+1}})$ are KLT, we may assume that $g$ is a very log resolution of $(Z_m, B_{Z_m})$ and $(Z_{m+1}, B_{Z_{m+1}})$. For each $m$, consider the number

$$d_m = \sum_{a \in \Sigma} \#\{c_i \mid c_i > a\} + \sum_{a \in \Sigma} \#\{c_i \mid c_i \geq a\}.$$

Here we consider all coefficients appearing in $\sum c_i \bar{C}_i$. Note that $d_m$ is a well-defined nonnegative integer since the sum only considers finitely many $a \in \Sigma \cap [0, \max\{c_i\}]$ as $c_i < 1$, and the definition does not depend on the choice of very log resolutions since we only consider nonnegative $c_i$. As $c_i \geq c_i'$, we know that $d_m \geq d_{m+1}$. If $Z_{m+1}$ contains a divisor $P$ which is not a divisor on $Z_m$, then $P$ comes from some $\bar{C}_i$, where $C_i$ is supported over the exceptional set of $X_m \dashrightarrow X_{m+1}$. So this means that $c_i > c_i' \in \Sigma$, and in this case we have $d_m > d_{m+1}$. If $Z_m$ contains a divisor $P$ which is not a divisor on $Z_{m+1}$, then again $P$ comes from some $\bar{C}_i$, where $C_i$ is supported over the exceptional set of $X_m \dashrightarrow X_{m+1}$. So this means that $\Sigma \ni c_i > c_i'$, and in this case we again have $d_m > d_{m+1}$.

In this way, we can see that $\alpha_m': Z_m \dashrightarrow Z_{m+1}$ is isomorphic in codimension 1 after removing finitely many steps.

Since $A, A'$ are general, their restrictions to $Z$ (for the strict transform of $Z$ after finitely many steps from the beginning) are relatively big. Since in finitely many steps from the beginning there might be new divisors appearing on $Z$, the assumption on the generality is necessary.

By the subadjunction formula, we may write $(K_X + B)|_Z = K_Z + B_Z$ and $(K_X + B')|_Z = K_Z + B_Z'$. Then the pairs $(Z, B_Z)$ and $(Z, B_Z')$ are KLT, $B_Z \geq A|_Z$, and $B_Z' \geq A'|_Z$. So they satisfy the BCHM condition, except that $Z$ is not necessarily $\mathbf{Q}$-factorial. Here we take $h: \tilde{Z} \to Z$ to be a $\mathbf{Q}$-factorialization, denote $\tilde{B} = h_*^{-1} B_Z$, $\tilde{B}' = h_*^{-1} B_Z'$, and $\tilde{f} = f \circ h$. Then we can consider the

196          3 *The Finite Generation Theorem*

MMP on $\tilde{f} \colon (\tilde{Z}, \tilde{B}) \to S$ with scaling of $\tilde{B}' - \tilde{B}$. By the inductive hypothesis, in this MMP, flips exist and terminate.

This MMP should match the pullback of the MMP for $f \colon (X, B) \to S$ to $\tilde{Z}$, so the original MMP terminates in a neighborhood of $Z$. However, the pullback of each step of the MMP for $f \colon (X, B) \to S$ on $\tilde{Z}$ corresponds to a composition of several steps of MMP, so we need a more detailed discussion. Let us explain this.

Suppose that the flip $\alpha_m \colon (X_m, B_m) \dashrightarrow (X_{m+1}, B_{m+1})$ is the composition of small birational maps $\phi_m \colon X_m \to S_m$ and $\phi_m^+ \colon X_{m+1} \to S_m$ and denote $t_m = \min\{t \mid K_{X_m} + B_m + t_m(B'_m - B_m) \text{ is nef over } S\}$, then $K_{X_m} + B_m + t_m(B'_m - B_m)$ is numerically trivial over $S_m$.

Consider the induced map $\alpha'_m \colon Z_m \dashrightarrow Z_{m+1}$, we can construct $\mathbf{Q}$-factorializations $h_m \colon \tilde{Z}_m \to Z_m$ and $h_{m+1} \colon \tilde{Z}_{m+1} \to Z_{m+1}$, and a decomposition into a sequence of flips over $S_m$:

$$\tilde{Z}_m = Z_{m,0} \dashrightarrow Z_{m,1} \dashrightarrow \cdots \dashrightarrow Z_{m,l} = \tilde{Z}_{m+1}$$

by induction on $m$.

Indeed, suppose that $h_m$ is given, take $\tilde{B}_{Z,m} = h_{m*}^{-1} B_{Z_m}$, and run an MMP with scaling for

$$\phi_m \circ h_m \colon (\tilde{Z}_m, \tilde{B}_{Z,m}) \to S_m.$$

Here the scale can be appropriately taken to be a relatively ample $\mathbf{R}$-divisor over $S_m$. By the inductive hypothesis, this MMP terminates to a minimal model $(\tilde{Z}_{m+1}, \tilde{B}_{Z,m+1}) \to S_m$. Since $K_{Z_{m+1}} + B_{Z_{m+1}}$ is ample over $S_m$, it is the canonical model of $\tilde{Z}_{m+1}$ and induces a morphism $h_{m+1} \colon \tilde{Z}_{m+1} \to Z_{m+1}$. Hence we get the required decomposition. Here as $Z_m$ and $Z_{m+1}$ are isomorphic in codimension 1, all models in the above MMP are isomorphic in codimension 1 with $Z_{m+1}$, and in particular, it consists of flips and $h_{m+1}$ is a $\mathbf{Q}$-factorialization.

Since $K_{X_m} + B_m + t_m(B'_m - B_m)$ is numerically trivial over $S_m$, the induced divisors $K_{Z_{m,i}} + B_{m,i} + t_m(B'_{m,i} - B_{m,i})$ are all numerically trivial over $S_m$. Hence it is easy to check that the sequence

$$\tilde{Z} = \tilde{Z}_0 = Z_{0,0} \dashrightarrow Z_{0,1} \dashrightarrow \cdots \dashrightarrow Z_{0,l} = \tilde{Z}_1 = Z_{1,0} \dashrightarrow Z_{1,1} \dashrightarrow \cdots$$

is indeed an MMP on $\tilde{f} \colon (\tilde{Z}, \tilde{B}) \to S$ with scaling of $\tilde{B}' - \tilde{B}$. Here $B_{m,i}, B'_{m,i}$ are the strict transforms of $\tilde{B}, \tilde{B}'$. By the inductive hypothesis, this MMP terminates, which implies that the original MMP terminates in a neighborhood of $Z$.

Indeed, after finitely many steps, $\tilde{Z}_m$ does not change, which means that $K_{\tilde{Z}_m} + \tilde{B}_{Z,m}$ is nef over $S_m$, and then $K_{Z_m} + B_{Z_m}$ is nef over $S_m$. On the other hand, $-(K_{Z_m} + B_{Z_m})$ is ample over $S_m$, so $-(K_{Z_m} + B_{Z_m})$ is both numerically trivial and ample over $S_m$, which implies that $Z_m \to S_m$ does not contract any curve on $Z_m$. Similarly, $K_{Z_{m+1}} + B_{Z_{m+1}}$ is both numerically trivial and ample over $S_m$, which implies that $Z_{m+1} \to S_m$ does not contract any curve on $Z_{m+1}$. If $Z_m$ intersects $\mathrm{Exc}(\alpha_m)$, then $Z_m$, as a divisor on $X_m$, is ample over $S_m$. Hence $-Z_{m+1}$ is ample over $S_m$, which contradicts the fact that $Z_{m+1} \to S_m$ does not contract any curve on $Z_{m+1}$. This implies that the original MMP terminates in a neighborhood of $Z$. $\qquad\square$

**Remark 3.3.6** Without assuming the BCHM condition, we have the following special termination: Suppose that the existence of flips in dimension $n - 1$ and the termination of MMP in dimension $n-1$ hold, then the special termination in dimension $n$ holds. That is, given a projective morphism $f : (X, B) \to S$ from a DLT pair, for an infinite sequence of flips over $S$ starting from $(X_0, B_0) = (X, B)$:

$$\alpha_m : (X_m, B_m) \dashrightarrow (X_{m+1}, B_{m+1}), \qquad m = 0, 1, 2, \ldots,$$

where $B_{m+1} = \alpha_{m*} B_m$, there exists a positive integer $m_0$ such that for any $m \geq m_0$, $\alpha_m$ is isomorphic in a neighborhood of the integral part $\llcorner B_m \lrcorner$ of the boundary. We will not use this fact in this book, please refer to [27].

## 3.4 The Existence and Finiteness of Minimal Models

In this section, we show the existence and finiteness of minimal models by induction on dimensions.

**Theorem 3.4.1** (*Existence of minimal models*) *Under the BCHM condition, suppose that the existence of PL flips in dimension n and the special termination of MMP with scaling in dimension n hold, and the nonvanishing theorem holds for a morphism $(X, B) \to S$ in dimension n, then the existence of minimal models holds for the morphism $(X, B) \to S$.*

**Remark 3.4.2** As can be seen from the proof below, the BCHM condition is not essential in this proof. To be more accurate, we can prove the following assertion:

Let $(X, B)$ be a KLT pair consisting of an $n$-dimensional $\mathbf{Q}$-factorial algebraic variety and an effective $\mathbf{R}$-divisor, and let $f : X \to S$ be a projective morphism to a quasi-projective algebraic variety. Assume the following:

(1) (Existence of PL flips) For any $n$-dimensional $\mathbf{Q}$-factorial DLT pair $(X', B')$ with a PL contraction (i.e. a small contraction with respect to

198                    3 The Finite Generation Theorem

which there is a prime divisor in $\llcorner B' \lrcorner$ which is relatively negative), the flip always exists.

(2) (Special termination) Any sequence of MMP with scaling starting from an $n$-dimensional **Q**-factorial DLT pair $(X', B')$ is isomorphic in neighborhoods of strict transforms of $\llcorner B' \lrcorner$ after finitely many steps.

(3) (Nonvanishing theorem) For the given pair $(X, B)$, there exists an effective **R**-divisor $D$ such that $K_X + B \equiv_S D$.

Then there exists a minimal model of the given morphism $f : (X, B) \to S$.

The existence of PL flips holds by summarizing all results proved in this chapter. The special termination can be proved assuming all statements in lower dimensional minimal model theory (Remark 3.3.6). Therefore, if one wants to try to prove the existence of minimal models without the BCHM condition, proving the nonvanishing theorem is a key point.

Here we will follow the proof of Birkar ([12]) which modifies that of [16]. First, recall the definition of minimal models:

**Definition 3.4.3** Let $(X, B)$ be a DLT pair consisting of a **Q**-factorial algebraic variety and an effective **R**-divisor, and let $f : X \to S$ be a projective morphism to a quasi-projective algebraic variety. A *minimal model* of $f : (X, B) \to S$ is given by a birational map $\alpha : (X, B) \dashrightarrow (Y, C)$ over $S$ to another **Q**-factorial algebraic variety projective over $S$ satisfying the following conditions:

(1) $\alpha$ is surjective in codimension 1, $C = \alpha_* B$, and the pair $(Y, C)$ is DLT.
(2) $K_Y + C$ is relatively nef over $S$.
(3) If we take birational projective morphisms $p : Z \to X$ and $q : Z \to Y$ from a third normal algebraic variety such that $\alpha = q \circ p^{-1}$, then the *discrepancy divisor* $G = p^*(K_X + B) - q^*(K_Y + C)$ is effective, and the support of $p_* G$ is the union of all prime divisors contracted by $\alpha$.

If in condition (3) we only assume that $G \geq 0$ while the support of $p_* G$ may not contain all prime divisors contracted by $\alpha$, then it is called a *weak minimal model*.

**Remark 3.4.4** Minimal models obtained by MMP satisfy the following condition stronger than (3): Denote by $\mathrm{Exc}(\alpha)$ the *exceptional set* of $\alpha$, that is, the complement set of the maximal open subset on which $\alpha$ is an isomorphism, then the support of $G$ coincides with $p^{-1}(\mathrm{Exc}(\alpha))$. Condition (3) only focuses on the phenomenon in codimension 1, but it is indeed sufficient.

*Proof* Since we only assume the existence of PL flips, we need to be careful on running MMP, that is, we can run the MMP as long as each small contraction

## 3.4 The Existence and Finiteness of Minimal Models 199

is a PL contraction. On the other hand, in order to show the termination of certain MMP after finitely many steps, the idea is to adjust the boundary $B$ and to apply the special termination. Therefore, it is necessary to consider DLT pairs instead of KLT pairs.

By assumption, $K_X + B \equiv_S D$ for some effective **R**-divisor $D$.

**Step 0** We reduce to the case that $X$ is smooth and the support of $B + D$ is a normal crossing divisor.

First, after replacing $B$, we may assume that $(X, B)$ satisfies the KLT version BCHM condition and $B \geq A$, where $A$ is a general effective relatively ample **R**-divisor. Next, take a log resolution $g: X' \to (X, B + D)$, we can construct an effective **R**-divisor $B'$ such that $(X', B')$ is KLT and

$$E = K_{X'} + B' - g^*(K_X + B)$$

is effective whose support coincides with the support of the exceptional set of $g$. Then a minimal model of $f \circ g: (X', B') \to S$ is also a minimal model of $f: (X, B) \to S$.

Indeed, take $\alpha: (X', B') \dashrightarrow (X'', B'')$ to be a minimal model, since $E$ is contained in the numerically fixed part of $K_{X'} + B'$ over $S$, it is contracted by $\alpha$ by Theorem 2.9.6, that is, $\alpha_* E = 0$. Hence $\alpha \circ g^{-1}: X \dashrightarrow X''$ is surjective in codimension 1, and the negativity can be checked easily. So it is a minimal model of $f: (X, B) \to S$.

Also, note that $B' \geq g_*^{-1} A$ is still relatively big since $A$ is general. Hence $f \circ g: (X', B') \to S$ satisfies the KLT version BCHM condition. After replacing $f: (X, B) \to S$ by $f \circ g: (X', B') \to S$, we may assume that $X$ is smooth and the support of $B + D$ is of normal crossing in the beginning.

In the following, we always assume that $f: (X, B) \to S$ satisfies the BCHM condition (while $(X, B)$ is not necessarily KLT), $K_X + B \equiv_S D \geq 0$, $X$ is smooth, the support of $B + D$ is of normal crossing, and $B \geq A$ is a general effective relatively ample **R**-divisor. In particular, $A$ has no common irreducible component with $D$ and $\llcorner B \lrcorner$.

**Step 1** Write $B = \sum b_i D_i$ and $D = \sum d_i D_i$ by distinct prime divisors $D_i$. We will do induction on

$$\theta = \theta(X, B, D) = \#\{i \mid b_i \neq 1, d_i \neq 0\}.$$

If $D = 0$, then $f: (X, B) \to S$ is already minimal, hence we assume that $D \neq 0$ in the following.

Suppose that $\theta = 0$. Take a general relatively ample **R**-divisor $H$ such that $K_X + B + H$ is relatively ample and $(X, B + H)$ is DLT, we can run an MMP

200             3 *The Finite Generation Theorem*

with scaling of $H$. Here since $\theta = 0$, the support of $D$ is contained in $\llcorner B \lrcorner$, hence in each step the contracted curves are contained in the support of $D$ and each small contraction is a PL contraction. By the existence of PL flips and the special termination of MMP with scaling, this MMP works and terminates.

Next, suppose that $\theta > 0$. Take

$$t = \min\{t' \in \mathbf{R}_{>0} \mid \mathrm{Supp}(\llcorner B \lrcorner) \neq \mathrm{Supp}(\llcorner B + t'D \lrcorner)\}.$$

Take $B + C$ to be the divisor obtained by cutting the coefficients of $B + tD$ by 1: $B + C = \sum \min\{b_i + td_i, 1\}D_i$. Here $C$ is effective. Then $K_X + B + C \equiv_S D + C$ and $\mathrm{Supp}(D + C) = \mathrm{Supp}\, D$. Note that $t$ is the smallest number making $\theta(X, B + C, D + C) < \theta(X, B, D)$.

Consider the morphism $f : (X, B + C) \to S$. Here we recall that $(X, B)$ satisfies the BCHM condition, where $\mathrm{Supp}(B + C)$ is simple normal crossing, $B \geq A$, and $A$ has no common irreducible component with $D$, hence $f : (X, B + C) \to S$ still satisfies the BCHM condition. Since $\theta(X, B + C, D + C)$ is smaller, by the inductive hypothesis on $\theta$, there exists a minimal model $f' : X' \to S$ over $S$: there exists a birational map $\alpha : (X, B + C) \dashrightarrow (X', B' + C')$ such that $K_{X'} + B' + C'$ is relatively nef. Here $B' = \alpha_* B$ and $C' = \alpha_* C$. Denote $D' = \alpha_* D$, then $D' \equiv_S K_{X'} + B'$.

We may run an MMP on $f' : (X', B') \to S$ with scaling of $C'$. As $(X, B)$ and $(X, B + C)$ satisfy the BCHM condition, $(X', B')$ and $(X', B' + C')$ also satisfy the BCHM condition. Moreover, $\llcorner B' + C' \lrcorner \geq \llcorner B' \lrcorner$.

By construction, there exists an effective $\mathbf{R}$-divisor $E'$ whose support is contained in $\llcorner B' \lrcorner$ such that $K_{X'} + B' + C' + E' = K_{X'} + B' + tD' \equiv_S (1 + t)(K_{X'} + B')$. For any extremal ray $R$ in this MMP, we have $((K_{X'} + B') \cdot R) < 0$ and $((K_{X'} + B' + C') \cdot R) \geq 0$, hence $(E' \cdot R) < 0$. Therefore, any curve in $R$ is contained in $\llcorner B' \lrcorner$ and each small contraction is a PL contraction. Here we keep the same symbol to denote strict transforms in the MMP. By the existence of PL flips and the special termination of MMP with scaling, this MMP works and terminates to a minimal model: a morphism $f'' : X'' \to S$ with a birational map $\beta : (X', B') \dashrightarrow (X'', B'')$ over $S$.

The composition map $\beta \circ \alpha : X \dashrightarrow X''$ is surjective in codimension 1 and $K_{X''} + B''$ is relatively nef. However, $K_X + B$ may not satisfy the negativity for the divisors contracted by $\beta \circ \alpha$, so we need a little more detailed discussions.

**Step 2** Consider the following set:

$$I = \{s \in [0, 1] \mid f : (X, B + sC) \to S \text{ has a minimal model}\}.$$

Note that $1 \in I$, and our goal is to show that $0 \in I$. By modifying the argument in Step 1, we will show that if $s \in I$ and $s > 0$, then there exists a sufficiently small $\epsilon > 0$ such that $s' \in I$ if $0 \leq s - s' \leq \epsilon$.

## 3.4 The Existence and Finiteness of Minimal Models

Take $\alpha_s : (X, B+sC) \dashrightarrow (X'_s, B'_s + sC'_s)$ to be a minimal model of $(X, B+sC)$ with natural morphism $f'_s : X'_s \to S$. Since negativity is an open condition, if $\epsilon$ is sufficiently small, $K_X + B + s'C$ is negative with respect to divisors contracted by $\alpha_s$.

We can run an MMP on $f'_s : (X'_s, B'_s + s'C'_s) \to S$ with scaling of $(s - s')C'_s$. As before, extremal rays in this MMP intersect negatively on an effective $\mathbf{R}$-divisor supported in $\llcorner B'_s \lrcorner$. Hence by the existence of PL flips and the special termination of MMP with scaling, this MMP works and terminates to a minimal model $\beta'_s : (X'_s, B'_s + s'C'_s) \dashrightarrow (X''_{s'}, B''_{s'} + s'C''_{s'})$ with morphism $f''_{s'} : X''_{s'} \to S$. The composition map $\beta_{s'} \circ \alpha_s : X \dashrightarrow X''_{s'}$ is surjective in codimension 1 and $K_{X''_{s'}} + B''_{s'} + s'C''_{s'}$ is relatively nef. Moreover, $K_X + B + s'C$ satisfies the negativity, hence $s' \in I$.

**Step 3** Take $s_0 = \inf I$. Note that $s_0 < 1$. We will show that $s_0 \in I$, then by Step 2, $s_0 = 0 \in I$. In order to show this, we first construct a weak minimal model.

Take a strictly decreasing sequence $I \ni s_k \to s_0$ and take minimal models $\alpha_k : (X, B + s_k C) \dashrightarrow (X'_k, B'_k + s_k C'_k)$ with morphisms $f'_k : X'_k \to S$.

We can run an MMP on $f'_k : (X'_k, B'_k + s_0 C'_k) \to S$ with scaling of $(s_k - s_0)C'_k$. As before, by the existence of PL flips and the special termination of MMP with scaling, this MMP works and terminates to a minimal model $\beta_k : (X'_k, B'_k + s_0 C'_k) \dashrightarrow (X''_k, B''_k + s_0 C''_k)$ with morphism $f''_k : X''_k \to S$.

The divisors contracted by $\beta_k \circ \alpha_k$ are all irreducible components of $D$, hence after replacing $\{s_k\}$ by a subsequence, we may assume that the contracted divisors are the same for all $k$. Then $X''_k$ are all isomorphic in codimension 1. Since $K_{X''_k} + B''_k + s_0 C''_k$ are all relatively nef, they are crepant to each other. That is, the pullbacks of any two of them coincide on a common resolution. Therefore, the discrepancy coefficients of $K_X + B + s_0 C$ for contracted divisors are independent of $k$.

Take an arbitrary prime divisor $P$ contracted by $\beta_k \circ \alpha_k$. Denote by $a_k \geq 0$ the discrepancy coefficient of $P$ for $K_X + B + s_k C$ with respect to $\alpha_k$ and denote by $b_k \geq 0$ the discrepancy coefficient of $P$ for $K_{X'_k} + B'_k + s_0 C'_k$ with respect to $\beta_k$. Denote by $a_P$ the coefficient of $P$ in the discrepancy divisor of $K_X + B + s_0 C$ with respect to $\beta_k \circ \alpha_k$, then $a_P \geq \lim_{k \to \infty}(a_k + b_k)$, which implies that $a_P \geq 0$. Indeed, take common resolutions $p : W \to X$, $p' : W \to X'_k$, $p'' : W \to X''_k$, we have

$$a_P = \operatorname{coeff}_P (p^*(K_X + B + s_0 C) - p''^*(K_{X''_k} + B''_k + s_0 C''_k))$$

$$\geq \operatorname{coeff}_P (p^*(K_X + B + s_0 C) - p'^*(K_{X'_k} + B'_k + s_k C'_k))$$

$$+ \operatorname{coeff}_P (p'^*(K_{X'_k} + B'_k + s_0 C'_k) - p''^*(K_{X''_k} + B''_k + s_0 C''_k))$$

$$= \operatorname{coeff}_P (p^*((s_0 - s_k)C)) + a_k + b_k.$$

202          3 *The Finite Generation Theorem*

Therefore, for any $k$, $\gamma = \beta_k \circ \alpha_k$ gives a weak minimal model $(X_k'', B_k'' + s_0 C_k'') = (X'', B'' + s_0 C'')$ of $(X, B + s_0 C)$. Here note that we only need to check that the discrepancy coefficients of contracted divisors are nonnegative, because in this case the effectivity of the discrepancy divisor is a consequence of the negativity lemma.

**Step 4** If $a_P > 0$ holds for all $P$, then $\gamma$ is a minimal model, and $s_0 \in I$, which concludes the proof.

If $a_P = 0$ for some prime divisors contracted by $\gamma$, by the following lemma, we can take a "crepant extraction" of these divisors from $(X'', B'' + s_0 C'')$ to make them not contracted, and this gives a minimal model of $(X, B + s_0 C)$.

Note that to apply the lemma, we need to check that for such $P$, the center of $P$ on $X''$ does not contain any LC center of $(X'', B'' + s_0 C'')$, but this is from the negativity in minimal models: If $P$ is contracted by $\alpha_k$, then the center of $P$ on $X_k'$ does not contain any LC center of $(X_k', B_k' + s_k C_k')$; if $P$ is contracted by $\beta_k$, then the center of $P$ on $X_k''$ does not contain any LC center of $(X_k'', B_k'' + s_0 C_k'')$.

$\square$

**Lemma 3.4.5** (*Crepant extraction*) *Let $(X, B)$ be an n-dimensional quasi-projective DLT pair and let $P$ be a discrete valuation on the function field of $X$ whose center is on $X$. Take a log resolution $f : Y \to (X, B)$ such that the center of $P$ on $Y$ is a prime divisor $E_P$ on $Y$, write $f^*(K_X + B) = K_Y + B_Y$, and suppose that the coefficient of $E_P$ in $B_Y$ is in the half-open interval $[0, 1)$. Moreover, we assume that the center of $P$ on $X$ does not contain any LC center of $(X, B)$. Suppose that the existence of PL flips in dimension $n$ and the special termination of MMP with scaling in dimension $n$ hold. Then there exists a birational projective morphism $g : (X', B') \to (X, B)$ from a $\mathbf{Q}$-factorial DLT pair such that*

(1) *$g$ is crepant: $g^*(K_X + B) = K_{X'} + B'$.*
(2) *The center of $P$ is a prime divisor on $X'$, and it is the only exceptional divisor of $g$.*

*Proof* Take a log resolution $f : Y \to (X, B)$ such that the coefficients of exceptional divisors in $B_Y$ are strictly smaller that 1. Such log resolution exists by the definition of DLT and the assumption on $P$. Take $E$ to be the divisor with all coefficients 1 whose support is the union of all exceptional divisors of $f$ except for $E_P$, and take $B_Y' = \max\{B_Y, E\}$. Then $(Y, B_Y')$ is a DLT pair and $K_Y + B_Y' - f^*(K_X + B) = F$ is effective whose support equals $E = \llcorner B_Y' \lrcorner$.

Take a general relatively ample effective divisor $H$ such that $K_Y + B_Y' + H$ is relatively nef over $X$. For a sufficiently small positive real number $t$, we can run an MMP on $f : (Y, B_Y' + tH) \to X$ with scaling of $(1 - t)H$. Here

## 3.4 The Existence and Finiteness of Minimal Models

note that $(Y, B'_Y + tH)$ satisfies the BCHM condition and all extremal rays intersect negatively with $F$. Hence by the existence of PL flips and the special termination of MMP with scaling, we get a minimal model $g \colon X' \to X$. As $K_Y + B'_Y + tH \equiv_X F + tH$, if $t$ is sufficiently small, then the support of the numerically fixed part of $K_Y + B'_Y + tH$ over $X$ coincides with the support of $F$. Therefore, this MMP over $X$ contracts all divisors in $E$ and obtains the desired extraction. $\qquad\square$

At the end of this section, we show the finiteness of minimal models.

**Theorem 3.4.6** (*Finiteness of minimal models*) *Under the BCHM condition, suppose that the existence of minimal models in dimension n holds, then the finiteness of minimal models in dimension n holds.*

Recall that in Theorem 2.10.3, we showed the finiteness of canonical models assuming the existence of minimal models and canonical models. On the other hand, in Remark 2.10.5 and Example 2.10.7, finiteness of minimal models does not hold in general. But if we assume the BCHM condition, then we can show the finiteness of minimal models, which indeed can be reduced to the finiteness of canonical models.

*Proof* Fix $f \colon X \to S$. Suppose that $P$ is a polytope spanned by effective **R**-divisors such that for any $B \in P$, $f \colon (X, B) \to S$ satisfies the BCHM condition. We may assume further that for any $B \in P$, $(X, B)$ is KLT and $B \geq A$, where $A$ is a fixed general relatively ample effective **Q**-divisor.

Take $H_1, \ldots, H_s$ to be relatively ample effective divisors whose classes form a basis of $N^1(X/S)$. Fix a sufficiently small real number $\epsilon > 0$, after changing $A$, we may assume that $A - \epsilon \sum_i H_i$ is effective and relatively ample. Consider a new polytope

$$Q' := \left\{ B + \sum_i h_i H_i \mid B \in P, -\epsilon \leq h_i \leq \epsilon \right\}.$$

If taking $\epsilon$ sufficiently small, we may assume that for any $B' \in Q'$, $f \colon (X, B') \to S$ satisfies the BCHM condition. Consider the polytope

$$Q := \{ B' \in Q' \mid [K_X + B'] \in \overline{\mathrm{Eff}}(X/S) \}.$$

Hence by the existence of minimal models and Theorem 2.10.3, there are finitely many canonical models corresponding to divisors in $Q$. Here note that for pairs with the BCHM condition, the existence of minimal models automatically implies the existence of canonical models by the basepoint-free theorem.

204                    *3 The Finite Generation Theorem*

In order to finish the proof, we only need to show that every minimal model for a divisor in $P$ is a canonical model for a divisor in $Q$.

Take any $B \in P$ and suppose that $\alpha \colon (X, B) \dashrightarrow (Y, C)$ is a minimal model of $f \colon (X, B) \to S$. Take a relatively ample divisor $H_Y$ on $Y$ and take $H = \alpha_*^{-1} H_Y$. We can write $H \equiv_S \sum d_i H_i$ for some real numbers $d_i$. Take a sufficiently small real number $\delta > 0$, we may assume that $B + \delta \sum d_i H_i \in Q$ and $\alpha \colon (X, B + \delta \sum d_i H_i) \dashrightarrow (Y, C + \delta \sum d_i \alpha_* H_i)$ is a minimal model of $f \colon (X, B + \delta \sum d_i H_i) \to S$ as negativity is an open condition, this is also a canonical model since $K_Y + C + \delta \sum d_i \alpha_* H_i \equiv_S K_Y + C + \delta H_Y$ is ample over $S$. $\qquad\square$

## 3.5 The Nonvanishing Theorem

Among a series of theorems deriving geometric consequences from numerical conditions, the nonvanishing theorem is one of the most difficult ones. It was proved in dimension 3 unconditionally ([95, 97]). Under the BCHM condition that the boundary is big, this difficult theorem can be proved by induction on dimensions.

Let us recall the statement of the nonvanishing theorem: If $K_X + B$ is relatively pseudo-effective, then there exists an effective **R**-divisor $D$ such that $D \equiv_S K_X + B$.

**Theorem 3.5.1** (*Nonvanishing theorem*) *Under the BCHM condition, suppose that the existence of PL flips in dimension n, the special termination of MMP with scaling in dimension n, the existence and finiteness of minimal models for pairs $(X, B)$ with $K_X + B$ relatively big in dimension n hold, then the nonvanishing theorem in dimension n holds.*

**Remark 3.5.2** In the following proof, we actually prove a slightly stronger statement: There exists an effective **R**-divisor $D$ such that $D \sim_{\mathbf{R}} K_X + B$. This is because we are going to derive the general result from the statement on generic fibers, but numerical equivalence does not work for this purpose.

On the other hand, by Theorem 3.4.1 and Remark 3.4.2, assuming the existence of PL flips in dimension $n$ and the special termination of MMP with scaling in dimension $n$, then the nonvanishing theorem stating that $K_X + B$ is "numerically equivalent" to an effective **R**-divisor is sufficient to show the existence of a minimal model $(Y, C)$. We can make $(X, B)$ satisfying the KLT version BCHM condition, then $(Y, C)$ is KLT and $C$ is big, then by the basepoint-free theorem, $K_Y + C$ is semi-ample, and in particular, it is **R**-linearly equivalent to an effective **R**-divisor. Hence $K_X + B$ is **R**-linearly

## 3.5 The Nonvanishing Theorem

equivalent to an effective **R**-divisor, as it is obtained by the sum of the strict transform pullback and exceptional divisors with positive coefficients.

*Proof* **Step 0** We may assume that $(X, B)$ is KLT and $B = 3A + E$ is big, where $A$ is a general effective ample **R**-divisor and $E$ is an effective **R**-divisor. Up to the end of Step 5, we suppose that $S$ is a point.

Take a log resolution $g: X' \to X$ of $(X, B)$, write $g^*(K_X + B) = K_{X'} + B'$, it suffices to show the theorem for the pair $(X', (B')^+)$. Therefore, from the beginning, we may assume that $X$ is smooth and the support of $B$ is a normal crossing divisor. Also, we may assume that $A$ is a **Q**-divisor and $kA$ is integral for a sufficiently large positive integer $k$.

**Step 1** Consider the divisorial Zariski decomposition $K_X + B = P + N$. First, we consider the case $P \equiv 0$. In this case, $K_X + B \equiv N$. From the above Remark 3.5.2, we can get the nonvanishing up to **R**-linear equivalence.

**Step 2** In the following, we assume that $P \not\equiv 0$. We will construct a PLT pair $(Y, C)$ by increasing the boundary.

Since $k(K_X + B)$ is pseudo-effective, by Theorem 2.9.8, after replacing $k$, for any sufficiently large $m$,

$$\dim H^0(X, \llcorner mk(K_X + B) \lrcorner + kA) > \binom{(k+1)n}{kn}.$$

Fix a general smooth point $x$ in $X$ and denote by $m_x$ the maximal ideal of the local ring of this point, since

$$\text{length}(\mathcal{O}_{X,x}/m_x^{kn+1}) = \binom{(k+1)n}{kn},$$

there exists an effective **R**-divisor $G \sim_{\mathbf{R}} m(K_X + B) + A$ such that $\text{mult}_x G > n$.

Then $(X, G)$ is not LC at $x$. Indeed, if we consider the blowup at $x$, and denote the exceptional divisor by $E_x$, then the coefficient of $E_x$ in the pullback of $K_X + G$ is larger than 1. Take a log resolution $g: Y \to X$ of $(X, B + G)$ in strong sense such that $E_x$ is a divisor on $Y$. Take an effective **R**-divisor $F$ with sufficiently small coefficients whose support is the exceptional set of $g$ such that $g^*A - F$ is ample and take a sufficiently general effective **R**-divisor $A' \sim_{\mathbf{R}} g^*A - F$.

Take $B_t = 2A + (1 - t/m)A + E + (t/m)G$. Note that $B_0 = B$, $B_m = 2A + E + G$, and $K_X + B_t \sim_{\mathbf{R}} (1 + t)(K_X + B)$.

206               *3 The Finite Generation Theorem*

Denote $g^*(K_X + B_t) = K_Y + \bar{C}_t$, take $C'_t = (\bar{C}_t)^+ - g^*A + A' + F$, and $E_t = (\bar{C}_t)^-$. Here $^+$, $^-$ mean the positive and negative parts. Then

$$K_Y + C'_t \equiv g^*(K_X + B_t) + E_t.$$

Take the divisorial Zariski decomposition $K_Y + C'_t = P_t + N_t$ and take $C_t = (C'_t - N_t)^+$.

Note that $C_t$ is continuous for $t \in (0, m)$. Indeed, from the construction, $N_t = (1 + t)g^*N + E_t$, where $N$ is the numerically fixed part of $K_X + B$. So $N_t$ is continuous and hence $C_t$ is continuous.

As $B_0 = B$, $(Y, C_0)$ is KLT. On the other hand, $B_m = 2A + E + G$ and as $x$ is a general point, the coefficient of $E_x$ in $N_m$ is 0. Here we use the fact that

$$K_Y + C'_m \equiv g^*(K_X + B_m) + E_m \equiv (1 + m)g^*(K_X + B) + (\bar{C}_m)^-$$

and $E_x$ is not in the support of $(\bar{C}_m)^-$. Therefore, $(Y, C_m)$ is not LC. We can consider the *LC threshold*

$$t_0 = \max\{t \mid (Y, C_t) \text{ is LC}\}.$$

Take $C = C_{t_0}$. As the support of $C$ is a normal crossing divisor, $(Y, C)$ is DLT.

By using the $A$ contained in $B_t$ to perturb the coefficients by the tiebreaking method, we may assume that $(Y, C)$ has a unique LC center. That is, $(Y, C)$ is PLT.

By the construction, $K_Y + C$ is pseudo-effective and $C - \lfloor C \rfloor \geq A'$. So, $(Y, C)$ satisfies the BCHM condition. Moreover, by the construction of $C_t$, $\mathrm{Supp}(C)$ does not contain any prime divisor in the numerically fixed part of $K_Y + C$, and in particular, $\lfloor C \rfloor$ is not contained in the numerically fixed part of $K_Y + C$.

Once we can show that $K_Y + C$ is numerically equivalent to an effective **R**-divisor, then $K_X + B_{t_0}$ is numerically equivalent to an effective **R**-divisor by taking the image under $g_*$. Since $K_X + B_{t_0} \sim_{\mathbf{R}} (1 + t_0)(K_X + B)$, we can conclude the proof.

**Step 3** Replacing $(X, B)$ by the PLT pair $(Y, C)$, we may reduce to the case that $X$ is smooth, $B = A + E + Z$, where $A$ is an effective ample **Q**-divisor, $E$ is an effective **R**-divisor, and $Z = \lfloor B \rfloor$ is irreducible. The subadjunction formula $(K_X + B)|_Z = K_Z + B_Z$ determines $B_Z$. By the construction in Step 2, $Z$ is not contained in the numerically fixed part of $K_X + B$.

Take $\{E_i\}$ to be the set of irreducible components of $E$ and consider the vector $v = \sum e_i E_i$. For a sufficiently small real number $\epsilon > 0$, suppose that $|e_i| \leq \epsilon$. For a sufficiently small real number $t > 0$, take $B_{t,v} = B + t(v + A)$. As $\epsilon$ is sufficiently small, we may assume that $v + A$ is ample.

## 3.5 The Nonvanishing Theorem

Since $K_X + B_{t,v}$ is big, by the assumption on the existence of minimal models for the big case, there exists a minimal model $\alpha_{t,v} : (X, B_{t,v}) \dashrightarrow (Y_{t,v}, C_{t,v})$.

**Step 4** We will show that if $\epsilon$ and $t$ are sufficiently small, then the birational map $\alpha_{t,v}$ induces a birational map $\alpha_Z : Z \dashrightarrow W$ in a neighborhood of $Z$ which is independent of the choice of $t, v$. Here note that $Z$ is not contained in the numerically fixed part of $K_X + B_{t,v}$ if $\epsilon$ and $t$ are sufficiently small, hence is not contracted by $\alpha_{t,v}$.

Fix a sufficiently small $t_1 > 0$, take $0 \le t < t_1$, consider the polytope

$$V_t = \{B + t'(v + A) \mid v = \sum e_i E_i, |e_i| \le \epsilon, t \le t' \le t_1\}$$

in the linear space of divisors which is not necessarily rational. Fix $t > 0$, for any $B' \in V_t$, $(X, B')$ is PLT and $K_X + B'$ is big. By the assumption on the existence and finiteness of minimal models, there are finitely many birational maps $\alpha_k : X \dashrightarrow X_k$ such that for any $B_{t',v} \in V_t$, $\alpha_{t',v}$ coincides with one of $\alpha_k$. So by taking the limit $t \to 0$, there are at most countably many minimal models for $V_0 \setminus \{B\}$.

Consider the restrictions of those countably many birational maps on $Z$. If all $\alpha_k|_Z$ are the same for some $\epsilon, t$ sufficiently small, then we can finish this step. So we may assume that there are at least two different $\alpha_k|_Z$ for any sufficiently small $\epsilon, t_1$. For a birational map $\alpha_k$, consider the subset

$$Q_k = \{B' \in V_0 \mid \alpha_k \text{ is a weak minimal model of } (X, B')\}.$$

If $(X, B)$ has a weak minimal model, then we can finish the proof. So we may assume that $Q_k$ does not contain $B$, which implies that $Q_k$ is a closed sub-polytope of $V_0 \setminus \{B\}$. Take $V'_k$ to be the smallest cone containing $Q_k$ with vertex at $B$.

Replacing $V_0$ by $V'_{k_1}$, we can do the same argument on $V'_{k_1}$ as above. If we could not get the conclusion of this step, then this process does not terminate, and we can get a decreasing sequence of cones $V_0 \supset V'_{k_1} \supset V'_{k_2} \supset \cdots$ with vertex $B$, where we can make proper choice of $\alpha_{k_i}$ in each step such that $\alpha_{k_{i+1}}|_Z$ is different from $\alpha_{k_i}|_Z$ for each $i$. By the compactness, we can find a ray $L$ starting from $B$ such that $L$ intersects all $Q_{k_i}$. Using this line, we can construct an MMP on $(X, B)$ with scaling which consists of an infinite sequence of flips and is not isomorphic in a neighborhood of $Z$ for infinitely many steps, which contradicts the special termination of MMP with scaling for $(X, B)$.

208          *3 The Finite Generation Theorem*

**Step 5** By using a similar argument as in the proof of the basepoint-free theorem, we will show certain extension theorem from $Z$ to $X$ by the vanishing theorem. The subadjunction formula

$$(K_{Y_{t,v}} + C_{t,v})|_W = K_W + C_{W,t,v}$$

defines $C_{W,t,v}$, denote $\lim_{t \to 0} C_{W,t,v} = C_W$, note that this limit does not depend on $v$. Since $K_W + C_{W,t,v}$ is nef, $K_W + C_W$ is also nef. Therefore, the birational map $\alpha_Z : (Z, B_Z) \dashrightarrow (W, C_W)$ is a weak minimal model except that $W$ might not be **Q**-factorial. Take $\bar{A} = \alpha_{Z*}(A|_Z)$.

Recall that $B = A + E + Z$ and $E_i$ are irreducible components of $E$. Consider $P$ to be a sufficiently small rational polytope containing $E$ in the linear space spanned by $E_i$ such that if $E' \in P$, then $(X, A + E' + Z)$ is PLT. Since $\bar{A}$ is big,

$$N_W = \{\bar{E}' = \alpha_{Z*}(E'|_Z) \mid E' \in P, K_W + \bar{A} + \bar{E}' \text{ is nef}\}$$

is a rational polytope in the linear space of divisors on $W$ by the cone theorem. Its pullback

$$N = \{E' \in P \mid \bar{E}' = \alpha_{Z*}(E'|_Z) \in N_W\}$$

is a rational polytope containing $E$.

Take rational points $F_j \in N$ and real numbers $r_j > 0$ such that $\sum r_j = 1$ and $E = \sum r_j F_j$. Correspondingly, we have **Q**-divisor $B_j = A + F_j + Z$ such that $B = \sum r_j B_j$. Take $C_{W,j} = \bar{A} + \alpha_{Z*}(F_j|_Z)$, then $C_W = \sum r_j C_{W,j}$.

Taking $t > 0$ sufficiently small and $F_j$ sufficiently close to $E$, denote

$$B_j = A + F_j + Z = B + tv_j,$$

we may assume that $B + t(v_j + A)$ satisfies conditions in Steps 3–4. Denote $Y_j = Y_{t,v_j}$, $\alpha_j = \alpha_{t,v_j}$, and $(\alpha_j)_* B_j = C_j$, note that

$$(K_{Y_j} + C_j)|_W = K_W + C_{W,j}.$$

Denote $A_j = \alpha_{j*} A$, then $K_{Y_j} + C_j + t A_j$ is nef and big.

Take $q$ to be the common multiple of the denominators of the coefficients of all $F_j$, since $K_W + C_{W,j}$ is nef and $C_{W,j}$ is big, by the effective basepoint-free theorem, there exists a positive integer $m$ independent of $q$ such that $|mq(K_W + C_{W,j})|$ is free. Consider the approximation of the coefficients of $E$ by those of $F_j$, the differences are bounded by order $\frac{1}{q^{1+\delta}}$ for some $\delta > 0$. As $F_j - E = tv_j$, if $q$ is sufficiently large, then we can make $tq$ sufficiently small. Note that

$$mq(K_{Y_j} + C_j) - W$$
$$= (mq - 1)(K_{Y_j} + C_j + t A_j) + K_{Y_j} + \alpha_{j*}((1 - (mq - 1)t)A + F_j),$$

## 3.5 The Nonvanishing Theorem

209

and we may assume that

$$(Y_j, \alpha_{j*}((1 - (mq - 1)t)A + F_j))$$

is KLT, by the vanishing theorem,

$$H^1(Y_j, mq(K_{Y_j} + C_j) - W) = 0.$$

Hence

$$H^0(Y_j, mq(K_{Y_j} + C_j)) \to H^0(W, mq(K_W + C_{W,j}))$$

is surjective. So, $H^0(Y_j, mq(K_{Y_j} + C_j)) \neq 0$. Recall that $(Y_j, C_j + tA_j)$ is a minimal model of $(X, B_j + tA)$ and take common resolutions $p_1 \colon X' \to X$ and $p_2 \colon X' \to Y_j$, we know that $p_1^*(K_X + B_j + tA) \geq p_2^*(K_{Y_j} + C_j + tA_j)$. On the other hand, $p_1^*A \leq p_2^*A_j$ by the negativity lemma, hence $p_1^*(K_X + B_j) \geq p_2^*(K_{Y_j} + C_j)$. This implies that $H^0(X, mq(K_X + B_j)) \neq 0$. As $B = \sum r_j B_j$, there exists an effective **R**-divisor $D$ such that $K_X + B \sim_{\mathbf{R}} D$.

**Step 6** Finally, we consider the case that $S$ is not a point. Restricting to the generic fiber $X_\eta$ of $f$, from the above argument, there exists an effective **R**-divisor $D_\eta$ such that $K_{X_\eta} + B_\eta \sim_{\mathbf{R}} D_\eta$. That is, there exist real numbers $r_i$ and rational functions $h_i$ on $X_\eta$ such that, $K_{X_\eta} + B_\eta - D_\eta = \sum r_i \mathrm{div}(h_i)$.

Denote by $D$ the closure of $D_\eta$ on $X$. As $h_i$ are also rational functions on $X$, $G = K_X + B - D - \sum r_i \mathrm{div}(h_i)$ defines an **R**-divisor $G$ on $X$. By construction, there exists an effective ample divisor $H$ on $S$ such that $f(\mathrm{Supp}(G)) \subset H$ and $f^*H + G \geq 0$. Hence $K_X + B \sim_{\mathbf{R},S} D + G + f^*H$ which proves the theorem. $\qquad\square$

**Remark 3.5.3** (1) The content of the nonvanishing theorem is to show the *weak effectivity* (numerically equivalent to an effective divisor) assuming the pseudo-effectivity. At first glance, the difference between effectivity and pseudo-effectivity seems small. But in fact this difference is the root of difficulty and fun in the minimal model theory. It asserts that some nature of mathematics is condensed in the boundary of the cone of big divisors. The basepoint-free theorem is also a statement of this type.

(2) If one wants to partially solve some conjectures in the minimal model theory, what immediately comes to mind is, for example, the case that $B = 0$, or the case that $K_X + B$ is big. However, such conditions are not compatible with the inductive argument. On the other hand, the condition that $B$ can be written as the form $B = A + E$ works very well in induction as we have already seen.

(3) If $B$ is a **Q**-divisor, then in the proof we can show that $D$ can be also taken to be a **Q**-divisor.

210          *3 The Finite Generation Theorem*

## 3.6 Summary

Summarizing all discussions so far, by complicated inductive arguments on dimensions, all the theorems have been proved at the same time. In conclusion, we get the following theorem:

**Theorem 3.6.1** (*Existence of minimal models*) *Let* $(X, B)$ *be a KLT pair consisting of a normal* **Q**-*factorial algebraic variety of arbitrary dimension and an effective* **R**-*divisor. Let* $f : X \to S$ *be a projective morphism to a quasi-projective variety. Assume the following conditions:*

*(1) B is relatively big, that is, there exists a relatively ample* **R**-*divisor A and an effective* **R**-*divisor E such that* $B = A + E$.
*(2)* $K_X + B$ *is relatively pseudo-effective:* $[K_X + B] \in \overline{\mathrm{Eff}}(X/S)$.

*Then there exists a minimal model of the morphism* $f : (X, B) \to S$.

This theorem has many important corollaries. First, combining with the basepoint-free theorem, the following corollary directly follows:

**Corollary 3.6.2** *Under the assumption of Theorem 3.6.1, assume further that B is a* **Q**-*divisor. Then the canonical ring*

$$R(X/S, K_X + B) = \bigoplus_{m=0}^{\infty} f_*(\mathcal{O}_X(\llcorner m(K_X + B) \lrcorner))$$

*is a finitely generated graded* $\mathcal{O}_S$-*algebra.*

The assumption that $B$ is big is not necessary in the above result (see Theorem 3.8.1).

The condition that the boundary is big works well with inductive arguments on dimensions, but we can also conclude the existence of minimal models when the log canonical divisor is big:

**Corollary 3.6.3** *In Theorem 3.6.1, if we replace conditions (1) and (2) with the following condition (3), we can get the same conclusion:*

*(3)* $K_X + B$ *is relatively big.*

*Proof* By assumption, there exists an effective **R**-divisor $B'$ such that $K_X + B \equiv_S B'$. For a sufficiently small $\epsilon > 0$, $(X, B + \epsilon B')$ is KLT and $B + \epsilon B'$ is relatively big. Therefore, there exists a minimal model of $f : (X, B + \epsilon B') \to S$, which is also a minimal model of $f : (X, B) \to S$. $\qquad\square$

**Corollary 3.6.4** *Let X be a normal* **Q**-*factorial algebraic variety with terminal singularities of arbitrary dimension and let* $f : X \to S$ *be a projective*

## 3.6 Summary

morphism to a quasi-projective variety. Assume that $K_X$ is relatively big. Then there exists a minimal model of the morphism $f : X \to S$ with **Q**-factorial terminal singularities.

*Proof* This follows from Corollary 3.6.3 for the case $B = 0$ and $X$ has terminal singularities. The resulting minimal model automatically has terminal singularities. $\qquad\square$

As a special case of the existence of minimal models, the existence of flips is proved:

**Theorem 3.6.5** (*Existence of flips*) *Let* $(X, B)$ *be a* **Q**-*factorial DLT pair and let* $f : X \to S$ *be a projective morphism to a quasi-projective variety. Then the flip of any small contraction of* $f : (X, B) \to S$ *exists.*

If $K_X + B$ is not relatively pseudo-effective, then the existence of Mori fiber spaces can be proved unconditionally:

**Theorem 3.6.6** (*Existence of Mori fiber spaces*) *Let* $(X, B)$ *be a KLT pair consisting of a normal* **Q**-*factorial algebraic variety of arbitrary dimension and an effective* **R**-*divisor. Let* $f : X \to S$ *be a projective morphism to a quasi-projective variety. Assume that* $K_X + B$ *is not relatively pseudo-effective, that is,*

$$[K_X + B] \notin \overline{\mathrm{Eff}}(X/S).$$

*Then there exists a birational model of* $f$ *admitting a Mori fiber space structure. That is, the following assertion holds. There exists a birational map* $\alpha : (X, B) \dashrightarrow (Y, C)$ *to a* **Q**-*factorial KLT pair projective over S, satisfying the following conditions:*

(1) $\alpha$ *is surjective in codimension 1 and* $C = \alpha_* B$.
(2) *If we take birational projective morphisms* $p : Z \to X$ *and* $q : Z \to Y$ *from a normal algebraic variety such that* $\alpha = q \circ p^{-1}$, *then* $G = p^*(K_X + B) - q^*(K_Y + C)$ *is effective and the support of* $p_* G$ *contains all prime divisors contracted by* $\alpha$.
(3) *There exists a Mori fiber space* $h : Y \to T$ *over S.*

Note that we do not assume that $B$ is relatively big.

*Proof* Take a sufficiently general relatively ample effective **Q**-divisor $H$ such that $(X, B + H)$ is KLT and $K_X + B + H$ is relatively ample. Consider the *pseudo-effective threshold*

$$t_0 = \min\{t' \mid K_X + B + t'H \text{ is relatively pseudo-effective}\}.$$

212    3 The Finite Generation Theorem

Since $t_0$ is a positive real number, $B + t_0 H$ is relatively big. Therefore, by running an MMP on $f: (X, B) \to S$ with scaling of $H$, we get a minimal model $\alpha: (X, B + t_0 H) \dashrightarrow (Y, C + t_0 \alpha_* H)$ of $f: (X, B + t_0 H) \to S$. Continue to run this MMP, note that all new models are again weak minimal models of $f: (X, B + t_0 H) \to S$, then by the finiteness of minimal models, after finitely many steps, the contraction morphism corresponding to the extremal ray is a Mori fiber space. $\qquad \square$

**Remark 3.6.7** The partial results we proved so far are strong conclusions derived from the strong assumption on the BCHM condition. It is expected that induction methods will be successful even if we drop the BCHM condition, and all problems in the minimal model theory can be settled in the future.

As an auxiliary result, the following theorem was already proved in the process of the proof:

**Theorem 3.6.8** (*Crepant blowups* or *crepant extraction*) *Let* $(X, B)$ *be a KLT pair consisting of a normal quasi-projective algebraic variety and an effective* **R**-*divisor* $B$. *Take a very log resolution* $f': Y' \to X$ *of the pair* $(X, B)$ *and write* $(f')^*(K_X + B) = K_{Y'} + C'$. *Choose a set of several* $f'$-*exceptional divisors on* $Y$ *with nonnegative coefficients in* $C'$. *Then there exists a* **Q**-*factorial KLT pair* $(Y, C)$ *and a birational projective morphism* $f: (Y, C) \to (X, B)$ *which is crepant (that is,* $f^*(K_X + B) = K_Y + C$*) and the set of exceptional divisors of* $f$ *coincides with the chosen set of prime divisors.*

*Proof* Take an effective **R**-divisor $F$ on $Y'$ whose support is the $f'$-exceptional divisors not contained in the given set of prime divisors. If $F$ is sufficiently small, then $(Y', (C')^+ + F)$ is KLT. Then a minimal model of the morphism $f': (Y', (C')^+ + F) \to X$ is the crepant blowup we want. Here as $f'$ is birational, every **R**-divisor on $Y'$ is relatively big. $\qquad \square$

As special cases, we get "**Q**-factorializations" and "**Q**-factorial terminalizations":

**Corollary 3.6.9** (**Q**-*factorialization*) *Let* $(X, B)$ *be a KLT pair consisting of a normal quasi-projective algebraic variety and an effective* **R**-*divisor* $B$. *Then there exists a* **Q**-*factorialization of* $X$, *that is, there exists a normal* **Q**-*factorial algebraic variety* $Y$ *and a birational projective morphism* $g: Y \to X$ *which is isomorphic in codimension* 1.

*Proof* Take the chosen set of prime divisors to be the empty set. $\qquad \square$

## 3.6 Summary 213

**Corollary 3.6.10** (Q-*factorial terminalization*) *Let* $(X, B)$ *be a KLT pair consisting of a normal quasi-projective algebraic variety and an effective* **R**-*divisor B. Then there exists a* **Q**-*factorial terminalization of* $(X, B)$, *that is, there exists a* **Q**-*factorial terminal pair* $(Y, C)$ *and a birational projective crepant morphism* $g: (Y, C) \rightarrow (X, B)$.

*Proof* Take the chosen set of prime divisors to be the set of all exceptional divisors with nonnegative coefficients in $C'$. □

Note that **Q**-factorial terminalizations are maximal among all crepant blowups. In particular, if $B = 0$ and $X$ has canonical singularities, then this is the crepant blowup considered in [61]. [65] applied crepant blowups to prove the termination of flips inductively.

**Example 3.6.11** As toric varieties are $\overline{\text{KLT}}$, they admit **Q**-factorializations.

Take a toric variety $(X, B)$. That is, $T \subset X$ is an open immersion of an algebraic torus into a normal algebraic variety and $B = X \setminus T$ is a divisor with all coefficients 1. Take $\Sigma = \{\sigma\}$ to be the corresponding fan. Take $\Sigma'$ to be a fan with the same vertices as $\Sigma$ in which each $\sigma$ is subdivided into simplicial cones. Take $(X', B')$ to be the corresponding toric variety. In this case, $X'$ is **Q**-factorial and there is a birational proper morphism $f: X' \rightarrow X$ isomorphic in codimension 1 such that $f^*(K_X + B) = K_{X'} + B'$.

The choice of subdivision is not necessarily unique. If taking the subdivision appropriately, then $f$ is projective and we get a **Q**-factorialization.

Similarly, we can prove the following result, which is useful for generalizing assertions for KLT or DLT pairs to LC pairs:

**Corollary 3.6.12** (*DLT blowups*) *Let* $(X, B)$ *be an LC pair consisting of a normal quasi-projective algebraic variety of arbitrary dimension and an effective* **R**-*divisor B. Then there exists a* **Q**-*factorial DLT pair* $(Y, C)$ *and a birational projective crepant morphism* $f: (Y, C) \rightarrow (X, B)$ *such that the exceptional divisors of f are contained in* $\llcorner C \lrcorner$.

*Proof* Take a log resolution $f': Y' \rightarrow X$ of the pair $(X, B)$, write $(f')^*(K_X + B) = K_{Y'} + C'$. As $(X, B)$ is LC, the coefficients of $C'$ are at most 1. Write $C'' = (f')_*^{-1}B + \text{Exc}(f')$ as the sum of the strict transform of $B$ and all exceptional divisors with coefficients 1. Then $C'' - C'$ is effective and its support is the union of all exceptional divisors of $f'$ with coefficients less than 1 in $C'$.

Take a general relatively ample effective divisor $A$ and a sufficiently small positive real number $t$, we can apply the MMP to the morphism $f: (Y', C'' + $

$tA) \rightarrow X$. Here $(Y', C'' + tA)$ is DLT and its boundary contains a relatively ample divisor. As $K_{Y'} + C'' + tA = (f')^*(K_X + B) + (C'' - C') + tA$, if $t$ is sufficiently small, then the support of the numerically fixed part of $K_{Y'} + C'' + tA$ over $X$ coincides with the support of $C'' - C'$. Therefore, the MMP over $X$ contracts all such divisors and obtains a DLT blowup. $\qquad\square$

If we further blow up a DLT blowup along the intersection of several irreducible components in the boundary with coefficients 1, then we again obtain a model which is log crepant. So there is no maximal model among DLT blowups in general.

## 3.7 Algebraic Fiber Spaces

In this section, we introduce the weak semistable reduction theorem ([1]) and the semipositivity theorem ([52]) for algebraic fiber spaces. We will just give outlines without proofs. There is a relatively simple proof for the latter ([75]).

Algebraic fiber spaces can be considered as the relative version of algebraic varieties. Birational equivalences between algebraic varieties are given by their function fields. The function fields of algebraic varieties are regular extensions of the base field. So birational equivalences between algebraic fiber spaces are given by regular extensions of function fields.

The weak semistable reduction theorem can be viewed as the desingularization theorem for algebraic fiber spaces. The semipositivity theorem, similar to the vanishing theorem, is an important consequence of the Hodge theory. Both theorems are proved when the base field is of characteristic 0, and in positive characteristics the latter has counterexamples.

### 3.7.1 Algebraic Fiber Spaces and Toroidal Geometry

A finite extension $L/K$ of fields is a *regular extension* if the following conditions are satisfied:

(1) (Separability) There exists a transcendence basis $t_1, \ldots, t_n$ over $K$ such that $L$ is a separable algebraic extension of $K(t_1, \ldots, t_n)$.
(2) (Relative algebraic closedness) The elements in $L$ which are algebraic over $K$ are exactly the elements in $K$.

If $K$ is an algebraically closed field, then the above two conditions automatically hold. In this case, there exists an algebraic variety $X$ defined over $K$ such that $L = K(X)$.

## 3.7 Algebraic Fiber Spaces 215

If $K$ is a regular extension of an algebraically closed field $k$, then there exist algebraic varieties $X, Y$ defined over $k$ such that $L = k(X), K = k(Y)$ and a morphism $f : X \to Y$ satisfying the following conditions:

(1) $f$ is dominant, that is, the generic point of $X$ is mapped to the generic point of $Y$.
(2) The geometric generic fiber of $f$ is irreducible and reduced.

The morphism $f : X \to Y$ satisfying conditions above is called an *algebraic fiber space*. As in this book, we are mainly interested in projective algebraic varieties, $X$ and $Y$ are usually assumed to be projective. We often work over a base field of characteristic 0, in which case the separability of field extensions automatically holds.

Next, we explain the language of toroidal geometry. A toroidal variety is a pair of a variety and a divisor locally isomorphic to a toric variety. Here "locally" means in the classical analytic topology or étale topology, and the base field is the complex number field.

A pair $(X, B)$ consisting of a normal algebraic variety and an effective divisor with all coefficients 1 is called a *toroidal variety*, if for each point $x_i \in X$, there exists an analytic neighborhood $U_i$ of $x_i$, a toric variety $Y_i$, and an analytic neighborhood $V_i$ of a point $y_i \in Y_i$ such that there is an analytic isomorphism between the triples $(U_i, B \cap U_i, x_i) \cong (V_i, C_i \cap V_i, y_i)$. Here $C_i = Y_i \setminus T_i$ is a divisor which is the complement of the torus on $Y_i$. Here we use the subscript $i$ to indicate that $Y_i, y_i$ depend on $x_i$. Here in addition we assume that irreducible components of $B$ are normal, that is, we only consider toroidal varieties *without self-intersection*.

A pair $(X, B)$ is called a *smooth toroidal variety*, if $X$ is a smooth algebraic variety and $B$ is a normal crossing divisor. A pair $(X, B)$ is called a *quasi-smooth toroidal variety*, if locally it is the quotient space of a smooth toroidal variety by a finite Abelian group: For any point $x_i \in X$, there exists an analytic neighborhood $x_i \in U_i$, and a finite Abelian group $G_i$ acting diagonally on an analytic neighborhood $\tilde{V}_i$ of a point $\tilde{y}_i$ in the affine space $\mathbf{C}^n$ such that there is an analytic isomorphism of pairs $(U_i, B \cap U_i, x_i) \cong (\tilde{V}_i/G_i, (\tilde{C} \cap \tilde{V}_i)/G_i, y_i)$. Here $\tilde{C}$ is the union of $n$ coordinate hyperplanes and $y_i$ is the image of $\tilde{y}_i$. Quasi-smooth toroidal varieties are $\mathbf{Q}$-factorial.

**Remark 3.7.1** Similar to toric varieties, a toroidal variety is also associated with a fan ([79]). While a toric variety is completely determined by its fan, in the case of a toroidal variety, the analytically local structure and the information of global gluing are determined by the fan.

216                    3 The Finite Generation Theorem

A toroidal variety is analytically locally isomorphic to a $\overline{\text{KLT}}$ pair, so it is also $\overline{\text{KLT}}$ itself, and admits a $\mathbf{Q}$-factorization. A toroidal variety is $\mathbf{Q}$-factorial if and only if the corresponding fan is simplicial, if and only if the toroidal variety is quasi-smooth.

For a toric variety $(X, B)$, the sheaf $\Omega_X^1(\log B)$ of all *logarithmic differential forms*, that is, rational differential forms on $X$ with at most logarithmic poles along $B$, is a locally free sheaf of rank $n = \dim X$. Indeed, the extension of regular differential forms $dz_i/z_i$ $(i = 1, \ldots, n)$ on the algebraic torus $T = X \setminus B$ form a basis. Here $z_i$ $(i = 1, \ldots, n)$ are coordinates of $T$. Hence as a toroidal variety $(X, B)$ is locally isomorphic to a toric variety, $\Omega_X^1(\log B)$ is also locally free.

Take $\Omega_X^\bullet(\log B)$ to be the exterior algebra of $\Omega_X^1(\log B)$. Using the exterior derivative $d$ on logarithmic differential forms, we can define the *log De Rham complex*

$$\Omega_X^\bullet(\log B) = \{0 \to \mathcal{O}_X \to \Omega_X^1(\log B) \to \Omega_X^2(\log B) \to \cdots \to \Omega_X^n(\log B) \to 0\}.$$

A dominant morphism $f \colon (X, B) \to (Y, C)$ between toroidal varieties is called a *toroidal morphism* if analytically locally it is isomorphic to a toric morphism. If $(X, B)$ and $(Y, C)$ are quasi-smooth, this is equivalent to the following: For any point $x_i \in X$, there exist analytic neighborhoods $x_i \in U_i$ and $y_i = f(x_i) \in U_i'$, and finite morphisms from open subsets of affine spaces $\pi_i \colon \tilde{V}_i \to U_i$ and $\pi_i' \colon \tilde{V}_i' \to U_i'$ such that $f$ is induced by $f_i \colon \tilde{V}_i \to \tilde{V}_i'$, where we may write

$$f_i^* w_j = \prod_k z_k^{c_{jk}}$$

for coordinates $(z_1, \ldots, z_n)$ and $(w_1, \ldots, w_m)$. Here $n = \dim X$, $m = \dim Y$, and $[c_{jk}]$ is an integer matrix whose entries are nonnegative.

For a toroidal morphism $f \colon (X, B) \to (Y, C)$, the sheaf of *relative logarithmic differential forms* $\Omega_{X/Y}^1(\log)$ is defined by

$$\Omega_{X/Y}^1(\log) = \Omega_X^1(\log B)/f^*\Omega_Y^1(\log C).$$

It is a locally free sheaf on $X$ of rank $\dim X - \dim Y$. In particular, if $f$ is finite, then $f^*\Omega_Y^1(\log C) \cong \Omega_X^1(\log B)$ and $\Omega_{X/Y}^1(\log) \cong 0$.

We can similarly define the *relative log De Rham complex* $\Omega_{X/Y}^\bullet(\log)$. Denote $d = \dim X - \dim Y$, then

$$\Omega_{X/Y}^d(\log) \cong \mathcal{O}_X(K_X + B - f^*(K_Y + C)).$$

This is denoted by $\omega_{X/Y}(\log)$ and called the *relative log canonical sheaf*.

### 3.7 Algebraic Fiber Spaces

## 3.7.2 The Weak Semistable Reduction Theorem and the Semipositivity Theorem

Assume that the base field is of characteristic 0.

The desingularization theorem is a fundamental theorem in birational geometry of algebraic varieties, while the "weak semistable reduction theorem" proved by Abramovich and Karu is a fundamental theorem in birational geometry of algebraic fiber spaces:

**Theorem 3.7.2** (*Weak semistable reduction theorem* [1]) *Let* $f_0: X_0 \to Y_0$ *be a surjective morphism between projective varieties with geometrically irreducible generic fiber and let* $Z \subset X_0$ *be a closed proper subset. Then we can construct the following algebraic fiber space models:*

*(1)* Well-prepared model: *There exists a quasi-smooth projective toroidal variety* $(X_1, B_1)$, *a smooth projective toroidal variety* $(Y_1, C_1)$, *a morphism* $f_1: X_1 \to Y_1$, *and birational morphisms* $g_1: X_1 \to X_0$, $h_1: Y_1 \to Y_0$ *such that* $f_0 \circ g_1 = h_1 \circ f_1$ *with the following properties:*

*(a)* $g_1^{-1}(Z) \subset B_1$.
*(b)* $f_1: (X_1, B_1) \to (Y_1, C_1)$ *is a toroidal morphism.*
*(c)* $f_1$ *is* equi-dimensional, *that is, every geometric fiber is of dimension* $\dim X_0 - \dim Y_0$.

*(2)* Weakly semistable model: *There exists a quasi-smooth projective toroidal variety* $(X_2, B_2)$, *a smooth projective toroidal variety* $(Y_2, C_2)$, *a morphism* $f_2: X_2 \to Y_2$, *a Galois finite morphism* $h_2: Y_2 \to Y_1$, *and a birational projective morphism* $\mu: X_2 \to (X_1 \times_{Y_1} Y_2)^{\nu}$ *isomorphic in codimension* 1 *with the following properties:*

*(a)* $g_2^{-1}(B_1) = B_2$, $h_2^{-1}(C_1) = C_2$.
*(b)* $f_2: (X_2, B_2) \to (Y_2, C_2)$ *is a toroidal morphism.*
*(c)* $f_2$ *is* equi-dimensional, *and every geometric fiber is reduced.*

*Here* $(X_1 \times_{Y_1} Y_2)^{\nu}$ *is the normalization of the fiber product and* $g_2: X_2 \to X_1$ *is the induced morphism. Moreover, by adding some reduced divisor to* $C_1$ *and replacing* $B_1$ *accordingly, we may assume that* $g_2: (X_2, B_2) \to (X_1, B_1)$ *and* $h_2: (Y_2, C_2) \to (Y_1, C_1)$ *are toroidal morphisms.*

**Remark 3.7.3** (1) A well-prepared model is a birational model, but a weakly semistable model is not as there is a base change. The birational morphism $\mu$ is a **Q**-factorialization.

218                     3 The Finite Generation Theorem

(2) The reason that a weakly semistable model is called "weak" is that the ambient space $X_2$ is not necessarily smooth. If the base space $Y_0$ is of dimension 1, then there exists a semistable model which is not "weak," that is, $X_2$ is smooth. However, $X_2 \to (X_1 \times_{Y_1} Y_2)^\nu$ is not isomorphic in codimension 1, but is a resolution of singularities ([79]). The base change $h_2$ is constructed by using Theorem 1.8.2.

In general, a locally free sheaf $F$ on a projective algebraic variety $X$ is said to be *numerically semipositive* or *nef* if the corresponding tautological quotient invertible sheaf $\mathcal{O}_{\mathbf{P}_X(F)}(1)$ on the projecive bundle $\mathbf{P}_X(F)$ over $X$ is nef. The following semipositivity theorem represents the geometric property of algebraic fiber spaces:

**Theorem 3.7.4** (*Semipositivity theorem* [52]) *For a well-prepared algebraic fiber space* $f : (X, B) \to (Y, C)$, *the following properties hold:*

*(1) For any integers* $p, q$, $R^q f_*(\Omega^p_{X/Y}(\log))$ *is a locally free sheaf on* $Y$.
*(2) For any integer* $q$, $R^q f_*(\omega_{X/Y}(\log))$ *is numerically semipositive.*

This result can be generalized by using the covering trick:

**Theorem 3.7.5** ([69, Theorem 2]) *Let* $f : (X, \bar{B}) \to (Y, \bar{C})$ *be a well-prepared algebraic fiber space and let* $B$ *be an effective* $\mathbf{Q}$-*divisor whose support is contained in* $\bar{B}$ *with coefficients in the interval* $[0, 1)$. *Assume that* $\kappa(X_y, (K_X + B)|_{X_y}) = 0$ *for the generic fiber* $X_y$ *of* $f$. *Write* $\bar{B} = \sum B_i$, $\bar{C} = \sum C_j$ *into irreducible components, write* $B = \sum b_i B_i$.
  *Assume that there exists a positive integer* $m$ *and an integral effective divisor* $D$ *on* $X$ *satisfying the following conditions, which determine effective* $\mathbf{Q}$-*divisors* $M$ *and* $C$ *on* $Y$:

*(1)* $m(K_X + B)$ *is Cartier,* $D \in |m(K_X + B)|$, *and the support of* $D$ *is contained in* $\bar{B}$. *Write* $D = \sum d_i B_i$.
*(2)* $M$ *is the largest* $\mathbf{Q}$-*divisor on* $Y$ *satisfying* $f^*M \le D$. *Write* $M = \sum m_j C_j$.
*(3) Take* $B^0 := B - D/m + f^*(M/m) = \sum b_i^0 B_i$, $f^*C_j = \sum b_{ij} B_i$ *and*

$$c_j = \max_i \{(b_i^0 + b_{ij} - 1)/b_{ij} \mid f(B_i) = C_j\}.$$

*Take* $C = \sum c_j C_j$.

*Then* $(Y, C)$ *is KLT and* $L := M/m - (K_Y + C)$ *is nef.*

*Proof* **Step 1** From the construction, $K_X + B^0 \sim_{\mathbf{Q}} f^*M/m \sim_{\mathbf{Q}} f^*(L + K_Y + C)$. Take $B^1 = B^0 - f^*C$, then $K_X + B^1 \sim_{\mathbf{Q}} f^*(L + K_Y)$.

$C$ is the smallest $\mathbf{Q}$-divisor satisfying $f^*(\bar{C} - C) \le \bar{B} - B^0$. Also,

$$f^* L \sim_{\mathbf{Q}} K_X + B^0 - f^*(K_Y + C) = K_X + \bar{B} - f^*(K_Y + \bar{C}) - (\bar{B} - B^0) + f^*(\bar{C} - C).$$

Therefore, $L$ is the largest $\mathbf{Q}$-divisor satisfying

$$f^* L \le K_X + \bar{B} - f^*(K_Y + \bar{C}).$$

In particular, $L$ is independent of the choice of the coefficients of $B$. However, the comparison of $K_X + \bar{B}$ and $K_Y + \bar{C}$ is given by $D$.

**Step 2** We show that $(Y, C)$ is KLT.

For each $j$, there exists $i_j$ such that $f(B_{i_j}) = C_j$ and the coefficient of $B_{i_j}$ in $-D/m + f^*(M/m)$ is 0. Therefore, $b_i^0 \le b_i$ and $b_{i_j}^0 = b_{i_j}$, and hence $0 \le c_j < 1$. So $(Y, C)$ is KLT. In the following we show that $L$ is nef.

**Step 3** We construct a covering space.

Take a rational function $h \ne 0$ such that $D - m(K_X + B) = \operatorname{div}(h)$ and denote by $X'$ the normalization of $X$ in the field $k(X)(h^{1/m})$. The field extension $k(X')/k(X)$ is a Galois extension with the cyclic group $G = \mathbf{Z}/(m_0)$ as the Galois group. Here the degree $m_0$ of the extension is a divisor of $m$.

As the support of $\operatorname{div}(h)$ is contained in $\bar{B}$, we get a finite toroidal morphism $\pi_X \colon (X', \bar{B}') \to (X, \bar{B})$. Take the Stein factorization of $f \circ \pi_X$, we get a finite toroidal morphism $\pi_Y \colon (Y', \bar{C}') \to (Y, \bar{C})$ and a toroidal algebraic fiber space $f' \colon (X', \bar{B}') \to (Y', \bar{C}')$.

We construct a Galois covering $\rho \colon Y_1 \to Y$ using Theorem 1.8.2 and take base changes of the above constructions: Let $Y_1'$ be the normalization of the fiber product $Y' \times_Y Y_1$ and let $X_1'$ be the normalization of $X' \times_Y Y_1$, et cetera, whereby if something is not irreducible, then take one of its irreducible components. By the following lemma, $Y_1'$ becomes smooth and $f_1' \colon (X_1', \bar{B}_1') \to (Y_1', \bar{C}_1')$ becomes a well-prepared algebraic fiber space.

**Lemma 3.7.6** *Let $X = \mathbf{C}^n$ be an affine space with coordinates $x_1, \ldots, x_n$ and let $\pi \colon Y \to X$ be a finite surjective morphism from a normal affine variety such that $\pi$ is etale outside a normal crossing divisor defined by $x_1 \cdots x_n = 0$. Then there exist positive integers $m_1, \ldots, m_n$ such that, if $X' = \mathbf{C}^n$ is another affine space with coordinates $x_1', \ldots, x_n'$ such that $(x_i')^{m_i} = x_i$, then the normalization of the fiber product $Y \times_X X'$ is smooth.*

*Proof* We can write $X = \operatorname{Spec}(\sigma \cap M)$ for a lattice $M = \mathbf{Z}^n$ and a simplicial cone $\sigma \subset M_{\mathbf{R}} := M \otimes \mathbf{R}$ generated by the fundamental vectors $e_i$ ($1 \le i \le n$). Then there is another lattice $M'$ containing $M$ which is contained in $M_{\mathbf{R}}$ such that $Y = \operatorname{Spec}(\sigma \cap M')$. If we take sufficiently divisible $m_i$, then the lattice

220                 *3 The Finite Generation Theorem*

$M''$ generated by $e_i/m_i$ contains $M'$. This means that there is a morphism $X' = \text{Spec}(\sigma \cap M'') \to Y$, hence the normalization of the fiber product is smooth. $\qquad\qquad\square$

In the following, we just use $f': (X', \bar{B}') \to (Y', \bar{C}')$ to denote the well-prepared algebraic fiber space.

Take the minimal positive integer $m_1$ such that $H^0(X_y, m_1(K_X + B)|_{X_y}) \neq 0$, which is a divisor of $m_0$. The induced morphism $X'_{\bar{y}'} \to X_{\bar{y}}$ between geometric generic fibers is a finite Galois morphism with the subgroup $G_1 = \mathbf{Z}/(m_1) < G$ as the Galois group. The morphism $\pi_Y$ between base varieties is a finite Galois morphism with Galois group $G/G_1$.

Write $K_{X'} + B' = \pi_X^*(K_X + B)$. As the coefficients of $B'$ are strictly smaller than 1, $D$ induces an effective divisor $D' \in |K_{X'}|$. The support of $D'$ is contained in $\bar{B}'$.

Further, take the weak semistable reduction by a base change $\pi'_Y : (Y'', \bar{C}'') \to (Y', \bar{C}')$, we get a weakly semistable algebraic fiber space $f'': (X'', \bar{B}'') \to (Y'', \bar{C}'')$. Denote the induced morphisms by $h_X: (X'', \bar{B}'') \to (X, \bar{B})$ and $h_Y: (Y'', \bar{C}'') \to (Y, \bar{C})$. $D'$ induces an effective divisor $D'' \in |K_{X''}|$. The support of $D''$ is contained in $\bar{B}''$.

**Step 4** Apply the semipositivity theorem to $f': (X'', \bar{B}'') \to (Y'', \bar{C}'')$. There exists a numerically semipositive locally free sheaf $F$ on $Y''$ such that $f''_* \mathcal{O}_{X''}(K_{X''}) = F \otimes \mathcal{O}_{Y''}(K_{Y''})$.

The Galois group $G_1$ acts on $F$. The generator $g$ of $G_1$ acts on the element $s \in H^0(X'', K_{X''})$ determining $D''$ by $g(s) = \zeta s$. Here $\zeta$ is a primitive $m_1$th root of 1. The corresponding invariant subsheaf of $F$ is a nef invertible sheaf $\mathcal{O}_{Y''}(L'')$.

Since $L'', D''$ are integral divisors and the fibers of $f''$ are reduced, $L''$ is the largest $\mathbf{Q}$-divisor satisfying $(f'')^* L'' \leq D'' + \bar{B}'' - (f'')^*(K_{Y''} + \bar{C}'')$. On the other hand, $h_X^*(K_X + \bar{B}) = K_{X''} + \bar{B}''$ and $h_Y^*(K_Y + \bar{C}) = K_{Y''} + \bar{C}''$, so $L'' = h_Y^* L$. Therefore, $L$ is nef. $\qquad\qquad\square$

## 3.8 The Finite Generation Theorem

In this section, we prove the main theorem of this book: The finite generation theorem, that is, "the canonical ring of any algebraic variety is finitely generated." This can be reduced to the general type case as in BCHM using the semipositivity theorem after simplifying the situation by the weak semistable reduction theorem. Here slightly generally, we introduce the proof for KLT pairs ([29]).

## 3.8 The Finite Generation Theorem

**Theorem 3.8.1** (*Finite generation of canonical rings*) *Let* $(X, B)$ *be a KLT pair consisting of a projective algebraic variety and a* **Q**-*divisor. Then the log canonical ring*

$$R(X, K_X + B) = \bigoplus_{m=0}^{\infty} H^0(X, \llcorner m(K_X + B) \lrcorner)$$

*is a finitely generated graded ring.*

*Proof* First, the right-hand side has a graded ring structure as

$$\llcorner m_1(K_X + B) \lrcorner + \llcorner m_2(K_X + B) \lrcorner \leq \llcorner (m_1 + m_2)(K_X + B) \lrcorner$$

for positive integers $m_1, m_2$. Also note that the log canonical ring is a birational invariant in the following sense: For a log resolution $\mu \colon X' \to (X, B)$, write $\mu^*(K_X + B) = K_{X'} + B'$ and decompose $B' = (B')^+ - (B')^-$ into positive and negative parts, then $R(X, K_X + B) \cong R(X', K_{X'} + (B')^+)$.

The degree 0 part of the quotient field of the canonical ring

$$L = \{s_1/s_2 \mid s_1, s_2 \in H^0(X, \llcorner m(K_X + B) \lrcorner) \text{ for some } m \geq 0, s_2 \neq 0\}$$

is a subfield of the function field $\mathbf{C}(X)$, and the field extension $\mathbf{C}(X)/L$ is a regular extension. Therefore, by taking birational models of $\mathbf{C}(X)$ and $L$ appropriately, we can construct an algebraic fiber space $f \colon X \to Y$. This is called the *Iitaka fibration*. By the construction, $\dim Y = \kappa(X, K_X + B)$ and $\kappa(X_y, (K_X + B)|_{X_y}) = 0$ for the generic fiber $X_y$ of $f$.

When $R(X, K_X + B) = \mathbf{C}$ the assertion is trivial, so in the following we assume that $R(X, K_X + B) \neq \mathbf{C}$, that is, $\dim Y > 0$. Take a sufficiently large and sufficiently divisible integer $m$ and fix an element $D \in |m(K_X + B)|$. Taking $Z = \mathrm{Supp}(D + B)$ and applying the weak semistable reduction theorem, we can take the birational model of the field extension $\mathbf{C}(X)/L$ to be a well-prepared algebraic fiber space $f \colon (X, \bar{B}) \to (Y, \bar{C})$. By the construction, $\bar{B}$ is a divisor contained in the support of the pullback of $D + B$ with coefficients 1. Take irreducible decompositions $\bar{B} = \sum B_i$ and $\bar{C} = \sum C_j$.

We use the same symbols $B, D$ for divisors on the new model $X$. That is, we denote again by $B$ the **Q**-divisor obtained by the original $B$ after replacing by the birational model and denote by $D \in |m(K_X + B)|$ the divisor corresponding to the original $D$. Write $B = \sum b_i B_i$ and $D = \sum d_i B_i$. Write $D = D^h + D^v$ into the horizontal and vertical irreducible components with respect to $f$. Each irreducible component of $D^h$ intersects $X_y$ and each irreducible component of $D^v$ is mapped by $f$ to a divisor of $Y$. As $\kappa(X_y, (K_X + B)|_{X_y}) = 0$, any multiple of $D^h$ is not movable.

222         *3 The Finite Generation Theorem*

Define $M$ to be the largest $\mathbf{Q}$-divisor on $Y$ such that $f^*M \leq D^v$. After replacing $m$ by its multiple, we may assume that $M$ is an integral divisor. As $f$ is equi-dimensional,

$$R(X, K_X + B)^{(m)} \cong R(Y, M) = \bigoplus_{m'=0}^{\infty} H^0(Y, m'M).$$

Therefore, it suffices to show that $R(Y, M)$ is finitely generated. By the construction, $M$ is big.

By Theorem 3.7.5, take $B - D/m + f^*(M/m) = \sum a_i B_i$, $f^*C_j = \sum b_{ij} B_i$,

$$c_j = \max_i \{(a_i + b_{ij} - 1)/b_{ij} \mid f(B_i) = C_j\},$$

and $C = \sum c_j C_j$. Then $(Y, C)$ is KLT and $M/m - (K_Y + C)$ is nef.

Since $M$ is big, there exists an ample $\mathbf{Q}$-divisor $A$ and an effective $\mathbf{Q}$-divisor $E$ such that $M/m = A + E$. Take a sufficiently small positive rational number $\epsilon$ such that $(Y, C + \epsilon E)$ is KLT and $M/m - (K_Y + C) + \epsilon A$ is ample. Take a general effective ample $\mathbf{Q}$-divisor $A' \sim_{\mathbf{Q}} M/m - (K_Y + C) + \epsilon A$ with sufficiently small coefficients such that $(Y, C + \epsilon E + A')$ is KLT. As

$$(1 + \epsilon)M/m \sim_{\mathbf{Q}} K_Y + C + \epsilon E + A',$$

the finite generation of $R(Y, M)$ follows from the finite generation of log canonical rings of pairs of log general type. $\qquad\square$

**Remark 3.8.2** (1) Even in the no-boundary case $B = 0$, the boundary divisor constructed on $Y$ may be not 0. This is because there are variations and degenerations of fibers in the algebraic fiber space $f$. Therefore, in order to prove theorems in the usual nonlog world, the theory in the log world is indispensable.

(2) When the coefficients of $B$ are not rational, the log canonical ring is not necessarily finitely generated. This is because the sheaf of graded rings

$$\bigoplus_{m=0}^{\infty} \mathcal{O}_X(\llcorner m(K_X + B) \lrcorner)$$

is not necessarily finitely generated over $\mathcal{O}_X$.

## 3.9 Generalizations of the Minimal Model Theory

So far in this book, we developed the minimal model theory for algebraic varieties over a base field which is an algebraically closed field of characteristic 0. This result can be easily generalized to algebraic varieties admitting finite

## 3.9 Generalizations of the Minimal Model Theory

group actions or over algebraically nonclosed fields. Results obtained over an algebraically closed field of characteristic 0 without a group action can be all generalized after appropriate modifications. Let us check these one by one.

### 3.9.1 The Case with a Group Action

Consider a pair $(X, B)$ with a morphism $f \colon X \to S$ admitting an action of a finite group $G$. That is, $G$ acts on $X$ and $S$, $f$ is $G$-equivariant, and $B$ is a $G$-invariant $\mathbf{R}$-divisor.

First, we extend definitions in numerical geometry. Linear spaces $N^1(X/S)$ and $N_1(X/S)$ admit actions of $G$. Take invariant subspaces $N^1(X/S)^G \subset N^1(X/S)$ and $N_1(X/S)^G \subset N_1(X/S)$ which are dual to each other. $N^1(X/S)^G$ is generated by $G$-invariant Cartier divisors and $N_1(X/S)^G$ is generated by $G$-invariant 1-cycles. The dimension is denoted by $\rho^G(X/S)$.

Since prime divisors and curves are not necessarily $G$-invariant, they do not necessarily generate $N^1(X/S)^G$ or $N_1(X/S)^G$. Since $G$ is finite, we can consider the trace maps, which are projections $p^1 \colon N^1(X/S) \to N^1(X/S)^G$ and $p_1 \colon N_1(X/S) \to N_1(X/S)^G$ defined by $p^1(D) = (1/\#G) \sum_{g \in G} g^* D$ and $p_1(C) = (1/\#G) \sum_{g \in G} g_* C$. The images of prime divisors and curves under these maps generate $N^1(X/S)^G$ and $N_1(X/S)^G$. If $D$ is an $f$-ample Cartier divisor, then $p^1(D)$ is a $G$-invariant $f$-ample $\mathbf{Q}$-Cartier divisor.

Denote by $\overline{NE}(X/S)^G = \overline{NE}(X/S) \cap N_1(X/S)^G$ the closed cone of $G$-invariant effective 1-cycles and by $\mathrm{Amp}(X/S)^G = \mathrm{Amp}(X/S) \cap N_1(X/S)^G$ the cone of $G$-invariant ample divisors.

A log resolution of $(X, B)$ is a $G$-equivariant birational projective morphism $f \colon Y \to X$ with a $G$-invariant normal crossing divisor on $Y$ satisfying prescribed conditions. Here a $G$-invariant normal crossing divisor is assumed to be a sum of smooth $G$-invariant divisors (which is not necessarily irreducible, but with irreducible components disjoint from each other).

The definitions of KLT, LC, et cetera for $(X, B)$ are the same as the case without the action of $G$. DLT is defined by requiring the existence of a $G$-equivariant log resolution satisfying prescribed conditions. Note that the irreducible components of $\llcorner B \lrcorner$ might be permuted by the action of $G$.

We introduce a new concept about $\mathbf{Q}$-factoriality. $X$ is said to be *$G$-equivariantly $\mathbf{Q}$-factorial* if any $G$-invariant integral divisor is $\mathbf{Q}$-Cartier. Even if there exists a prime divisor $D$ which is not $\mathbf{Q}$-Cartier, $\sum_{g \in G} g^* D$ could be $\mathbf{Q}$-Cartier and $X$ could be $G$-equivariantly $\mathbf{Q}$-factorial. Therefore, $G$-equivariantly $\mathbf{Q}$-factoriality does not necessarily imply $\mathbf{Q}$-factoriality.

In the $G$-equivariant MMP, we assume that the pair $(X, B)$ is KLT or DLT and $X$ is $G$-equivariantly $\mathbf{Q}$-factorial.

224          *3 The Finite Generation Theorem*

**Lemma 3.9.1** *If a G-invariant Cartier divisor $D$ is positive on $\overline{\mathrm{NE}}(X/S)^G \setminus \{0\}$, then it is $f$-ample.*

*Proof* Take any $z \in \overline{\mathrm{NE}}(X/S) \setminus \{0\}$. If $(D \cdot z) \leq 0$, then as $(D \cdot g_* z) = (g^* D \cdot z) = (D \cdot z) \leq 0$, we get $(D \cdot p_1(z)) \leq 0$ which is a contradiction. $\square$

The vanishing theorems can be used exactly the same way and the basepoint-free theorem can be applied. If $D$ is $G$-invariant, then the morphism corresponding to $|mD|$ is $G$-equivariant. By the rationality theorem, we derive the cone theorem. The proof is by replacing everything by $G$-invariant or $G$-equivariant one in the proof of the case without $G$ action. Here an extremal ray is an extremal ray of $\overline{\mathrm{NE}}(X/S)^G$, and the corresponding morphism $g \colon X \to Y$ has $\rho(X/Y)^G = 1$. However, $\rho(X/Y)$ might be greater than 1.

The contraction morphism associated to an extremal ray can be classified into divisorial contractions, small contractions, and Mori fiber spaces. However, the exceptional locus of a divisorial contraction is not necessarily a prime divisor:

**Lemma 3.9.2** *Let $g \colon X \to Y$ be a contraction morphism associated to an extremal ray. Suppose that $g$ is birational and its exceptional set contains a prime divisor. Then the exceptional set is of pure codimension 1 and its irreducible components are transitive under the action of $G$.*

*Proof* Take a prime divisor $E_0$ contained in the exceptional set, write $E = \sum_{g \in G} g^* E_0$. Then $E$ is a $G$-invariant $\mathbf{Q}$-Cartier divisor. By the negativity lemma, there is a $G$-invariant effective 1-cycle $C$ which is contracted by $g$ to finitely many points such that $(E \cdot C) < 0$. If the exceptional set does not coincide with the support of $E$, then there is a $G$-invariant effective 1-cycle $C'$ which is contracted by $g$ to finitely many points such that $(E \cdot C') \geq 0$, which contradicts the fact that $\rho(X/Y)^G = 1$. $\square$

In this way, the framework of the minimal model theory in Chapter 2 can be extended to morphisms from pairs admitting $G$-actions. Also arguments in Chapter 3 can be extended to DLT pairs $(X, B)$ satisfying the $G$-equivariant BCHM condition. Here the $G$-equivariant version of the BCHM condition requires that the $\mathbf{R}$-divisors $A, E$ are $G$-invariant. Hence when $B$ is a $\mathbf{Q}$-divisor, the canonical ring is a finitely generated graded $\mathcal{O}_S$-algebra with an action of $G$. In particular, the existence of $G$-equivariant flips follows.

### 3.9.2 The Case when the Base Field Is Not Algebraically Closed

Let us consider the generalization to a base field $k$ of characteristic 0, which is not algebraically closed.

As the characteristic of $k$ is 0, the algebraic closure $\bar{k}$ is a Galois extension of $k$. Take base changes $\bar{X} = X \times_k \bar{k}$ and $\bar{S} = S \times_k \bar{k}$. The Galois group $\bar{G} = \mathrm{Gal}(\bar{k}/k)$ is infinite in general, but an algebraic cycle on $\bar{X}$ can be defined on an algebraic extension of $k$, so the action of $\bar{G}$ passes through a finite quotient group. Since $N^1(\bar{X}/\bar{S})$ and $N_1(\bar{X}/\bar{S})$ are finite-dimensional, we can take the composition of algebraic extension fields of the generators and take one finite quotient group $G$ acting on $N^1(\bar{X}/\bar{S})$ and $N_1(\bar{X}/\bar{S})$. Then $N^1(X/S) = N^1(\bar{X}/\bar{S})^G$.

Although $G$ might not act on $\bar{X}$, we can replace "$G$-invariant" and "$G$-equivariant" by "defined over $k$" in Section 3.9.1 to get the same generalizations of results over algebraically closed fields.

## 3.10 Remaining Problems

Although there has been great progress in the minimal model theory, there are still two big open problems. One is the abundance conjecture, including the nonvanishing conjecture, and the other is the problem of generalization to positive characteristics.

The existence of minimal models in the general case is also a major open problem, but there are two approaches to it. One is to prove the termination of flips. The other is to show the nonvanishing conjecture, which is a part of the abundance conjecture, and to apply the framework of inductive arguments discussed in this chapter (Remark 3.4.2).

### 3.10.1 The Abundance Conjecture

The abundance conjecture has been miraculously solved in dimension 3, but the method cannot be generalized to higher dimensions, so it is necessary to develop a completely new type of inductive argument on dimensions.

A minimal model discussed in this book has the numerical property that the canonical divisor $K_X + B$ is nef. So we can call it a *numerically minimal model*. In contrast, a birational model with a stronger condition that the canonical divisor $K_X + B$ is semi-ample, is called a *geometric minimal model*. This is the same with a *good minimal model* in [76].

Let us give the accurate definition. A projective morphism $f : (X, B) \to S$ from a $\mathbf{Q}$-factorial DLT pair is called a *geometric minimal model* if the following conditions are satisfied: There is a decomposition $f = h \circ g$ into

226                    3 *The Finite Generation Theorem*

projective morphisms $g: X \to Z$ and $h: Z \to S$ such that there is an $h$-ample $\mathbf{R}$-divisor $H$ on $Z$ with $K_X + B \sim_{\mathbf{R}} g^*H$.

It is conjectured that a numerically minimal model is automatically a geometric minimal model:

**Conjecture 3.10.1** *A projective morphism $f: (X, B) \to S$ from a DLT pair is a numerically minimal model if and only if it is a geometric minimal model.*

This conjecture can be proved to follow from the following abundance conjecture ([57]). The proof is a generalization of the basepoint-free theorem:

**Conjecture 3.10.2** (*Abundance conjecture*) *For a projective dominant morphism $f: (X, B) \to S$ from a DLT pair with $\mathbf{Q}$-divisor $B$, the following assertions hold:*

*(1) (Nonvanishing conjecture) If $\nu(X/S, K_X + B) \geq 0$, then $\kappa(X/S, K_X + B) \geq 0$.*

*(2) If $\nu(X/S, K_X + B) > 0$, then $\kappa(X/S, K_X + B) > 0$.*

The Kodaira dimension $\kappa$ and its numerical version $\nu$ have been defined in Chapter 2, here we only recall the definitions when $K_X + B$ is $f$-nef. These invariants are defined by the values on the generic fiber. First, the relative Kodaira dimension $\kappa(X/S, K_X + B)$ is the Kodaira dimension $\kappa(X_\eta, (K_X+B)|_{X_\eta})$ of the generic fiber $X_\eta$. The *numerical Kodaira dimension* $\nu(X/S, K_X + B)$ in this case is defined as the following (which is the original definition):

$$\nu(X/S, K_X + B) = \max\{r \mid ((K_X + B)^r \cdot H^{n-r})_{X_\eta} > 0\}.$$

Here $H$ is an $f$-ample divisor and $n = \dim X_\eta$. The value of $\nu(X/S, K_X + B)$ is among $0, 1, \ldots, \dim X_\eta$. Therefore, the assumption of the nonvanishing conjecture is always satisfied (when $K_X + B$ is $f$-nef).

The abundance conjecture is the last remaining big open problem in the minimal model theory. The conjecture is automatically true when $\nu = \dim X_\eta$. Indeed, in this case the assertion that numerically minimal models are geometric minimal models follows from the basepoint-free theorem.

The conjecture can be also proved when $\nu = 0$. When $K_X + B$ is $f$-nef, it can be proved by using additivity of Kodaira dimensions ([58]). We will give a proof without assuming the minimality in the next subsection.

The abundance conjecture was proved in dimension 3 ([63, 78, 95–97]). The proof of the nonvanishing part ([95, 97]) made full use of special properties in dimension 3 and the generalization to higher dimensions seems difficult.

## 3.10 Remaining Problems

As a related topic, it has been proved that if the generic fiber has a geometric minimal model, then the ambient space also has a geometric minimal model:

**Theorem 3.10.3** (Hacon–Xu [40]) *Let* $f: (X, B) \to S$ *be a projective morphism from a DLT pair. Suppose that $B$ is a $\mathbf{Q}$-divisor. Take a non-empty open subset $S^o$ of $S$, denote $X^o = f^{-1}(S^o)$, $B^o = B|_{X^o}$, and $f^o = f|_{X^o}$. Assume that the following conditions hold:*

*(1) The morphism $f^o: (X^o, B^o) \to S^o$ has a geometric minimal model.*
*(2) Any LC center of $(X, B)$ intersects with $X^o$.*

*Then $f: (X, B) \to S$ has a geometric minimal model.*

### 3.10.2 The Case of Numerical Kodaira Dimension 0

The existence of minimal models in general is still an open problem as the termination of flips is not yet proved. However, if we consider the problem only in codimension 1, we already know that the prime divisors contracted by the MMP are irreducible components of the numerically fixed part of the divisorial Zariski decomposition of the log canonical divisor. Indeed, during the process of the MMP, we can contract all prime divisors that should be contracted:

**Theorem 3.10.4** *Let $f: (X, B) \to S$ be a projective morphism from a $\mathbf{Q}$-factorial DLT pair to a quasi-projective algebraic variety. Suppose that $K_X + B$ is relatively pseudo-effective. Run an MMP with scaling of a relatively ample divisor $H$. Then after finitely many steps, all irreducible components of the numerically fixed part of $K_X + B$ are contracted.*

*Proof* Take $(X, B) = (X_0, B_0)$ and denote by $\alpha_i: (X_i, B_i) \dashrightarrow (X_{i+1}, B_{i+1})$ $(i = 0, 1, \ldots)$ the sequence of MMP with scaling. After removing finitely many steps in the beginning, we may assume that $\alpha_i$ are all isomorphic in codimension 1. Denote by $H_i$ the strict transform of $H$ on $X_i$, we have a sequence $1 \geq t_0 \geq t_1 \geq \cdots$ determined by

$$t_i = \inf\{t \mid K_{X_i} + B_i + t H_i \text{ is relatively nef}\}.$$

Take $t_\infty = \lim t_i$.

If $t_\infty > 0$, take $t_\infty > t' > 0$, then this MMP is an MMP of $f: (X, B + t'H) \to S$ with scaling of $(1 - t')H$. Since $H$ is $f$-ample, this MMP terminate by the theorem of BCHM.

228                    3 The Finite Generation Theorem

Since $K_{X_i} + B_i + t_i H_i$ is relatively nef, $K_X + B + t_i H$ is numerically movable. So, if $t_\infty = 0$, then the limit $K_X + B$ is also numerically movable. Therefore, the numerically fixed part is contracted. □

As a corollary, we can show the existence of geometric minimal models when $\nu = 0$:

**Corollary 3.10.5** (Druel, Gongyo [24, 32]) *Let $(X, B)$ be a **Q**-factorial projective DLT pair. Suppose that $K_X + B$ is pseudo-effective and $\nu(X, K_X + B) = 0$, then there exists a minimal model $\alpha \colon (X, B) \dashrightarrow (Y, C)$ such that $K_Y + C \sim_{\mathbf{R}} 0$.*

*Proof* We construct $\alpha$ by Theorem 3.10.4. By Theorem 2.9.8, $K_Y + C \equiv 0$. By Theorem 2.10.1, there are positive real numbers $r_i$ with $\sum r_i = 1$ and **Q**-divisors $C_i$ such that $(Y, C_i)$ are DLT, $C = \sum r_i C_i$, and $K_Y + C_i \equiv 0$. By [74], $K_Y + C_i \sim_{\mathbf{Q}} 0$. This concludes the assertion. □

**Remark 3.10.6** We will explain later in Section 3.11.5 that when $B$ is a **Q**-divisor, there are also proofs dealing only with smooth algebraic varieties which do not use the minimal model theory.


### 3.10.3 Generalization to Positive Characteristics

In dimension 2, the minimal model theory and the classification theory of algebraic surfaces work for all characteristics. Mumford discovered several "pathological" phenomena in algebraic surface theory in positive characteristics ([105, 108, 109]), but from a higher point of view, algebraic surface theory can be formulated uniformly and independently of characteristics ([110]).

In contrast, the minimal model theory in dimensions 3 and higher deeply relies on vanishing theorems of Kodaira type, so it is hard to be generalized to positive characteristics. For this issue, there are only some partial results so far.

The idea is to use Frobenius morphisms instead of vanishing theorems, which is a method specific in positive characteristics. For example, although far from perfect, there is an alternative to the basepoint-free theorem in [77]. In singularity theory, there are developments on $F$-*singularity* by using *test ideals* instead of multiplier ideals.

In dimension 3, when the Kodaira dimension is nonnegative and the characteristic is bigger than 5, the existence of minimal models is proved ([41]). Also, the existence of minimal models for semistable families of surfaces (the ambient space is of dimension 3) is proved ([66]).

In characteristic 0, results derived from analytic methods such as vanishing theorems and extension theorems are crucial. Also, the general theory of deformation invariance of plurigenera has only an analytic proof. In order to make such analytic theory algebraic, it is necessary to go beyond the algebraic theory in characteristic 0 and study algebraic geometry in positive characteristics. That is why the theory for positive characteristics is very important.

## 3.11 Related Topics

In this section, we discuss related topics without proof.

### 3.11.1 Boundedness Results

For a given class of varieties, we discuss about what kind of boundedness we should expect.

Algebraic varieties $X$ can be roughly classified by the Kodaira dimensions $\kappa(X)$. For different Kodaira dimenisons, their geometric properties are completely different. There are three distinct classes, $\kappa(X) = -\infty, 0, \dim X$, and their representatives are varieties with negative, 0, or positive canonical divisors. Here we consider KLT pairs $(X, B)$ and discuss the boundedness of certain classes according to the positivity of $K_X + B$.

A *Fano variety* is a variety whose anti-canonical divisor is ample. First, we discuss the boundedness of Fano varieties. Fano varieties are expected to have strong boundedness.

For the boundedness of Fano varieties, the ultimate problem is the following BAB conjecture proposed by Alexeev and the Borisov brothers. For a positive real number $\epsilon$, a KLT pair $(X, B)$ is $\epsilon$-*KLT* or $\epsilon$-*LC* if it satisfies the following condition: Take a log resolution $f : Y \to (X, B)$ and write $f^*(K_X + B) = K_Y + C$ and $C = \sum c_i C_i$, then for each $i$, $c_i < 1 - \epsilon$ or $c_i \leq 1 - \epsilon$. Here $C_i$ are distinct prime divisors.

**Conjecture 3.11.1** (*BAB conjecture*) *Fix a positive integer n and a positive real number $\epsilon$. Then there is a scheme S of finite type and a flat projective morphism $f : \mathcal{X} \to S$ with the following property: For any $\epsilon$-KLT pair $(X, B)$ consisting of a projective algebraic variety of dimension n and an $\mathbf{R}$-divisor such that $-(K_X + B)$ is ample, there is a closed point $s \in S$ such that there is an isomorphism $X \cong f^{-1}(s)$.*

This conjecture is known when $n = 2$ ([3]). Also it is known for toric varieties in arbitrary dimensions ([20]). The same boundedness result holds

230                           3 The Finite Generation Theorem

when $X$ is smooth and $B = 0$ ([87]). There are also some partial results when $n = 3$ ([64, 90]). (Birkar ([13, 14]) solved completely the BAB conjecture.)

In order to show the BAB conjecture, we need to show the following two conjectures:

- (*Boundedness of volumes*) Fix a positive integer $n$ and a positive real number $\epsilon$. Then there exists a real number $M$ with the following property: For any $\epsilon$-KLT pair $(X, B)$ consisting of a projective algebraic variety of dimension $n$ and an **R**-divisor such that $-(K_X + B)$ is ample, $(-(K_X + B))^n \leq M$.
- (*Boundedness of Cartier indices*) Fix a positive integer $n$, a positive real number, and a finite set of rational numbers $I \subset (0, 1)$. Then there exists a positive integer $m$ with the following property: For any $\epsilon$-KLT pair $(X, B)$ consisting of a projective algebraic variety of dimension $n$ and a **Q**-divisor such that the coefficients of $B$ are in $I$ and $-(K_X + B)$ is ample, $m(K_X + B)$ is Cartier.

The boundedness of Cartier indices is a difficult conjecture, currently the only known strategy is in [64], which depends on special phenomenon in dimension 3 and is hard to be generalized to higher dimensions.

A *Calabi–Yau variety* is a variety whose canonical divisor is 0. It is expected that certain boundedness can be also established for Calabi–Yau varieties. However, simple boundedness as in the case of Fano varieties does not hold.

On a K3 surface $X$ which is a smooth Calabi–Yau variety of dimension 2, an ample divisor $H$ is said to be *primitive* if it cannot be written as $H \sim mH'$ for some $m \geq 2$ and some divisor $H'$. The pair $(X, H)$ is called a *polarized K3 surface* and $(H^2)$ is called the *degree*. Polarized K3 surfaces of the same degree form a deformation family of dimension 19. Since there are infinitely many possibilities of degrees, algebraic K3 surfaces are not bounded as a family of algebraic varieties. However, if we extend the set to all complex K3 surfaces, then they are in one deformation family of dimension 20. Although there is no boundedness in the category of algebraic varieties, it becomes a bounded family once we enlarge the category to complex manifolds. The same thing happens to Abelian varieties.

A *Calabi–Yau 3-fold* is a simply connected smooth projective algebraic variety of dimension 3 with $K \sim 0$. Finiteness of deformation families of Calabi–Yau 3-folds is a big problem. At least, there are mountains of (although finitely many) known examples of Calabi–Yau 3-folds with distinct Euler numbers ([81]). But it is still expected that some kind of unified theory exists ([123]).

*3.11 Related Topics* 231

By theorems proved in this chapter, an algebraic variety of general type is birationally equivalent to a variety with canonical singularities and ample canonical divisor. Varieties of general type are general, as the name implies, so it is difficult to control the whole family with great diversity.

The entire family of varieties of general type is too general, but we can prove the boundedness once we fix the range of volumes defined as the following. The *volume* $\mathrm{vol}(X)$ of a smooth projective variety $X$ is defined by

$$\mathrm{vol}(X) = \limsup_{m \to \infty} \frac{n!}{m^n} \dim H^0(X, mK_X).$$

Here $n = \dim X$. Being of general type is equivalent to $\mathrm{vol}(X) > 0$. Take the canonical model $X'$ of $X$, then for any integer $m > 0$, $H^0(X, mK_X) \cong H^0(X', mK_{X'})$. By the Riemann–Roch theorem, $\mathrm{vol}(X) = (K_{X'}^n)$. Note however that the volume is a birational invariant which can be defined without passing to the canonical model.

**Theorem 3.11.2** ([35, 42, 137]) *(1) For any fixed positive integer $n$, there exists a positive integer $m_n$ satisfying the following property: If $X$ is a smooth projective variety of general type of dimension $n$, then for any integer $m \geq m_n$, the linear system $|mK_X|$ defines a birational map onto its image.*

*(2) Fix a positive integer $n$ and a positive real number $M$. There exists a scheme $S = S_M$ of finite type and a flat projective morphism $f : \mathcal{X} \to S$ with the following property: For any projective algebraic variety $X$ of dimension $n$ with canonical singularities such that $K_X$ is ample and $\mathrm{vol}(X) \leq M$, there is a closed point $s \in S$ and an isomorphism $X \cong f^{-1}(s)$.*

*(3) For any fixed positive integer $n$, there exists a positive integer $r_n$ satisfying the following property: If $X$ is a smooth projective variety of general type of dimension $n$, then $\mathrm{vol}(X) \geq r_n$.*

*(4) For any fixed positive integer $n$, there exists a positive real number $s_n$ satisfying the following property: If $X$ is a smooth projective variety of general type of dimension $n$, then the order of the birational automorphism group of $X$ is bounded from above by $s_n \mathrm{vol}(X)$.*

### 3.11.2 Minimal Log Discrepancies

Section 3.11.1 discussed global boundedness, and we consider local boundedness in this subsection.

We define the *MLD* $\mathrm{mld}_Z(X, B)$ in order to quantitatively measure the singularity of a pair $(X, B)$. Here $X$ is a normal algebraic variety, $B$ is an effective **R**-divisor, $Z$ is a closed subset of $X$, and $K_X + B$ is **R**-Cartier.

For a log resolution $f : Y \to (X, B)$, write $f^*(K_X + B) = K_Y + C$ with $C = \sum c_i C_i$. Here $C_i$ are disctinct prime divisors. Then define

$$\mathrm{mld}_Z(X, B) = \inf_{f, C_i} \{1 - c_i \mid f(C_i) \subset Z\}.$$

Here the infimum runs over all log resolutions and all prime divisors $C_i$. It is not hard to see that the value of $\mathrm{mld}_Z(X, B)$ is automatically $-\infty$ if it is negative. If $\mathrm{mld}_Z(X, B)$ is nonnegative, then $(X, B)$ is LC in a neighborhood of $Z$, and in this case, its value can be computed by the minimum value of $1 - c_i$ running over all $C_i$ with $f(C_i) \subset Z$ for a fixed log resolution $f$ on which $f^{-1}(Z)$ is a divisor and $C + f^{-1}(Z)$ has normal crossing support. For example, $(X, B)$ is KLT, LC, $\epsilon$-KLT, $\epsilon$-LC is equivalent to $\mathrm{mld}_X(X, B) > 0$, $\geq 0$, $> \epsilon$, $\geq \epsilon$, respectively.

For MLD, Shokurov proposed the following two conjectures and proved that they imply the termination of flips ([131]):

**Conjecture 3.11.3** *(1)* ACC conjecture: *Fix a positive integer $n$ and a set $I \subset [0, 1]$ satisfying the DCC. Then the set of all minimal log discrepancies $\mathrm{mld}_x(X, B)$, where $\dim X = n$, the coefficients of $B$ are in $I$, and $x \in X$ is a closed point, satisfies the ACC.*

*(2)* LSC conjecture: *The function $\mathrm{mld}_x(X, B)$ for all closed points $x \in X$ is lower semicontinuous (= LSC).*

These conjectures predict that the set of all algebraic varieties has some kind of boundedness.

For an LC pair $(X, B)$ and an effective **R**-Cartier divisor $M$ on $X$, the *LC threshold* $\mathrm{lct}(X, B; M)$ is defined by

$$\mathrm{lct}(X, B; M) = \max\{t \mid (X, B + tM) \text{ is LC}\}.$$

LC thresholds are simpler than MLD, for which the ACC conjecture is proved:

**Theorem 3.11.4** ([43]) *Fix a positive integer $n$ and DCC sets $I \subset [0, 1]$, $J \subset [0, \infty)$. Then the set of values of LC thresholds $\mathrm{lct}(X, B; M)$, where $(X, B)$ is an LC pair, the coefficients of $B$ are in $I$, and $M$ is an effective **R**-Cartier with coefficients in $J$, satisfies the ACC.*

### 3.11.3 The Sarkisov Program

The outcomes of the MMP are minimal models and Mori fiber spaces. In either case, the outcome is not unique, and there are birationally equivalent minimal models and Mori fiber spaces. In the former case, we have seen that minimal models are connected to each other by a sequence of flops (Corollary 2.10.10). In the latter case, they can be connected by fundamental transformations called *Sarkisov links* ([39, 132]).

In this subsection, a *Mori fiber space* $f : (X, B) \to Y$ satisfies the following conditions:

(1) $(X, B)$ is KLT and $X$ is **Q**-factorial.
(2) $f$ is a projective surjective morphism over $S$ with connected geometric fibers.
(3) $\rho(X/Y) = 1$ and $-(K_X + B)$ is $f$-ample.

In this case, we may take ample **R**-divisors $H_X, H_Y$ appropriately such that $K_X + B + H_X = f^* H_Y$. So $(X, B + H_X)$ is a KLT minimal model and $f$ is the canonical model. The idea of [39] is to express Sarkisov links using the decomposition of the space of divisors with respect to canonical models.

A *Sarkisov link* between Mori fiber spaces $f_i : (X_i, B_i) \to Y_i$ ($i = 1, 2$) is as the first diagram in Figure 3.1. Indeed, it belongs to one of the four types I, II, III, and IV in Figure 3.1 and some of $E_i, F_i, G$ coincide with $X_i, Y_i$. Moreover, the following conditions are satisfied:

(1) There exists an **R**-divisor $B_{E_1}$ on $E_1$ such that $(E_1, B_{E_1})$ is a **Q**-factorial KLT pair and $K_{E_1} + B_{E_1} \equiv_G 0$.
(2) The horizontal dotted arrow $\alpha : E_1 \dashrightarrow E_2$ is an isomorphism or a composition of $(K_{E_1} + B_{E_1})$-flops.
(3) The nonhorizontal arrows $g_i, h_i$ ($i = 1, 2$) are contraction morphisms associated to extremal rays of KLT pairs which coincides with $f_i$ if the source is $X_i$, and is a divisorial contraction if the target is $X_i$. Also $h_1 \neq h_2$.

By assumption, $\rho(E_1/G) = 2$, so these diagrams are a kind of 2-ray game. When $f_1 = f_2$, the Sarkisov link becomes a birational self-map of a Mori fiber space. When $f_1 \neq f_2$, we automatically have $h_1 \neq h_2$.

Mori fiber spaces are connected by a sequence of Sarkisov links:

**Theorem 3.11.5** ([39]) *Let $f_i : (X_i, B_i) \to Y_i$ ($i = 1, 2$) be Mori fiber spaces obtained by MMP over $S$ starting from a **Q**-factorial KLT pair $(X_0, B_0) \to S$. Then the induced birational map $\alpha$ can be decomposed into a sequence of Sarkisov links.*

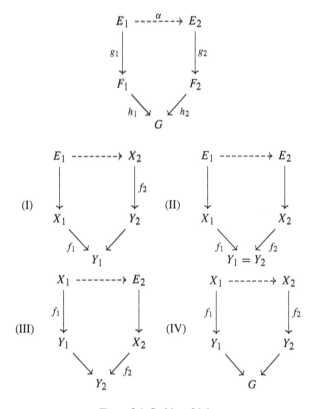

Figure 3.1 Sarkisov Links.

### 3.11.4 Rationally Connected Varieties

A *rational variety* is a variety birational to a projective space and a *unirational variety* is a variety which admits a dominant map from a projective space. Those are interesting varieties but many things are unknown in dimensions 3 and higher. In contrast, *rationally connected varieties*, on which any two general points are contained in a rational curve, are easier to handle ([87–89]). These three concepts coincide in dimension no greater than 2, which is not the case for dimensions 3 and higher. Rationality implies unirationality, and unirationality implies rational connectedness. But there are unirational varieties which are not rational ([8, 23, 50]). Also it is expected that there are rationally connected varieties which are not unirational.

It can be proved that fibers of contraction morphisms in the minimal model theory are rationally connected:

## 3.11 Related Topics

235

**Theorem 3.11.6** ([37]) *Let $f : (X, B) \to S$ be a surjective morphism from an LC pair and let $s \in f(X)$ be a point. Suppose that $-(K_X + B)$ is $f$-ample. Then any connected component of $f^{-1}(s)$ is* rationally chain connected, *that is, for any two points $x, y$ in a connected component of $f^{-1}(s)$, there are rational curves $C_1, \ldots, C_r$ containing in $f^{-1}(s)$ such that $x \in C_1$, $y \in C_r$, and $C_i \cap C_{i+1} \neq \emptyset$ $(i = 1, \ldots, r - 1)$.*

The concepts of rationality, unirationality, and rational connectedness are birationally invariant properties in the category of algebraic varieties. Also, for smooth projective algebraic varieties, rational connectedness and rational chain connectedness are equivalent. However, for singular algebraic varieties, rational chain connectedness cannot imply rational connectedness. Therefore, rational chain connectedness is not a birationally invariant property. For example, in a divisorial contraction of a 3-dimensional algebraic variety, the exceptional divisor can be a cone over a curve of degree 3. In this case, the exceptional divisor is rationally chain connected but not rationally connected.

### 3.11.5 The Category of Smooth Algebraic Varieties

Traditionally, algebraic geometry usually deals with smooth varieties. However, in the minimal model theory, the new point of view of considering varieties with mild singularities led to numerous successes. Fundamental conjectures such as the subadditivity of Kodaira dimensions and the deformation invariance of plurigenera had also been reduced to propositions in the minimal model theory ([58, 114]).

Later, Siu proved the deformation invariance of plurigenera without assuming the minimal model theory ([133, 134]). This was the beginning of the recent development of the minimal model theory discussed in Chapter 3. It was a revival of the point of view of focusing on smooth varieties. When dealing with smooth varieties, one advantage is that one can apply analytic methods. In this subsection, we recall results related to smooth algebraic varieties.

In the minimal model theory, it is expected that any algebraic variety has a good birational model with mild singularities, which is a minimal model or a Mori fiber space. Two birationally equivalent algebraic varieties have birationally equivalent minimal models or Mori fiber spaces, and the latter are connected by special birational transformations. Hence any birational map can be decomposed into a decomposition of standard birational maps as divisorial contractions and flips and their inverse maps, flops, and Sarkisov links.

236                    3 The Finite Generation Theorem

In contrast, in the traditional point of view, smooth varieties are in the special position. From this point of view, the following "weak factorization theorem" is sufficient. According to this theorem, any birational map can be decomposed into blowups and inverses of blowups ([2, 141, 143]):

**Theorem 3.11.7** (*Weak factorization theorem*) *Let $\alpha: X \dashrightarrow Y$ be a birational map between two smooth projective varieties. Then there exists a sequence of birational maps $\alpha_i: X_{i-1} \dashrightarrow X_i$ ($i = 1, \ldots, r$) with $X_0 = X$, $X_r = Y$, and $\alpha = \alpha_r \circ \ldots \circ \alpha_1$ such that for any $i$,*

*(1) $X_i$ is smooth;*
*(2) $\alpha_i$ or $\alpha_i^{-1}$ is a blowup along a smooth subvariety.*

We say that the *strong factorization theorem* holds if there is a decomposition such that for some integer $0 \le s \le t$, the first $s$ $\alpha_i^{-1}$ ($i = 1, \ldots, s$) and the last $t - s$ $\alpha_i$ ($i = s + 1, \ldots, t$) are morphisms. The strong factorization theorem holds in dimension 2 and remains open in dimensions 3 and higher.

The following conjecture is closely related to the basepoint-free theorem. It seems easy at the first glance, but it is a surprisingly deep and difficult conjecture:

**Conjecture 3.11.8** (*Fujita's conjecture*) *Let $X$ be a smooth projective algebraic variety of dimension n and let $H$ be an ample divisor on $X$. Then the following assertions hold.*

*(1) (Freeness) For any integer $m \ge n + 1$, $|K_X + mH|$ is basepoint free.*
*(2) (Very ampleness) For any integer $m \ge n + 2$, $K_X + mH$ is very ample.*

As the lengths of extremal rays of smooth varieties are bounded from above by $n + 1$, $K_X + mH$ is nef in (1) and ample in (2) ([100]). This conjecture predicts that it does not only have numerical properties but also geometric properties. The freeness conjecture (1) is true when dim $X \le 4$ ([68]). Also, in any dimension, for $m \ge n(n + 1)/2 + 1$, $|K_X + mH|$ is basepoint free ([4]). The very ampleness part is not yet well understood.

As Kodaira dimensions and numerical Kodaira dimensions are birational invariants, the abundance conjecture can be also considered in the category of smooth varieties. It suffices to consider a smooth projective variety $X$ and an **R**-divisor $B$ on $X$ with normal crossing support whose coefficients are in the interval $[0, 1]$. The conjecture says that $\kappa(X, K_X + B) = \nu(X, K_X + B)$.

The conjecture is trivial when $\nu(X, K_X + B) = \dim X$. When $\nu(X, K_X + B) = 0$ and $B$ is a **Q**-divisor, by Theorem 2.9.8, we can show that $K_X + B$ is **Q**-linearly equivalent to an effective **Q**-divisor ([21, 22, 74, 135]).

# References

[1] Abramovich, Dan and Karu, Kalle. 2000. Weak semistable reduction in characteristic 0. *Invent. Math.*, **139**(2), 241–273.

[2] Abramovich, Dan, Karu, Kalle, Matsuki, Kenji, and Włodarczyk, Jarosław. 2002. Torification and factorization of birational maps. *J. Amer. Math. Soc.*, **15**(3), 531–572.

[3] Alexeev, Valery. 1994. Boundedness and $K^2$ for log surfaces. *Internat. J. Math.*, **5**(6), 779–810.

[4] Angehrn, Urban and Siu, Yum-Tong. 1995. Effective freeness and point separation for adjoint bundles. *Invent. Math.*, **122**(2), 291–308.

[5] Arbarello, Enrico, Cornalba, Maurizio, Griffiths, Phillip A., and Harris, Joe. 1985. *Geometry of algebraic curves. Vol. I.* Grundlehren der mathematischen Wissenschaften [Fundamental Principles of Mathematical Sciences], vol. 267. Springer-Verlag, New York.

[6] Artin, Michael. 1962. Some numerical criteria for contractability of curves on algebraic surfaces. *Amer. J. Math.*, **84**, 485–496.

[7] Artin, Michael. 1966. On isolated rational singularities of surfaces. *Amer. J. Math.*, **88**, 129–136.

[8] Artin, Michael and Mumford, David. 1972. Some elementary examples of unirational varieties which are not rational. *Proc. London Math. Soc. (3)*, **25**, 75–95.

[9] Barth, Wolf P., Hulek, Klaus, Peters, Chris A. M., and Van de Ven, Antonius. 1984. *Compact complex surfaces*. Ergebnisse der Mathematik und ihrer Grenzgebiete (3) [Results in Mathematics and Related Areas (3)], vol. 4. Springer-Verlag, Berlin.

[10] Beauville, Arnaud. 1983. *Complex algebraic surfaces*. London Mathematical Society Lecture Note Series, vol. 68. Cambridge University Press, Cambridge. Translated from the French by Richard Barlow, Nicholas I. Shepherd-Barron, and Miles Reid.

[11] Bierstone, Edward and Milman, Pierre D. 1997. Canonical desingularization in characteristic zero by blowing up the maximum strata of a local invariant. *Invent. Math.*, **128**(2), 207–302.

238  *References*

[12] Birkar, Caucher. 2010. On existence of log minimal models. *Compos. Math.*, **146**(4), 919–928.

[13] Birkar, Caucher. 2019. Anti-pluricanonical systems on Fano varieties. *Ann. Math. (2)*, **190**(2), 345–463.

[14] Birkar, Caucher. 2021. Singularities of linear systems and boundedness of Fano varieties. *Ann. Math. (2)*, **193**(2), 347–405.

[15] Birkar, Caucher and Shokurov, Vyacheslav V. 2010. Mld's vs thresholds and flips. *J. Reine Angew. Math.*, **638**, 209–234.

[16] Birkar, Caucher, Cascini, Paolo, Hacon, Christopher D., and M\u1d9cKernan, James. 2010. Existence of minimal models for varieties of log general type. *J. Amer. Math. Soc.*, **23**(2), 405–468.

[17] Bombieri, Enrico. 1973. Canonical models of surfaces of general type. *Inst. Hautes Études Sci. Publ. Math.*, 171–219.

[18] Bombieri, Enrico and Mumford, David. 1976. Enriques' classification of surfaces in char. *p*. III. *Invent. Math.*, **35**, 197–232.

[19] Bombieri, Enrico and Mumford, David. 1977. Enriques' classification of surfaces in char. *p*. II. In Walter L. Baily, Jr. and Tetsuji Shioda (eds.), *Complex analysis and algebraic geometry*. Cambridge University Press, Cambridge, pp. 23–42.

[20] Borisov, Alexander A. and Borisov, Lev A. 1992. Singular toric Fano three-folds. *Mat. Sb.*, **183**(2), 134–141.

[21] Campana, Frédéric and Peternell, Thomas. 2011. Geometric stability of the cotangent bundle and the universal cover of a projective manifold. *Bull. Soc. Math. France*, **139**(1), 41–74. With an appendix by Matei Toma.

[22] Campana, Frédéric, Koziarz, Vincent, and Păun, Mihai. 2012. Numerical character of the effectivity of adjoint line bundles. *Ann. Inst. Fourier (Grenoble)*, **62**(1), 107–119.

[23] Clemens, C. Herbert and Griffiths, Phillip A. 1972. The intermediate Jacobian of the cubic threefold. *Ann. Math. (2)*, **95**, 281–356.

[24] Druel, Stéphane. 2011. Quelques remarques sur la décomposition de Zariski divisorielle sur les variétés dont la première classe de Chern est nulle. *Math. Z.*, **267**(1–2), 413–423.

[25] Francia, Paolo. 1980. Some remarks on minimal models. I. *Compositio Math.*, **40**(3), 301–313.

[26] Fujiki, Akira. 1980. On the minimal models of complex manifolds. *Math. Ann.*, **253**(2), 111–128.

[27] Fujino, Osamu. 2007. Special termination and reduction to pl flips. In Alessio Corti (ed.), *Flips for 3-folds and 4-folds*. Oxford Lecture Ser. Math. Appl., vol. 35. Oxford University Press, Oxford, pp. 63–75.

[28] Fujino, Osamu. 2011. On Kawamata's theorem. In Carel Faber, Gerard van der Geer, and Eduard Looijenga (eds.), *Classification of algebraic varieties*. EMS Ser. Congr. Rep. Eur. Math. Soc., Zürich, pp. 305–315.

[29] Fujino, Osamu and Mori, Shigefumi. 2000. A canonical bundle formula. *J. Differential Geom.*, **56**(1), 167–188.

[30] Fujita, Takao. 1979. On Zariski problem. *Proc. Japan Acad. Ser. A Math. Sci.*, **55**(3), 106–110.

# References 239

[31] Fulton, William. 1993. *Introduction to toric varieties*. Annals of Mathematics Studies, vol. 131. Roever Lectures in Geometry. Princeton University Press, Princeton, NJ.

[32] Gongyo, Yoshinori. 2011. On the minimal model theory for dlt pairs of numerical log Kodaira dimension zero. *Math. Res. Lett.*, **18**(5), 991–1000.

[33] Grauert, Hans. 1962. Über Modifikationen und exzeptionelle analytische Mengen. *Math. Ann.*, **146**, 331–368.

[34] Grothendieck, Alexander. 1962. *Fondements de la géométrie algébrique [Extraits du Séminaire Bourbaki, 1957–1962.]* Secrétariat mathématique, Paris.

[35] Hacon, Christopher D. and M$^c$Kernan, James. 2006. Boundedness of pluricanonical maps of varieties of general type. *Invent. Math.*, **166**(1), 1–25.

[36] Hacon, Christopher D. and M$^c$Kernan, James. 2007a. Extension theorems and the existence of flips. In Alessio Corti (ed.), *Flips for 3-folds and 4-folds*. Oxford Lecture Ser. Math. Appl., vol. 35. Oxford University Press, Oxford, pp. 76–110.

[37] Hacon, Christopher D. and M$^c$Kernan, James. 2007b. On Shokurov's rational connectedness conjecture. *Duke Math. J.*, **138**(1), 119–136.

[38] Hacon, Christopher D. and M$^c$Kernan, James. 2010. Existence of minimal models for varieties of log general type. II. *J. Amer. Math. Soc.*, **23**(2), 469–490.

[39] Hacon, Christopher D. and M$^c$Kernan, James. 2013. The Sarkisov program. *J. Algebraic Geom.*, **22**(2), 389–405.

[40] Hacon, Christopher D. and Xu, Chenyang. 2013. Existence of log canonical closures. *Invent. Math.*, **192**(1), 161–195.

[41] Hacon, Christopher D. and Xu, Chenyang. 2015. On the three dimensional minimal model program in positive characteristic. *J. Amer. Math. Soc.*, **28**(3), 711–744.

[42] Hacon, Christopher D., M$^c$Kernan, James, and Xu, Chenyang. 2013. On the birational automorphisms of varieties of general type. *Ann. Math. (2)*, **177**(3), 1077–1111.

[43] Hacon, Christopher D., M$^c$Kernan, James, and Xu, Chenyang. 2014. ACC for log canonical thresholds. *Ann. Math. (2)*, **180**(2), 523–571.

[44] Hartshorne, Robin. 1977. *Algebraic geometry*. Graduate Texts in Mathematics, No. 52. Springer-Verlag, New York.

[45] Hironaka, Heisuke. 1964. Resolution of singularities of an algebraic variety over a field of characteristic zero. I, II. *Ann. Math. (2)*, **79**, 109–203; *(2)*, **79**, 205–326.

[46] Hironaka, Heisuke. 1977. Idealistic exponents of singularity. In Jun-Ichi Igusa (ed.), *Algebraic geometry*. (J. J. Sylvester Sympos., Johns Hopkins University, Baltimore, MD, 1976), pp. 52–125.

[47] Iitaka, Shigeru. 1982. *Algebraic geometry: An introduction to birational geometry of algebraic varieties*. North-Holland Mathematical Library, vol. 24. Springer-Verlag, New York.

[48] Iskovskih, Vasily A. 1977. Fano threefolds. I. *Izv. Akad. Nauk SSSR Ser. Mat.*, **41**(3), 516–562, 717.

[49] Iskovskih, Vasily A. 1978. Fano threefolds. II. *Izv. Akad. Nauk SSSR Ser. Mat.*, **42**(3), 506–549.

[50] Iskovskih, Vasily A. and Manin, Yuri I. 1971. Three-dimensional quartics and counterexamples to the Lüroth problem. *Mat. Sb. (N.S.)*, **86(128)**, 140–166.

240                          *References*

[51] Kawamata, Yujiro. 1979. On the classification of noncomplete algebraic surfaces. In Knud Lønsted (ed.), *Algebraic geometry (Proc. Summer Meeting, Univ. Copenhagen, Copenhagen, 1978)*. Lecture Notes in Math., vol. 732. Springer, Berlin, pp. 215–232.

[52] Kawamata, Yujiro. 1981. Characterization of abelian varieties. *Compositio Math.*, **43**(2), 253–276.

[53] Kawamata, Yujiro. 1982. A generalization of Kodaira-Ramanujam's vanishing theorem. *Math. Ann.*, **261**(1), 43–46.

[54] Kawamata, Yujiro. 1984a. The cone of curves of algebraic varieties. *Ann. Math. (2)*, **119**(3), 603–633.

[55] Kawamata, Yujiro. 1984b. Elementary contractions of algebraic 3-folds. *Ann. Math. (2)*, **119**(1), 95–110.

[56] Kawamata, Yujiro. 1984c. On the finiteness of generators of a pluricanonical ring for a 3-fold of general type. *Amer. J. Math.*, **106**(6), 1503–1512.

[57] Kawamata, Yujiro 1985a. Pluricanonical systems on minimal algebraic varieties. *Invent. Math.*, **79**(3), 567–588.

[58] Kawamata, Yujiro. 1985b. Minimal models and the Kodaira dimension of algebraic fiber spaces. *J. Reine Angew. Math.*, **363**, 1–46.

[59] Kawamata, Yujiro. 1986. On the plurigenera of minimal algebraic 3-folds with $K \equiv 0$. *Math. Ann.*, **275**(4), 539–546.

[60] Kawamata, Yujiro. 1987. The Zariski decomposition of log-canonical divisors. In Spencer J. Bloch (ed.), *Algebraic geometry, Bowdoin, 1985 (Brunswick, Maine, 1985)*. Proc. Sympos. Pure Math., vol. 46. Amer. Math. Soc., Providence, RI, pp. 425–433.

[61] Kawamata, Yujiro. 1988. Crepant blowing-up of 3-dimensional canonical singularities and its application to degenerations of surfaces. *Ann. Math. (2)*, **127**(1), 93–163.

[62] Kawamata, Yujiro. 1989. Small contractions of four-dimensional algebraic manifolds. *Math. Ann.*, **284**(4), 595–600.

[63] Kawamata, Yujiro. 1992a. Abundance theorem for minimal threefolds. *Invent. Math.*, **108**(2), 229–246.

[64] Kawamata, Yujiro. 1992b. Boundedness of **Q**-Fano threefolds. In Leonid A. Bokut, Yuri L. Ershov, and Alexei I. Kostrikin (eds.), *Proceedings of the International Conference on Algebra, Part 3 (Novosibirsk, 1989)*. Contemp. Math., vol. 131. Amer. Math. Soc., Providence, RI, pp. 439–445.

[65] Kawamata, Yujiro. 1992c. Termination of log flips for algebraic 3-folds. *Internat. J. Math.*, **3**(5), 653–659.

[66] Kawamata, Yujiro. 1994. Semistable minimal models of threefolds in positive or mixed characteristic. *J. Algebraic Geom.*, **3**(3), 463–491.

[67] Kawamata, Yujiro. 1997a. *Algebraic varieties (in Japanese)*. Kyoritsu Shuppan.

[68] Kawamata, Yujiro. 1997b. On Fujita's freeness conjecture for 3-folds and 4-folds. *Math. Ann.*, **308**(3), 491–505.

[69] Kawamata, Yujiro. 1998. Subadjunction of log canonical divisors. II. *Amer. J. Math.*, **120**(5), 893–899.

[70] Kawamata, Yujiro. 1999a. Deformations of canonical singularities. *J. Amer. Math. Soc.*, **12**(1), 85–92.

## References

[71] Kawamata, Yujiro. 1999b. On the extension problem of pluricanonical forms. In Piotr Pragacz, Michał Szurek, and Jaroslaw Wiśniewski (eds.), *Algebraic geometry: Hirzebruch 70 (Warsaw, 1998)*. Contemp. Math., vol. 241. Amer. Math. Soc., Providence, RI, pp. 193–207.

[72] Kawamata, Yujiro. 2008. Flops connect minimal models. *Publ. Res. Inst. Math. Sci.*, **44**(2), 419–423.

[73] Kawamata, Yujiro. 2011. Remarks on the cone of divisors. In Carel Faber, Gerard van der Geer, and Eduard Looijenga (eds.), *Classification of algebraic varieties*. EMS Ser. Congr. Rep. Eur. Math. Soc., Zürich, pp. 317–325.

[74] Kawamata, Yujiro. 2013. On the abundance theorem in the case of numerical Kodaira dimension zero. *Amer. J. Math.*, **135**(1), 115–124.

[75] Kawamata, Yujiro. 2015. Variation of mixed Hodge structures and the positivity for algebraic fiber spaces. In Jungkai Alfred Chen, Meng Chen, Yujiro Kawamata, and JongHae Keum (eds.), *Algebraic geometry in east Asia – Taipei 2011*. Adv. Stud. Pure Math., vol. 65. Math. Soc. Japan, Tokyo, pp. 27–57.

[76] Kawamata, Yujiro, Matsuda, Katsumi, and Matsuki, Kenji. 1987. Introduction to the minimal model problem. In Tadao Oda (ed.), *Algebraic geometry, Sendai, 1985*. Adv. Stud. Pure Math., vol. 10. North-Holland, Amsterdam, pp. 283–360.

[77] Keel, Seán. 1999. Basepoint freeness for nef and big line bundles in positive characteristic. *Ann. Math. (2)*, **149**(1), 253–286.

[78] Keel, Sean, Matsuki, Kenji, and M$^c$Kernan, James. 1994. Log abundance theorem for threefolds. *Duke Math. J.*, **75**(1), 99–119.

[79] Kempf, George, Knudsen, Finn F., Mumford, David, and Saint-Donat, Bernard. 1973. *Toroidal embeddings. I*. Lecture Notes in Mathematics, vol. 339. Springer-Verlag, Berlin-New York.

[80] Kleiman, Steven L. 1966. Toward a numerical theory of ampleness. *Ann. Math. (2)*, **84**, 293–344.

[81] Klemm, Albrecht and Schimmrigk, Rolf. 1994. Landau-Ginzburg string vacua. *Nuclear Phys. B*, **411**(2–3), 559–583.

[82] Kodaira, Kunihiko. 1953. On a differential-geometric method in the theory of analytic stacks. *Proc. Nat. Acad. Sci. USA*, **39**, 1268–1273.

[83] Kodaira, Kunihiko. 1954. On Kähler varieties of restricted type (an intrinsic characterization of algebraic varieties). *Ann. Math. (2)*, **60**, 28–48.

[84] Kollár, János. 1993. Effective base point freeness. *Math. Ann.*, **296**(4), 595–605.

[85] Kollár, János. 1996. *Rational curves on algebraic varieties*. Ergebnisse der Mathematik und ihrer Grenzgebiete. 3. Folge. A Series of Modern Surveys in Mathematics [Results in Mathematics and Related Areas. 3rd Series. A Series of Modern Surveys in Mathematics], vol. 32. Springer-Verlag, Berlin.

[86] Kollár, János and Mori, Shigefumi. 1998. *Birational geometry of algebraic varieties*. Cambridge Tracts in Mathematics, vol. 134. Cambridge University Press, Cambridge. With the collaboration of C. Herbert Clemens and Alessio Corti. Translated from the 1998 Japanese original.

[87] Kollár, János, Miyaoka, Yoichi, and Mori, Shigefumi. 1992a. Rational connectedness and boundedness of Fano manifolds. *J. Differential Geom.*, **36**(3), 765–779.

242 *References*

[88] Kollár, János, Miyaoka, Yoichi, and Mori, Shigefumi. 1992b. Rational curves on Fano varieties. In Edoardo Ballico, Fabrizio Cantanese, and Ciro Ciliberto (eds.), *Classification of irregular varieties (Trento, 1990)*. Lecture Notes in Math., vol. 1515. Springer, Berlin, pp. 100–105.

[89] Kollár, János, Miyaoka, Yoichi, and Mori, Shigefumi. 1992c. Rationally connected varieties. *J. Algebraic Geom.*, **1**(3), 429–448.

[90] Kollár, János, Miyaoka, Yoichi, Mori, Shigefumi, and Takagi, Hiromichi. 2000. Boundedness of canonical **Q**-Fano 3-folds. *Proc. Japan Acad. Ser. A Math. Sci.*, **76**(5), 73–77.

[91] Kollár, János et al. 1992. *Flips and abundance for algebraic threefolds*. Société Mathématique de France, Paris. Papers from the Second Summer Seminar on Algebraic Geometry held at the University of Utah, Salt Lake City, Utah, August 1991, Astérisque No. 211 (1992) (1992).

[92] Lazarsfeld, Robert. 2004. *Positivity in algebraic geometry. I*. Ergebnisse der Mathematik und ihrer Grenzgebiete. 3. Folge. A Series of Modern Surveys in Mathematics [Results in Mathematics and Related Areas. 3rd Series. A Series of Modern Surveys in Mathematics], vol. 48. Springer-Verlag, Berlin. Classical setting: line bundles and linear series.

[93] Lesieutre, John. 2016. A pathology of asymptotic multiplicity in the relative setting. *Math. Res. Lett.*, **23**(5), 1433–1451.

[94] Matsumura, Hideyuki. 1970. *Commutative algebra*. W. A. Benjamin, Inc., New York.

[95] Miyaoka, Yoichi. 1987. The Chern classes and Kodaira dimension of a minimal variety. In Tadao Oda (ed.), *Algebraic geometry, Sendai, 1985*. Adv. Stud. Pure Math., vol. 10. North-Holland, Amsterdam, pp. 449–476.

[96] Miyaoka, Yoichi. 1988a. Abundance conjecture for 3-folds: case $\nu = 1$. *Compositio Math.*, **68**(2), 203–220.

[97] Miyaoka, Yoichi. 1988b. On the Kodaira dimension of minimal threefolds. *Math. Ann.*, **281**(2), 325–332.

[98] Miyaoka, Yoichi and Mori, Shigefumi. 1986. A numerical criterion for uniruledness. *Ann. Math. (2)*, **124**(1), 65–69.

[99] Mori, Shigefumi. 1979. Projective manifolds with ample tangent bundles. *Ann. Math. (2)*, **110**(3), 593–606.

[100] Mori, Shigefumi. 1982. Threefolds whose canonical bundles are not numerically effective. *Ann. Math. (2)*, **116**(1), 133–176.

[101] Mori, Shigefumi. 1985. On 3-dimensional terminal singularities. *Nagoya Math. J.*, **98**, 43–66.

[102] Mori, Shigefumi. 1988. Flip theorem and the existence of minimal models for 3-folds. *J. Amer. Math. Soc.*, **1**(1), 117–253.

[103] Mori, Shigefumi and Mukai, Shigeru. 2004. Extremal rays and Fano 3-folds. In Alberto Collino, Alberto Conte, and Gino Fano (eds.), *The fano conference*. University of Torino, Turin, pp. 37–50.

[104] Mukai, Shigeru. 1995. New developments in Fano manifold theory related to the vector bundle method and moduli problems. *Sūgaku*, **47**(2), 125–144.

[105] Mumford, David. 1961a. Pathologies of modular algebraic surfaces. *Amer. J. Math.*, **83**, 339–342.

## References 243

[106] Mumford, David. 1961b. The topology of normal singularities of an algebraic surface and a criterion for simplicity. *Inst. Hautes Études Sci. Publ. Math.*, **9**, 5–22.

[107] Mumford, David. 1962. The canonical ring of an algebraic surface. Appendix to [144].

[108] Mumford, David. 1962. Further pathologies in algebraic geometry. *Amer. J. Math.*, **84**, 642–648.

[109] Mumford, David. 1967. Pathologies. III. *Amer. J. Math.*, **89**, 94–104.

[110] Mumford, David. 1969. Enriques' classification of surfaces in char $p$. I. In *Global analysis (Papers in Honor of K. Kodaira)*. University of Tokyo Press, Tokyo, pp. 325–339.

[111] Nadel, Alan M. 1990. Multiplier ideal sheaves and Kähler-Einstein metrics of positive scalar curvature. *Ann. Math. (2)*, **132**(3), 549–596.

[112] Nakano, Shigeo. 1973. Vanishing theorems for weakly 1-complete manifolds. In Yusuke Kusunoki, Sigeru Mizohata, Masayoshi Nagata, Hiroshi Toda, Masaya Yamaguti, and Hiroki Yoshizawa (eds.), *Number theory, algebraic geometry and commutative algebra, in honor of Yasuo Akizuki*. Kinokuniya Book Store Co. Ltd., Tokyo, pp. 169–179.

[113] Nakano, Shigeo. 1974/75. Vanishing theorems for weakly 1-complete manifolds. II. *Publ. Res. Inst. Math. Sci.*, **10**(1), 101–110.

[114] Nakayama, Noboru. 1986. Invariance of the plurigenera of algebraic varieties under minimal model conjectures. *Topology*, **25**(2), 237–251.

[115] Nakayama, Noboru. 1987. The lower semicontinuity of the plurigenera of complex varieties. In Tadao Oda (ed.), *Algebraic geometry, Sendai, 1985*. Adv. Stud. Pure Math., vol. 10. North-Holland, Amsterdam, pp. 551–590.

[116] Nakayama, Noboru. 2004. *Zariski-decomposition and abundance*. MSJ Memoirs, vol. 14. Mathematical Society of Japan, Tokyo.

[117] Norimatsu, Yoshiki. 1978. Kodaira vanishing theorem and Chern classes for $\partial$-manifolds. *Proc. Japan Acad. Ser. A Math. Sci.*, **54**(4), 107–108.

[118] Raynaud, M. 1978. Contre-exemple au "vanishing theorem" en caractéristique $p > 0$. In Kollagunta G. Ramanathan (ed.), *C. P. Ramanujam – a tribute*. Tata Inst. Fund. Res. Studies in Math., vol. 8. Springer, Berlin-New York, pp. 273–278.

[119] Reid, Miles. *Projective morphisms according to Kawamata*. Warwick preprint, 1983 (unpublished), www.maths.warwick.ac.uk/~miles/3folds/Ka.pdf

[120] Reid, Miles. 1983a. Decomposition of toric morphisms. In Michael Artin and John Tate (eds.), *Arithmetic and geometry, Vol. II*. Progr. Math., vol. 36. Birkhäuser Boston, Boston, MA, pp. 395–418.

[121] Reid, Miles. 1983b. Minimal models of canonical 3-folds. In Shigeru Iitaka (ed.), *Algebraic varieties and analytic varieties (Tokyo, 1981)*. Adv. Stud. Pure Math., vol. 1. North-Holland, Amsterdam, pp. 131–180.

[122] Reid, Miles. 1986. Surfaces of small degree. *Math. Ann.*, **275**(1), 71–80.

[123] Reid, Miles. 1987a. The moduli space of 3-folds with $K = 0$ may nevertheless be irreducible. *Math. Ann.*, **278**(1–4), 329–334.

[124] Reid, Miles. 1987b. Young person's guide to canonical singularities. In Spencer J. Bloch (ed.), *Algebraic geometry, Bowdoin, 1985 (Brunswick, Maine, 1985)*.

Proc. Sympos. Pure Math., vol. 46. Amer. Math. Soc., Providence, RI, pp. 345–414.

[125] Sakai, Fumio. 1982. Anti-Kodaira dimension of ruled surfaces. *Sci. Rep. Saitama Univ. Ser. A*, **10**(2), 1–7.

[126] Serre, Jean-Pierre. 1955. Faisceaux algébriques cohérents. *Ann. Math. (2)*, **61**, 197–278.

[127] Shokurov, Vyacheslav V. 1985. A nonvanishing theorem. *Izv. Akad. Nauk SSSR Ser. Mat.*, **49**(3), 635–651.

[128] Shokurov, Vyacheslav V. 1992. Three-dimensional log perestroikas. *Izv. Ross. Akad. Nauk Ser. Mat.*, **56**(1), 105–203.

[129] Shokurov, Vyacheslav V. 1996. 3-fold log models. Algebraic geometry, 4. *J. Math. Sci.*, **81**(3), 2667–2699.

[130] Shokurov, Vyacheslav V. 2003. Prelimiting flips. *Tr. Mat. Inst. Steklova*, **240**(Biratsion. Geom. Lineĭn. Sist. Konechno Porozhdennye Algebry), 82–219.

[131] Shokurov, Vyacheslav V. 2004. Letters of a bi-rationalist. V. Minimal log discrepancies and termination of log flips. *Tr. Mat. Inst. Steklova*, **246**(Algebr. Geom. Metody, Svyazi i Prilozh.), 328–351.

[132] Shokurov, Vyacheslav and Choi, Sung Rak. 2011. Geography of log models: theory and applications. *Cent. Eur. J. Math.*, **9**(3), 489–534.

[133] Siu, Yum-Tong. 1998. Invariance of plurigenera. *Invent. Math.*, **134**(3), 661–673.

[134] Siu, Yum-Tong. 2002. Extension of twisted pluricanonical sections with plurisubharmonic weight and invariance of semipositively twisted plurigenera for manifolds not necessarily of general type. In Ingrid Bauer, Fabrizio Catanese, Yujiro Kawamata, Thomas Peternell, and Yum-Tong Siu (eds.), *Complex geometry (Göttingen, 2000)*. Springer, Berlin, pp. 223–277.

[135] Siu, Yum-Tong. 2011. Abundance conjecture. In Lizhen Ji (ed.), *Geometry and analysis. No. 2*. Adv. Lect. Math. (ALM), vol. 18. Int. Press, Somerville, MA, pp. 271–317.

[136] Szabó, Endre. 1994. Divisorial log terminal singularities. *J. Math. Sci. Univ. Tokyo*, **1**(3), 631–639.

[137] Takayama, Shigeharu. 2006. Pluricanonical systems on algebraic varieties of general type. *Invent. Math.*, **165**(3), 551–587.

[138] Tsuji, Hajime. 1992. Analytic Zariski decomposition. *Proc. Japan Acad. Ser. A Math. Sci.*, **68**(7), 161–163.

[139] Viehweg, Eckart. 1982. Vanishing theorems. *J. Reine Angew. Math.*, **335**, 1–8.

[140] Villamayor, Orlando. 1989. Constructiveness of Hironaka's resolution. *Ann. Sci. École Norm. Sup. (4)*, **22**(1), 1–32.

[141] Włodarczyk, Jarosław. 2003. Toroidal varieties and the weak factorization theorem. *Invent. Math.*, **154**(2), 223–331.

[142] Włodarczyk, Jarosław. 2005. Simple Hironaka resolution in characteristic zero. *J. Amer. Math. Soc.*, **18**(4), 779–822.

[143] Włodarczyk, Jarosław. 2009. Simple constructive weak factorization. In Dan Abramovich, Aaron Bertram, Ludmil Katzarkov, Rahul Pandharipande, and Michael Thaddeus (eds.), *Algebraic geometry – Seattle 2005. Part 2*. Proc. Sympos. Pure Math., vol. 80. Amer. Math. Soc., Providence, RI, pp. 957–1004.

[144] Zariski, Oscar. 1962. The theorem of Riemann–Roch for high multiples of an effective divisor on an algebraic surface. *Ann. Math. (2)*, **76**, 560–615.

# Index

$(-1)$-curve, 66
$(-2)$-curve, 69
2-ray game, 129
$F$-singularity, 228
$G$-equivariant log resolution, 75
$G$-equivariantly $\mathbf{Q}$-factorial, 223
$K$-equivalent, 64
$\epsilon$-KLT, 229
$\epsilon$-LC, 229
$\mathbf{Q}$-factorial terminalization, 213
$\mathbf{Q}$-factorialization, 212
$\mathbf{Q}$-Cartier divisor, 10
$\mathbf{Q}$-divisor, 2, 9
$\mathbf{Q}$-factorial, 10
   $G$-equivariantly $\mathbf{Q}$-factorial, 223
   analytically $\mathbf{Q}$-factorial, 76
$\mathbf{Q}$-factorial DLT minimal model, 126
$\mathbf{Q}$-factorial terminal minimal model, 126
$\mathbf{R}$-Cartier divisor, 10
$\mathbf{R}$-divisor, 2, 9
$\mathbf{R}$-linearly equivalent, 10
$\overline{\mathrm{KLT}}$, 54
$b$-divisor, 184
$f$-ample, 25, 26
$f$-big, 17, 26
$f$-nef, 26
$m$-genus, 18

Abelian variety, 28
abundance conjecture, 226
ACC, 194
ACC conjecture, 232
ADE singularity, 70
adjoint ideal sheaf, 164
adjunction formula, 19
algebraic fiber space, 104, 215

algebraic surface, 65
algebraic variety, 6
ample, 25
   $f$-ample, 25, 26
   ample over $S$, 25
   relatively ample, 25, 26
ample model, 126, 127
ample over $S$, 25
analytically $\mathbf{Q}$-factorial, 76
anti-canonical divisor, 16
anti-canonical ring, 16
Artin's contraction theorem, 69
ascending chain condition, 194
asymptotic multiplier ideal sheaf,
   167
Atiyah's flop, 122

BAB conjecture, 229
base locus, 14
basepoint-free theorem, 2, 88, 95
   effective basepoint-free theorem, 96
basket, 85
BCHM condition, 179
bend and break, 139
big, 16
   $f$-big, 17, 26
   relatively big, 17, 26
birational invariant, 1
birational map, 11
birational model, 1, 11
birational morphism, 11
birational transform, 12
birationally equivalent, 1, 11
blowup, 13
   crepant blowup, 156, 212
   DLT blowup, 213

246 *Index*

blowup (*cont*)
  maximal crepant blowup, 156
  weighted blowup, 85
boundary, 46
boundary divisor, 2
boundedness of Cartier indices, 230
boundedness of volumes, 230

Calabi–Yau 3-fold, 230
Calabi–Yau variety, 230
canonical cover, 40
canonical divisor, 17
canonical model, 69, 85, 115, 126, 127
canonical ring, 1, 18
canonical sheaf, 18
canonical singularity, 69
Cartier divisor, 8
Cartier index, 83
Castelnuovo's contraction theorem, 66
Castelnuovo–Mumford regularity, 167
center, 13
Chern class
  first Chern class, 68
  second Chern class, 68
closed cone of relative curves, 26
closed convex cone, 25
Cohen–Macaulay, 50, 51
complete linear system, 14
complex algebraic surface, 27
concentration method, 92, 93
cone, 104
  closed cone of relative curves, 26
  closed convex cone, 25
  convex cone, 25
  dual closed convex cone, 25
  numerically movable cone, 142
  relative ample cone, 26
  relative big cone, 26
  relative nef cone, 26
  relative pseudo-effective cone, 25
cone theorem, 107
conic bundle, 113
conic surface, 111
conjecture
  abundance conjecture, 226
  ACC conjecture, 232
  BAB conjecture, 229
  boundedness of Cartier indices, 230
  boundedness of volumes, 230
  Fujita's conjecture, 236
  LSC conjecture, 232

termination of flips, 125
contracted, 105
contraction morphism, 104
contraction morphism associated to $F$, 105
contraction theorem, 105
convex, 104
convex cone, 25
cover
  canonical cover, 40
  covering trick, 38
  cyclic covering, 38
  index 1 cover, 40, 76
  Kummer covering, 39
covering trick, 38
crepant, 64
crepant blowup, 156, 212
crepant extraction, 202, 212
curve, 22
  $(-1)$-curve, 66
  $(-2)$-curve, 69
  hyperelliptic curve, 65
  rational curve, 22
cyclic covering, 38

DCC, 49, 194
deformation family, 112
degree, 230
del Pezzo surface, 111
descending chain condition, 49, 194
dimension
  Iitaka–Kodaira dimension, 16
  Kodaira dimension, 18
  log Kodaira dimension, 18
  numerical Iitaka–Kodaira dimension, 147
  numerical Kodaira dimension, 226
  relative Iitaka–Kodaira dimension, 17
direct image, 23
directed MMP, 130
discrepancy, 144
discrepancy coefficient, 46
discrepancy divisor, 198
discrete valuation ring, 7
discreteness, 108
divisor, 7
  $\mathbf{Q}$-divisor, 2, 9
  $\mathbf{R}$-divisor, 2, 9
  $b$-divisor, 184
  boundary divisor, 2
  canonical divisor, 17
  Cartier divisor, 8
  discrepancy divisor, 198

# Index

247

exceptional divisor, 11
integral divisor, 8
log canonical divisor, 2, 18
normal crossing divisor, 9
prime divisor, 7
principal divisor, 8
simple normal crossing divisor, 9
Weil divisor, 8
divisorial contraction, 117
divisorial sheaf, 8
divisorial Zariski decomposition, 10, 142
divisorially log terminal, 51
DLT, 51
DLT blowup, 213
domain of definition, 11
Du Val singularity, 70
dual closed convex cone, 25
dual graph, 66
Dynkin diagram, 70

effective, 7, 9
effective basepoint-free theorem, 96
equi-dimensional, 217
Euler–Poincaré characteristic, 21
exceptional divisor, 11
exceptional set, 11, 198
exceptional type, 84
existence of flips, 125, 211
existence of minimal models, 197, 210
existence of Mori fiber spaces, 211
existence of PL flips, 183
exponential exact sequence, 74
extension theorem, 168, 175
extremal ray, 27, 104

face, 104
factorial, 10
fan, 54
Fano fibration, 118
Fano variety, 113, 229
finite generation of canonical rings, 2, 221
finite morphism, 27
finiteness of minimal models, 203
first Chern class, 68
fixed part, 14, 189
numerically fixed part, 142
flat morphism, 39
flip, 119
flop, 122, 158
Atiyah's flop, 122
flop decomposition, 158

formula
adjunction formula, 19
genus formula, 67
Kodaira's canonical bundle formula, 69
Noether's formula, 68
projection formula, 24
ramification formula, 19
Reid's plurigenus formula, 85
subadjunction formula, 20
free, 14
relatively free, 17
Frobenius morphism, 133
Fujita's conjecture, 236
fundamental cycle, 71

general, 112
general type, 2, 18, 84
generated by global sections, 14
generated by relative global sections, 16
generic fiber, 17
geometric generic fiber, 17
genus
virtual genus, 67
genus formula, 67
geometric generic fiber, 17
geometric invariant theory, 147
geometric minimal model, 225
GIT, 147
global section
relative global section, 16
good minimal model, 225
graph, 11

Hilbert polynomial, 133
Hilbert scheme, 133
Hironaka desingularization theorem, 33
Hodge index theorem, 65
hyperelliptic curve, 65

Iitaka fibration, 221
Iitaka–Kodaira dimension, 16
index 1 cover, 40, 76
integral divisor, 8
intersection number, 21
inverse image, 12
isomorphic in codimension 1, 13

Kawamata log terminal, 46
Kawamata–Viehweg vanishing theorem, 46
Kleiman's criterion, 28
Klein singularity, 70
KLT, 46

# Index

Kodaira dimension, 18
Kodaira embedding theorem, 36
Kodaira vanishing theorem, 36
Kodaira's canonical bundle formula, 69
Kodaira's lemma, 29
Kodaira–Enriques classification theory, 68
Kummer covering, 39

LC, 50
LC threshold, 89, 206, 232
length, 140
linear system, 14
  complete linear system, 14
linearly equivalent, 8
  $\mathbf{R}$-linearly equivalent, 10
  relatively linearly equivalent, 8
local coordinates, 7
locally closed, 25
locus
  base locus, 14
  non-KLT locus, 50
  non-PLT locus, 164
  numerical base locus, 146
  singular locus, 7, 9
log canonical, 50, 51
log canonical divisor, 2, 18
log canonical model, 127
log canonical ring, 3, 18
log crepant, 64
log De Rham complex, 216
log discrepancy coefficient, 46
log Kodaira dimension, 18
log minimal, 61
  log minimal in weak sense, 61
log minimal model, 126
log pair, 2
log resolution, 34
log resolution in strong sense, 34
log resolution in weak sense, 34
log terminal, 47
log version, 2
logarithmic differential form, 216
logarithmic multiplier ideal sheaf, 166
logarithmic poles, 18
LSC conjecture, 232

maximal crepant blowup, 156
minimal, 61
  minimal in the classical sense, 66
  minimal in weak sense, 61
minimal log discrepancy, 232

minimal log resolution of singularities, 67
minimal model, 67, 126, 198
minimal model program, 2, 128
minimal resolution of singularities, 67
MLD, 232
MMP, 2, 128
  directed MMP, 130
  MMP with scaling, 130
MMP with scaling, 130
model
  $\mathbf{Q}$-factorial DLT minimal model, 126
  $\mathbf{Q}$-factorial terminal minimal model, 126
  ample model, 126, 127
  birational model, 1, 11
  canonical model, 69, 85, 115, 126, 127
  geometric minimal model, 225
  good minimal model, 225
  log canonical model, 127
  log minimal model, 126
  minimal model, 67, 126, 198
  Mori fiber space, 67
  numerically minimal model, 225
  terminal model, 126
  weak minimal model, 198
  weakly semistable model, 217
  well-prepared model, 217
moduli space, 133
moduli space of morphisms, 134
Mori fiber space, 67, 117, 233
morphism, 8
movable
  numerically movable, 142
movable part, 15, 189
  numerically movable part, 142
multiplier ideal sheaf, 160, 163
  asymptotic multiplier ideal sheaf, 167
  logarithmic multiplier ideal sheaf, 166
Mumford's numerical pullback, 73

Nadel vanishing theorem, 162
nef, 218
  $f$-nef, 26
  relatively nef, 26
negative part, 10
Neron–Severi group
  relative Neron–Severi group, 23
Noether's formula, 68
Noetherian induction, 91
non-KLT locus, 50
non-PLT locus, 164
nonsingular, 6

# Index 249

nonvanishing theorem, 89, 204
normal, 7
normal crossing divisor, 9
normalization, 7
numerical base locus, 146
numerical geometry, 20
numerical Iitaka–Kodaira dimension, 147
numerical Kodaira dimension, 226
numerically equivalent, 23
   relatively numerically equivalent, 23
numerically fixed part, 142
numerically minimal model, 225
numerically movable, 142
numerically movable cone, 142
numerically movable part, 142
numerically semipositive, 218

obstruction space, 134

permissible blowup, 36
permissible center, 35
perturbation, 46
Picard number, 23, 111
pinch point, 34
PL contraction, 183
PLT, 51
polarization, 26
polarized K3 surface, 230
polytope, 147
   rational polytope, 147
positive part, 10
prime divisor, 7
primitive, 230
principal divisor, 8
projection formula, 24
projective, 25
pseudo-effective
   relatively pseudo-effective, 26
pseudo-effective threshold, 211
pullback, 12, 15
   Mumford's numerical pullback, 73
purely log terminal, 51

quasi-projective, 25
quasi-smooth toroidal variety, 215
quotient singularity, 47

ramification formula, 19
rational curve, 22
rational double point, 70
rational map, 11
rational polytope, 147

rational singularity, 50, 51, 71
rational variety, 234
rationality theorem, 100
rationally chain connected, 235
rationally connected, 118
rationally connected variety, 234
reduced, 7, 9
reduction, 133
reflexive sheaf, 8
reflexive sheaf of rank one, 8
regular extension, 214
regular system of parameters, 7
Reid's plurigenus formula, 85
relative $t$-cycle, 23
relative ample cone, 26
relative big cone, 26
relative curve, 21
relative global section, 16
relative Iitaka–Kodaira dimension, 17
relative log canonical sheaf, 216
relative log De Rham complex, 216
relative logarithmic differential form, 216
relative nef cone, 26
relative Neron–Severi group, 23
relative Picard number, 23
relative pseudo-effective cone, 25
relative section ring, 17
relative subvariety, 21
relative version, 2
relatively ample, 25, 26
relatively big, 17, 26
relatively free, 17
relatively linearly equivalent, 8
relatively nef, 26
relatively numerically effective, 26
relatively numerically equivalent, 23
relatively pseudo-effective, 26
relatively semi-ample, 17
resolution, 34
   $G$-equivariant log resolution, 75
   log resolution, 34
   log resolution in strong sense, 34
   log resolution in weak sense, 34
   minimal log resolution of singularities, 67
   minimal resolution of singularities, 67
   very log resolution, 48
Riemann–Roch theorem, 93
round down, 3, 10
round up, 10
ruled surface, 111

250          *Index*

Sarkisov link, 233
saturation, 192
scale, 130
second Chern class, 68
section ring, 15
  relative section ring, 17
self-intersection number, 22
semi-ample, 14
  relatively semi-ample, 17
semipositivity theorem, 2, 218
Serre duality theorem, 141
Serre vanishing theorem, 37
simple normal crossing divisor, 9
simple singularity, 70
singular Hermitian metric, 163
singular locus, 7, 9, 17
singularity
  $F$-singularity, 228
  $\epsilon$-KLT, 229
  $\epsilon$-LC, 229
  $\overline{\text{KLT}}$, 54
  ADE, 70
  canonical, 69
  divisorially log terminal, 51
  DLT, 51
  Du Val, 70
  Kawamata log terminal, 46
  Klein, 70
  KLT, 46
  LC, 50
  log canonical, 50, 51
  log terminal, 47
  non-KLT locus, 50
  pinch point, 34
  PLT, 51
  purely log terminal, 51
  quotient, 47
  rational, 50, 51, 71
  rational double point, 70
  simple, 70
  terminal, 60
  toric, 38
  type $\frac{1}{r}(a_1, \ldots, a_n)$, 48
  weak log terminal, 52
  WLT, 52
small contraction, 117
smooth, 6
smooth toroidal variety, 215
special termination theorem, 193
special type, 84
stable, 29

Stein factorization, 94, 106
strict transform, 12
strong factorization theorem, 236
subadjunction formula, 20
support, 9
supporting function, 104
surface
  algebraic surface, 65
  complex algebraic surface, 27
  conic surface, 111
  del Pezzo surface, 111
  polarized K3 surface, 230
  ruled surface, 111
surjective in codimension 1, 13

terminal, 60
terminal model, 126
termination of flips, 125
termination of MMP with scaling, 157
test ideal, 228
theorem
  Artin's contraction theorem, 69
  basepoint-free theorem, 2, 88, 95
  Castelnuovo's contraction theorem, 66
  cone theorem, 107
  contraction theorem, 105
  effective basepoint-free theorem, 96
  existence of flips, 125, 211
  existence of minimal models, 197, 210
  existence of Mori fiber spaces, 211
  existence of PL flips, 183
  extension theorem, 168, 175
  finite generation of canonical rings, 2, 221
  finiteness of minimal models, 203
  Hironaka desingularization theorem, 33
  Hodge index theorem, 65
  Kawamata–Viehweg vanishing theorem, 46
  Kleiman's criterion, 28
  Kodaira embedding theorem, 36
  Kodaira vanishing theorem, 36
  Nadel vanishing theorem, 162
  nonvanishing theorem, 89, 204
  rationality theorem, 100
  Riemann–Roch theorem, 93
  semipositivity theorem, 2, 218
  Serre duality theorem, 141
  Serre vanishing theorem, 37
  special termination theorem, 193
  strong factorization theorem, 236
  weak factorization theorem, 236
  weak semistable reduction theorem, 217

# Index     251

Zariski's main theorem, 105
threshold, 89
    LC threshold, 89, 206, 232
    pseudo-effective threshold, 211
tiebreaking, 89
toric singularity, 38
toric variety, 54
toroidal morphism, 216
toroidal variety, 215
    quasi-smooth toroidal variety, 215
    smooth toroidal variety, 215
total transform, 12
transform
    birational transform, 12
    strict transform, 12
    total transform, 12
tree, 72
type $\frac{1}{r}(a_1, \ldots, a_n)$, 48

unirational variety, 234
uniruled variety, 118
universal family, 133

variety
    Abelian variety, 28
    algebraic variety, 6
    Calabi–Yau variety, 230
    Fano variety, 229
    quasi-smooth toroidal variety, 215
    rational variety, 234
    rationally connected variety, 234

smooth toroidal variety, 215
toric variety, 54
toroidal variety, 215
unirational variety, 234
uniruled variety, 118
very general, 112
very log resolution, 48
volume, 231

wall crossing, 113, 132
weak effectivity, 209
weak factorization theorem, 236
weak log terminal, 52
weak minimal model, 198
weak semistable reduction theorem, 217
weakly 1-complete, 140
weakly semistable model, 217
weighted blowup, 85
Weil divisor, 8
well-prepared model, 217
without self-intersection, 215
WLT, 52

X-method, 93

Zariski decomposition, 80
    divisorial Zariski decomposition,
       10, 142
Zariski's main theorem, 105
Zariski tangent space, 134
Zariski's Riemann space, 184

Milton Keynes UK
Ingram Content Group UK Ltd.
UKHW021824271124
451464UK00008B/112